Professional Pilot's Career Guide

About the Author

Robert P. Mark is a former airline and corporate pilot, as well as a current Certified Flight Instructor. He has logged over 6,000 hours in aircraft ranging from a Cessna 150 to the Boeing BBJ. Ten years of his career was also devoted to the FAA as an air traffic controller and supervisor.

A contributing editor to *Aviation International News*, Mr. Mark was, in 2004, named the Airbus Aerospace Journalist of the Year. Author of four aviation books for McGraw-Hill, he has also written for *Business & Commercial Aviation*, *Professional Pilot*, *Flying*, *FLTops.com*, and *Air Line Pilot*, as well as the *Chicago Tribune*. He is a professional speaker on aviation and communications topics and is read around the world through his aviation industry blog at www.jetwhine.com. He has appeared as a source of expert aviation opinion on NBC, CBS, and Fox News television, as well as numerous national radio outlets.

Mr. Mark received his master's degree from the Integrated Marketing Communications program at Northwestern University's Medill School of Journalism. He received a bachelor's degree in English from Northeastern Illinois University.

You may contact Robert Mark through this book's web site at www.propilotbook.com, via e-mail at rob@propilotbook.com, or by joining in the discussion at www.jetwhine.com.

Professional Pilot's Career Guide

Robert P. Mark

Second Edition

McGraw-Hill

New York   Chicago   San Francisco   Lisbon   London   Madrid
Mexico City   Milan   New Delhi   San Juan   Seoul
Singapore   Sydney   Toronto

1 2 3 4 5 6 7 8 9 0   DOC/DOC   0 1 3 2 1 0 9 8 7

ISBN-13: 978-0-07-148553-1
ISBN-10: 0-07-148553-8

Sponsoring Editor: Stephen S. Chapman
Production Supervisor: Richard C. Ruzycka
Editing Supervisor: Stephen M. Smith
Project Manager: Patricia Wallenburg
Copy Editor: Marcia Baker
Proofreader: Paul Tyler
Indexer: Karin Arrigoni
Art Director, Cover: Jeff Weeks
Composition: TypeWriting

Printed and bound by RR Donnelley.

McGraw-Hill books are available at special quantity discounts to use as premiums and sales promotions, or for use in corporate training programs. For more information, please write to the Director of Special Sales, McGraw-Hill Professional, Two Penn Plaza, New York, NY 10121-2298. Or contact your local bookstore.

This book is printed on acid-free paper.

For Nancy, Abigail, and E. T. W. D.

Contents

Foreword

When Robert Mark asked me to write this foreword, my mind raced back 30 years to the beginning of my civilian flying career. I vividly remember the first day in my new hire class at a small "local service" airline known as Hughes Airwest. There were 26 pilots in that class (13 civilians and 13 ex-military), and all had three common traits in addition to their piloting skills: optimism, knowledge of the aviation industry, and an effective job-hunting strategy. All these pilots had successfully landed an airline job in 1976, a year during which U.S.-based airlines hired less than 500 pilots.

As the airline industry and corporate flight departments expand and contract, thousands of professional pilots are affected, including those who

Figure F-1 The future is bright. (Courtesy John Absolon: www.jetphotos.net.)

aspire to fly for a living. Career decisions are necessary at a maddening pace, all in the framework of uncertainty in a profession frequently described as "feast or famine."

Pilots at many passenger airlines have suffered significant reductions in pay and benefits the past few years (Figure F-1). Some believe these reductions are permanent and suggest aspiring pilots avoid the profession altogether. I am one who believes the professional pilot career will recover from the doldrums and provide those who want it badly enough a rewarding and satisfying professional life. Planning and preparation are the key ingredients to making it work for you.

It takes staying power and the proper tools to make it as a professional pilot. This excellent book is one of those tools. You need to read it once to understand the scope of the task ahead, and then reread it with an eye toward developing your own personal career plan.

Fly safely and plan your pilot career with passion!

Louis Smith
President
FLTops.com

Acknowledgments

While this author has played a role in the creation of a career guide like the one you hold in your hands, I represent only a portion of the talent necessary to see a volume like this through to completion. The *Professional Pilot's Career Guide* could not have been written without the help of many other aviation and business professionals whom I've recognized below. I sincerely hope I have not forgotten or misspelled a name along the way.

My thanks to Bill Traub, Louis Smith, Judy Tarver, David Ball, Victor Veltze, Emmett Johnson, Vic Lipsey, Randy Johnson, Greg Brown, Mike Collins, Steve Phillips, Rich Morris, Dick Skovgaard, Bree Cox, Bob Stangarone, Mark Fairchild, Pedro Ferraz, Randy Padfield, Nigel Moll, Sean Reilly, Scott Spangler, Jan Barden, Ruth Elliott, Peter Moll, Paul Berliner, Allan Greene, Liz Clark, Andrew Ponzoni, Mark Phelps, Annmarie Yannaco, Robert Baugniet, Mike Overly, Scott Randell, David Jones, Nancy Molitor, Derek Martin, David Manning, Sark Boyajian, John Basuerman, Craig Washka, Diane Powell, Jim North, Neal Schwartz, Pete Beckmeyer, Steve Mayer, Greg Watts, Joe Siok, Warren Cleveland, and Sandy Anderson.

A special thanks to the astounding photographers from the United States, Europe, Asia, and Australia who provided some of the shots in this volume. They include Nik French, Dale Boeru, Daniel Havlik, Ian Schofield, Peter Egglestone, Dean Heald, Tim Wagenknecht, Kevin Wachter, Seth Jaworski, Dan Valentine, Pan Jun, Stephen Toernblom, Curt Jans, and Chris Starnes.

Corporations that provided photographs and information include Cessna Aircraft Company, Gulfstream Aerospace, Embraer, FlightSafety International, Delta Connection Academy, FLTops.com, Avcrew.com, Auburn University, the Aircraft Owners and Pilots Association, Jetwhine.com, the U.S. Air Force, Frasca International Inc., the Boeing Company, Dassault Falcon, Sino Swearingen Corp., and Northwestern University.

Thanks to Bossard Publications/*Flying Careers* magazine for granting permission to reprint some of the many articles I wrote for them over the years, as well as *Aviation International News* for agreeing to share a few of the stories I wrote for that magazine. Thanks to Scott Spangler for his editorial on the CFI shortage, reprinted by permission (*Flight Training*, 1998).

Certainly this book would never have been finished without the infinite patience of my wife Nancy and daughter Abigail, who tolerated my crankiness during the seemingly endless writing and editing process.

And, finally, a big thanks to Amy Odwarka's seventh-grade writing class at Haven Middle School in Evanston, Illinois, for its editing expertise, with a special thanks to Molly, Sarah, and Lucy.

Robert P. Mark

Introduction

"The flight was short, but it was nevertheless the first flight in the history of the world in which a machine carrying a man raised itself by its own power into the air in full flight, had sailed forward without a reduction of speed, and had finally landed at a point as high as that from which it started."
—Orville Wright in a statement to the press, December 1903

As my high school graduation drew close in 1966, I was convinced my future was waiting in the cockpit of some airplane, somewhere. I spent so much of my free time riding my motorcycle to Chicagoland airports, such as Midway and O'Hare, where I would plant myself alongside the road to watch airplanes take off and land, that some of my friends gave up on me as a social lost cause. Little did I know that same pastime was responsible for adding thousands and thousands of other pilots to the industry ranks worldwide. Never let anyone tell you such an afternoon is wasted.

The '60s were days when the opportunities were legion for kids like me to see the latest propeller-driven airplanes. No one could have told me that a Douglas DC-3 or a Lockheed Constellation, Convair 580 (Figure I-1) or even the Electra were anything other than the most beautiful, the most fantastic, the most incredible pieces of flying machinery that existed on the face of the Earth or ever would. Just about the time I became really excited about these great old birds, they marched into history, replaced by the Boeing 707, the Douglas DC-8 and DC-9, and the Convair 880s and 990s. The term "business aviation" had not caught on yet, although twin-engine propeller-driven aircraft, such as the Queen Air, the Cessna 310, and the Beech 18, were common.

Cranking my head around every time an airplane flew seemed pretty normal as I wondered how I'd feel behind the controls of some big airliner

NORTH CENTRAL AIRLINES

Convair 580 prop-jet

Figure I-1 North Central Airlines Convair 580 (ca. 1960s).

or business airplane. I never thought much then about a plan focused on how I'd get into the cockpit. At the time, the dream seemed enough.

I took hundreds of photos, too, many of which have survived to this day in albums stashed in the closet. I built models and read any book that had either the word "airplane" in the title or offered the promise of a story about something that flew. For me, being immersed in aviation with books, pictures, model airplanes, and every air show I could find around my Midwestern home helped me stay motivated and focused on my goal—to make it to the cockpit any way I could. I was 17 years old.

Thinking back on it all now, some 40 years later, I wish I'd known someone in the industry, someone who could have shown me the ropes and helped me avoid some of the pits I managed to fall into along the way. I'm convinced now that, if a mentor had been part of my life, I would never have spent as much time as I did moving in so many different directions over the years, directions that, at times, seemed to take me further from my goal instead of closer to it. As a result, I did not wear my first airline uniform until I was in my 30s.

I grew up assuming that finding a job would be easy because I remembered seeing ads in the Chicago newspapers searching for new cockpit crewmembers in the '60s. One ad said the only requirement to be considered for an interview was a private pilot's license. The airline would pay for the remainder of the needed training, if only respondents would be kind enough to come and talk. Many pilots did. And, many of those pilots who took those jobs in the 1960s are on their way out to retirement as they reach the magic age of 60, the current limit to how long a pilot can fly for a scheduled carrier in America.

Because our audience has become significantly more international for this edition of the *Professional Pilot's Career Guide*, readers should be aware that, as I write this in late 2006, the retirement guidelines issued by the International Civil Aviation Organizations (ICAO) will be updated by raising the age limit to 65. Right now, it is unclear whether the U.S. will comply with the new international guidelines.

When the first edition of this volume was penned, the airline industry was moving through one of its greatest expansions in history, with carriers worldwide reporting record profits as new aircraft orders set unbelievable, almost unsustainable levels. The deliveries of general aviation aircraft, including business machines, had not reached nearly the levels of the airlines, however.

But the old maxim warning us about something that looks too good probably is, could have been written for the aviation industry.

Early in the twenty-first century, with an economy alternating between dramatic surges and anemic coughs, people began to revolt against the high-fare economic model of the airlines that had sustained the industry for 50 years. Their solution emerged as low-cost carriers that offered travelers a choice of airlines, some that offered full service and some, like pioneer low-cost Southwest Airlines, that simply offered low fares and no frills—no food, prereserved seats, and few hassles when it came to changes. The major airlines were slow to respond. By 2007, low-cost carriers such as Southwest, joined by JetBlue, Spirit, and AirTran, accounted for nearly 25 percent of U.S. domestic capacity. In Europe, despite the constraints of a much more complex airspace structure, low-cost carriers also began to emerge. Companies such as Ryan Air, Easyjet, and Air Berlin became household names. India and Asia are also witnessing dramatic demands for airline service.

In the United States, the stock market fell apart in 2000 as thousands of dot-com companies folded. And, just as airlines were reeling from the reduced demand fallout from the stock market crash came the horrendous 9/11 attacks in New York and Washington, which grounded almost every aircraft in North America. The threat of additional terrorist actions sent the U.S. airline industry into a cataclysmic tailspin that triggered tens of thou-

sands of airline employee layoffs as companies ran for cover trying to figure out their next move. Since 9/11, United, Delta, Northwest, America
West, ATA, and US Airways have filed bankruptcy here in the U.S. In
Europe, KLM and Air France have merged, while the Swissair millions had
known for decades has all but disappeared. Alitalia also stands on the
brink of bankruptcy if the possible deal with Air France/KLM falls apart.
Australia's former flag carrier, Qantas, is being purchased by a private
equity group and plans a major expansion of service.

For airline employees worldwide, the economic fallout from 9/11 meant
economic sanctions in the form of pay and benefit cuts that left many feeling they'd simply lost their way in an industry to which they'd given their
lives. At one point, experts released a sobering claim that the airline industry had collectively lost more money by 2005 than it had earned in total
since airlines first began to fly. ·

If the ups and downs of this industry have not deterred you yet, read on
because a career-focusing light beam is taking shape at the end of the tunnel, even though it might not always appear clearly to everyone. More
changes are in store for the aviation industry and pilots with career aspira-
tions. The Aerospace Industries Association said, in late December 2006,
that total civil aircraft sales increased by $14 billion, reaching $184.4 billion,
a record level for the third year in a row. The number represented a jump
of 8.4 percent over 2005's final sales total. AIA CEO and President John
Douglass said the outlook for next year remains strong, with sales forecast
to reach $195 billion, an increase of 6 percent. Aerospace as a whole also
logged a $52 billion positive trade balance, while the industry added about
23,000 new jobs, reaching a total of 635,000.

As in other parts of our business world, change has become a part of the
landscape to expect during your entire flying career. The key to turning
what may appear to be the turmoil of the day into a successful job search
focuses on how you react to any given situation. Some people, for example,
might see a box of smashed lemons as a damaged Vitamin C tablet, while
others view the same mess as an opportunity to make lemonade.

Someone asked me recently when the scales of hiring would again put
pilots in the driver's seat. While I would not say pilots are in the driver's
seat, by any stretch of the imagination, the job-hiring atmosphere is on the
upswing. But we all need a little reality check. It was not long ago that
many pilots were receiving multiple job offers. Regional airline pilots
would hire on and leave not long after they'd begun flying the line, leaving
the airline with huge, underfunded annual training bills. On the corporate
side, pilots would often stay only long enough to successfully complete aircraft specific training and pick up a type rating. While employee/employ-
er loyalty is a hot topic, I can certainly say this: loyalty is in the eyes of the
beholder. That can be a dangerous stance for anyone, in any profession. Be

careful of who you step on as you climb the ladder because you're sure to see them again when you fall from the top, as often happens.

This same concept also plays out in the entire flight education arena. Over the years, I have read the same e-mail written again and again. "Flight training is so expensive; shouldn't I just wait until the job market picks up before I start?" The simple answer is No. Wait until the market improves and you'll find the jobs will have evaporated by the time you're ready, that is, if you ever are ready (Figure I-2).

The good news and perhaps the bad news is the aviation industry moves in great cycles. What this means to you as a potential applicant is your chances of finding a cockpit position improve on a daily basis, certainly in proportion to the changes in the industry but, to an even greater degree in relation to your own personal job-networking efforts. One of the important aspects of the aviation profession is the trickle-down effect.

As someone is hired at the major airlines or into a senior business aviation cockpit, they leave their job at a smaller airline or, perhaps, a charter company. If business remains essentially good, those companies need to replace those workers, which results in pilots from smaller organizations being hired and that, in turn, opens up their old jobs. Add expansion into

Figure I-2 An Auburn University student preflights a school Cessna 172.

the equation and positions begin to emerge laterally, as well as vertically, within the industry.

Despite an ever-increasing need for pilots over the next ten years, the competition is always going to be stiff for the best jobs. It's not unusual for an airline to hire as few as 6 pilots for every 100 it interviews. At a place such as Southwest Airlines, the ratio is more like hundreds of interviews for every pilot hired because everyone wants to work there. A good corporate flying position might interview ten people, or more, before they offer someone the job. So, doesn't this kind of thing fly in the face of supply and demand? While a shortage of supply will certainly exist in the coming years, there will almost always be a shortage of qualified pilots. What a company sets as its minimums and the qualifications of the people they bring on board can be vastly different. These are facts you should track during your networking efforts.

If you're reading this volume for the first time, welcome. I believe you'll find data and stories here to help you make better career decisions along the way, simply by being better informed. You need to know the tools available to you that will enhance your job search efforts and the process of getting that nod to come aboard. We discuss many of them here. This information ranges from an understanding of how to write and deliver the letter that dramatizes your initial interest in a company to getting your foot in the door for that first big interview, how to prepare once you are invited to compete, and how to better understand not only the training process when you get the job, but also what you might see when you begin flying for money (Figure I-3).

One of the most important aspects of finding that first job, or the second, or the tenth is not to lose sight of how much work it took to climb that next rung on the career ladder. I guarantee you'll meet pilots along the way who are also on the way up, but who might not have made it quite as far as you . . . yet. Give them a hand up. This book and this industry should be about sharing the information where we can to help a fellow aviator.

At times during the job search, you'll feel like giving up. You might even think a well-paying office job is precisely where you *should* be. I did for a while in a beautiful office just off Michigan Avenue in Chicago—hustle, bustle, and even great shopping, according to my wife. The problem was that a nice cushy office job was not where I *wanted* to be. Take some advice from a guy who fell into the trappings of an office, but who escaped to fly again. The money and the benefits may look appealing at first, but a few months of sitting in a chair, staring out the window at the sky, and you'll be kicking yourself for having sold out.

You might learn that the coveted path to your cockpit career is not be quite as direct as you first expected. Don't be deterred. Occasionally, those diversions may not be as far off the path as you think. A move out of state might not be what you'd planned at the beginning, but it might evolve into

Figure I-3 Many airlines will soon fly this Boeing 787. (Courtesy The Boeing Company.)

the best career direction when the opportunity presents itself. That could mean modifying your goals from time to time. Of course, this also assumes you have specific goals. Few pilots do have specific goals, believe it or not.

This book is a guide, much like sitting down over a cup of coffee with a friend who has taken a significant personal interest in your career as a professional pilot. We called this book a "guide" because it represents a significant extension of the personal philosophy I've come to believe throughout my career: people should help other people climb the career ladder. This book is an index of resources. Some are ordinary people, others are industry experts, and others are storehouses of information. The list is also constantly evolving and represents an asset to be used by all.

In this text, you can find a number of relevant articles of mine, some originally penned for other publications, as well as interviews and remembrances from other pilots within the industry who share insights into the hiring game. What also makes this text quite different from competing volumes is we look at more than the airlines. As I mentioned earlier, airline hiring leaves a considerable void in other areas of the industry. But hiring on with a corporate, fractional, or charter company is different from hiring on with an airline. It's not necessarily more or less difficult. It's just different.

Before you think about a new flying position, I'll try and walk you though some of the decision processes and the soul searching you need to determine which direction within this industry best fits your lifestyle requirements and your attitude.

Here, my goal is to teach you to think differently than you did when you first purchased this book. One of the true marvels of this world is the ability for all of us to converse, no matter what part of the world we live in. For the *Professional Pilot's Career Guide*, this means using some of the newest technologies to engage in that two-way dialogue we all find so valuable. I included some articles originally published in other magazines. Some are about flying techniques and a few are pilot reports for some of the airplanes you might encounter in the next few years. If you have a question about something you read here, try these options:

- E-mail the author at rob@propilotbook.com

- Visit the web site at www.propilotbook.com

Want to read about or participate in discussions of some of the most formidable news issues in the aviation industry? Stop by our aviation industry blog at www.jetwhine.com.

A wise aviator, Bill Traub, a man who spent many years hiring other pilots for one of the top airlines in the world, offered some advice in an earlier edition of this book. Traub's advice still holds true today, especially during a time some might call chaotic. "This book could be the starting point for a very exciting, professionally rewarding, and challenging career," Traub said.

"Becoming a professional pilot takes a lot of work, but the rewards—both personal and financial—are well worth the investment. Our astronauts talked about having the 'right stuff.' But what is the right stuff to become a professional pilot? I've spent 17 years heavily involved in the hiring process for a major airline. And, to me, what makes a professional pilot can be summed up in one word—attitude.

"Before you launch your career as a professional pilot, assess your attitude toward work and life in general. Are you committed to excellence in everything you do? If you can honestly say you strive to be the best—in everything from your academic preparation to your performance in the cockpit—and that you take every opportunity to learn and improve your skills, then in my opinion, you're on the right track.

"Too often, people measure excellence in terms of flight hours. I don't believe that's an accurate gauge. I've hired pilots with 350 hours and pilots with 20,000 hours. I believe these pilots can perform equally well if they first have the proper attitude.

"After 35 years of professional flying, I still personally critique every flight I make, whether I'm flying a Boeing 747 or a single-engine aircraft. I

Figure I-4 Boeing 777. Is one in your future?

write down everything I can think of to improve the flight next time and I review those comments from time to time. I'm still waiting to see a perfect flight.

"Always strive to be the best pilot and good luck. There's a great opportunity out there for those who are willing to make the commitment. William Traub, Captain B-727, B-737, B-747, DC-6, DC-7, DC-8."

Finally I'd like to offer this stipulation. Unless otherwise stated, everything contained within this book represents my opinion alone. I encourage you not to accept these opinions as the only gospel possible to reach your goal as a professional pilot, but more as a series of suggestions to consider.

Allow the information you learn here to help shape the way you approach your search for that perfect flying job. And, when you find that perfect flying job or learn a technique that might be even more effective than something we've suggested here, share it with the rest of us. It's simply the way things are done (Figure I-4).

Good luck.

Robert P. Mark

1

Your Career Begins Here

A cartoon-like poster hangs on the south-facing wall of my Evanston, Illinois, office in a place I pass each and every morning. The caricature is a tall, animated fellow in a flying suit, wearing a traditional leather cap and goggles, and piloting a machine that, like a bumblebee, should not be capable of flying at all. The caption is a simple one: "Flying is the second greatest experience known to man . . . landing is the first." Looking back on that, I'd say, it depends. To me, flying airplanes and being paid for the experience is certainly one of the greatest experiences of my life (Figure 1-1).

As a naïve kid of 17 I never realized that little in my life to that point would prepare me for the experience of a job search, especially in aviation. Flying was, after all, not the same as a real job, I thought at the time. Many of my friends in other careers had spent weeks and months after college organizing themselves to be ready when that first big job interview came along. Flying was not like that, I thought. Flying airplanes meant challenging nature, overcoming fears, winning accolades from passengers, and walking around trying to decide where to put those big fat paychecks. Flying wasn't work.

OK, so perhaps I was a little green about what it might take to become a professional pilot. I know I'm not the only one, though, who sat around an airport watching airplanes with my ear pressed closely to a portable airband radio, trying to figure it all out. More than a few of you, I bet, still look up every time something flies over. And why do you look? Because you must! You may not even clearly understand the connection between airplanes and that desire to be a part of it all. But you know it exists. And that's a good thing.

What you also learn in your flying career is this: many people are flying airplanes who are not quite as enamored with flying as you are. It's not to say they don't enjoy banking around storm cells at 39,000 feet, or that they don't feel a sense of enormous satisfaction when they break out of a 200-foot overcast at night and find the runway exactly where they knew it would be. To some aviators, flying airplanes for a living is more of a job. They're good

Figure 1-1 Flying is the greatest of experiences!

at it, and they enjoy it, and they are often well paid for their efforts. They just don't necessarily come home and fly small airplanes in their spare time, or care about whether their kids follow in their footsteps as a pilot. They are more, well, pragmatic about the entire flying experience.

Years ago, I believed these people couldn't possibly become real pilots. They simply did not have the love, the passion, that total immersion in everything airplane-like that was a daily part of my life. No way could they possibly become a better, more successful pilot than I am. Guess what? I was wrong. Many thousands of astounding pilots around the world do not live, sleep, and breathe airplanes, and they are truly *almost* as good as the rest of us. What I've come to better understand from those more practical pilots over the years was a lesson I wish I'd known much earlier on.

Passion and enthusiasm for any profession are important, critical actually. But I don't believe emotions are enough to see you through to your goal. No matter how much you love airplanes, or flying, or the objective of flying and being paid, without some kind of plan that lays out the necessary steps to translate your dreams into reality, only a slim chance of success exists. I hope the reason you found a book like this of interest is because you inherently realize that someone looking over your shoulder a bit as you design and execute your plans is a good idea.

Allow me to offer up what will probably become the most important element in your search for the best flying job available . . . great information. It's true. Only one way exists to locate a job in aviation—networking—and that's impossible without good information. Whether you're exploring the outer reaches of the Internet, wandering around an airport with a fistful of resumes, or trying to absorb a bucket of information from an online or print publication, it's all networking. Today, the wealth of information online may make this all seem easy, but too much information can be almost as much trouble as too little. Almost as critical to good decision making as the information itself is verifying that the data you've collected is both timely and correct.

Word of Mouse—Networking and Technology

If networking is the preeminent method of finding a flying job, then a computer, a cell phone, and an intense curiosity about employers and their needs are the primary tools for success. Understanding the intricacies of the Web has, since the previous edition of this book was released, become the primary means of information gathering. But the search needs to be integrated with a sense of marketing that ensures that you present yourself as the answer to a potential employer's search for the right pilot.

Early on in my career, many of us were overjoyed at the thought of flying for a living because it meant we'd avoid the need to spend our lives selling things. Little did we know what some of those more pragmatic pilots already understood. As pilots in an ever-changing world, we'd always be selling ourselves—our personalities, our talents, and our abilities—to accomplish the job at hand, whoever our employer might be. The ability to analyze, and later synthesize, what you've learned into a specifically targeted resume and cover letter or e-mail is crucial. Technology makes the job-hunting work a little less tedious these days, but it's no substitute for smart thinking.

During most of the interviews I gathered for this book, I noticed that pilots seem to easily fit into one of two categories related to their perspectives on life and career. The first is those who would be happy with any job they could land, and the second is those who are completely focused on flying as a career. How these people seemed to gather the information they need also split the groups again: those who were rather sketchy and those who never seem to end their search.

My suggestion is to lean toward the side of having more information, rather than less. Weeding out the useless material is easier than you might think. And, today, the wealth of available information services specific to aviation employment is almost impossible to quantify. FLTops.com (Figure 1-2) and Aviation Information Resources (AIR Inc.) are two of the largest players with smaller firms, such as Avcrew.com, the job board of the National Business Aviation Association, or Flightcrews.com, serving other more niche markets. Most of these information sources are not free—and you shouldn't expect them to be. Some airlines and corporate operators have taken advantage of new technologies as well and only accept applications online. Take a look at airlineapps.com to get a better idea of this technique.

If that first, second, or fifteenth flying job is only to be found by combining solid networking with relevant information, I'd be remiss not to mention how differently aviation fits into the world now than it did in the latter part of the twentieth century. Many middle class Americans viewed the dot.com stock market failures as the end of hope, as the end to their dreams of success. And, for many, if they bet everything they had on the market as their best means to success, that may have been true.

And just as people in the U.S. were getting a handle on the market crisis in early 2001, the first terrorist attacks of the twenty-first century came in the United States, in Spain, Africa, England, the Middle East, and Asia. These attacks added a dimension to flying, travel, and life in general, which is often spoken of only peripherally by any of the aviation information services.

Prior to 9/11, the aviation industry was principally concerned with hijackers taking control of an airplane and releasing hostages once their

from first flight to last flight

Pilot Hiring Bulletin

Wednesday, August 09, 2006

Below is a listing of last month's pilot hiring activity. Visit the Pilot Hiring section at
FLTops.com for more details.

ABX Air
ABX Air has 17 pilots and 4 PFEs on furlough status.

Air Cargo Carriers
Air Cargo Carriers is currently interviewing and accepting resumes.

Air Group, Inc.
Air Group, Inc., is a corporate flight department recruiting G-IV SICs and PICs for Van
Nuys, CA and Teterboro, NJ bases.

Air Midwest
Air Midwest is accepting resumes.

Figure 1-2 Typical FLTops.com hiring bulletin.

demands were met. No one expected anarchists to use airplanes as
weapons. But they did. And, following the attacks here in the United States,
the world began to reassess aviation security and how it might affect airline, business aviation, charter, and personal aviation operations.

While commentary about new security systems and bureaucracies that
began around the world in the 21st century is beyond the scope of this
book, it is relevant to say that security has become an issue for job hunting
because it has now a significant operational element of the aviation industry. More on this later.

The Aviation Industry Today

With the terror attacks in the U.S. following so closely behind the market
crash in America, the aviation industry saw steep declines, not only in the
numbers of airline passengers purchasing tickets, but also in the overall
health of most of the traditional, called "legacy," airlines around the world.
One airline analyst I heard speak in early 2002 expected all of the top six
airlines to file for bankruptcy protection. Those predictions proved to be
pretty much correct, with the exception of American Airlines. As those carriers filed bankruptcy, they took many thousands of pilot jobs with them.
By early 2005, nearly 9,000 airline pilots alone were on furlough. Airline

bankruptcies also occurred in other parts of the world. Fifty thousand aviation jobs have been lost since 2000.

Despite the chaos within the airline industry, some bright spots were also beginning to emerge in the marketplace. If the bedlam of the early twenty-first century has confirmed anything about the aviation industry, it's this: like other sectors of the worldwide economy, this business experiences regular high- and low-expansion cycles. What made the current economic expansion and contraction different from some past episodes was the reaction to the airline's woes by other parts of the industry.

As many of the legacy carriers imploded, the regional airline systems around the world experienced unbelievable expansions as their major partners searched for cost-cutting ideas. In some portions of the U.S., regionals grew in the double-digit range for two to three years. This meant many routes once served by at least narrow-body airplanes such as an MD80 or an Airbus A320 were now seeing service with regional jets (RJs) that offered jet service, but reduced cabin comfort.

Major airlines also came to the realization that the economic model they'd used for decades was crumbling in front of their eyes. They'd always priced tickets so that those who had the greatest need, typically, the last-minute business traveler, paid the highest fare. In the past, travelers of all shapes and sizes had few alternatives, and they paid the going rate, whatever the ticket price because there was no alternative.

Then, in the 1990s, and early into the twenty-first century, a new, financially stable breed of airlines emerged, the low-cost carriers (LCCs). One well-known airline, Southwest Airlines, continued years of solid growth. At press time, Southwest has experienced nearly 26 years of profitable quarters, which made it clear its model beat the alternatives hands-down. Travelers disenchanted began avoiding the large airlines, the companies who rewarded their best customers with the highest prices, and, often, some of the toughest travel restrictions. In the late 1990s, LCCs accounted for less than 10 percent of the marketplace. By late 2006, that number had climbed to 25 percent, a number that made the major carriers woozy.

With the growth of the LCCs came more jobs, as did the expansion of the regional airlines. This time around, business aviation also saw their aircraft deliveries rise as airline-weary executives decided it was time to better control their own destinies. Another reason for a significant uptick in business aviation deliveries was this: more and more companies were opting out of the airline market due to the delays and security concerns the post-9/11 U.S. Transportation Security Administration has instituted. During the early twenty-first century, another sector of business aviation, fractionals, also benefited significantly from the need for business airplanes.

Fractionals, or more precisely, fractional ownership programs, offer businesses an opportunity to own a portion of a business jet or turboprop without taking the plunge of opening and managing a complete flight department. That translates into the chance to travel in a business airplane for more than the price of a charter, but much less than full ownership. Finally, as the airline sector contracted, traditional business aircraft charters also experienced significant growth, all of which translates into an ever-increasing need for more pilots. In fact, many pilots have looked at the upheaval in the airline industry and turned their attention to business aviation as their final destination. They never looked back because of the significant pay and benefits available.

This is not as grim as it might appear on the surface. The answers to what is happening to the industry are out there. You just need to decide whether you'll search them out yourself, subscribe to a job information service and hope you'll learn everything, or decide you need some combination of all these things (hint: the last suggestion is probably the best).

The key issue now is to realize that each time the industry works its way through one of these hi-lo cycles, it emerges in a different form. Whether or not the new version is bad or good truly depends on the eyes of the person conducting the evaluation. Most people don't particularly enjoy too much change in their lives. But change is simply the way of the world. This is not the first major transformation of the aviation business and it won't be the last. The key is to position yourself not only with the experience and ratings for a flying position, but also to build an educational portfolio to see you through the good and the bad times, which are bound to occur. This also means establishing flexible-enough goals to bend with the times, which means simply finding a well-paying pilot position is not a specific goal if you intend to enjoy your career.

Here's a look at how the industry is shaping up as we go to press. US Airways recently tendered an offer to purchase Delta Airlines in the United States which the Atlanta carrier ultimately rejected, while rumors abound about which airline United Airlines might tap. The European Union decided against the purchase of Aer Lingus by Ryanair. One industry expert there said 90 low-cost airlines were operating in Europe by the end of 2006. All need pilots.

A number of major aviation-related organizations also released their ten-year forecasts in late 2006. Honeywell predicts the delivery of 12,000 new aircraft by 2016, worth nearly $200 billion. Fractional fleets are still growing here in the United States, as well as in Europe. In December 2006, Cessna delivered the first of its Citation Mustangs (Figure 1-3), part of an emerging new arm of Very Light Jets (VLJs) capable of carrying two to three passengers for a thousand miles in jet comfort, while operating from shorter, nonhub airports.

Another major player in the VLJ market, *Eclipse* is expected to soon begin deliveries of its model 500 and claims orders for over 2,000 of the air-

Figure 1-3 Cessna's new Citation Mustang. (Courtesy Cessna Aircraft Company.)

craft. All the VLJs are certified as single-pilot machines, but most are expected to be operated with two pilots on board. As if this new VLJ segment, where none like it existed before, were not significant enough, smart business people have also developed the algorithms to make a true air-taxi fleet operate between nonhub airports. One of the major players in this realm is Day Jet of Delray Beach, Florida.

Embraer, Boeing, and Airbus—together representing the vast majority of large, commercial airframe manufacturing in the world—also released industry-delivery forecasts. Before we dive into any more numbers, a word of caution. These are forecasts produced by companies that often have a vested interest in building a tidal wave–like interest for their products. That doesn't mean these numbers are not accurate. They could well turn out to be conservative. But some international issue in the world might also send them plummeting.

Here's a look at the industry through Embraer's crystal ball. The Brazilian manufacturer expects to deliver approximately 370 aircraft through the end of 2008, including 15 to 20 of its Phenom 100 Very Light Jets. Embraer believes the industry overall will demand just over 11,000 business jets over the next ten years, with the blossoming air-taxi market requiring another 2,500 to 3,000 VLJs during that time period. Embraer also builds airliners that seat up to 120 people and sees a need for 7,500 new aircraft in the 30 to 120 seat category by 2026. The deliveries of airline category aircraft show just over half are headed for North America, 17 percent to Europe, 9 percent to China, and the rest spread around the globe.

The General Aviation Manufacturers Association (GAMA), the trade association for companies such as Cessna, Raytheon, Piper, Gulfstream, Eclipse, and a host of other smaller airframe builders, also shared some dramatic results in late 2006. Pete Bunce, GAMA's president and CEO, said in a news release, "Shipments of general aviation (GA) airplanes for the first nine months of this year totaled 2,842 units, an increase of 18.9 percent over the same period last year, while industry-wide billings were $13.2 billion, up 28.6 percent. Piston-engine aircraft shipments totaled 1,957 units compared to 1,653 units delivered in 2005, a 18.4 percent increase. Turboprop shipments increased from 228 units last year to 256 units this year. Business jet shipments were 629, a 23.3 percent increase over the 510 units delivered through the first nine months in 2005. Our manufacturers have seen growth in all airplane segments, part of which we attribute to strengthening sales in Europe, and into Russia, China, and India . . . we expect this trend to continue."

Boeing and Airbus, the other two large aircraft manufacturers in the world, also believe the market shows significant promise over the next 20 years. In a November 24, 2006, article in the *Wall Street Journal (WSJ)*, Airbus believes the world need for airliners will number 22,700 aircraft by 2026, while Boeing believes the number will run closer to 27,200 machines. Adding to these new numbers are the airplanes expected to remain in service worldwide and the numbers continue to climb. Airbus believes the world fleet will total about 27,000 aircraft, while Boeing believes the numbers will run closer to 36,000 airframes. The big differences highlighted in the *WSJ* were how each manufacturer breaks down the numbers. Airbus sees the need for more wide bodies, such as its A380, while Boeing believes the market is headed for more narrow- and medium-body machines, such as the Boeing 787.

All this data represents possibilities—better than you might have first thought when you began reading this section. Whether you see this as strong evidence of a positive upswing in the industry or yet another moment of industry uncertainty surely depends on your education and your perspective on life. And that's the point. You cannot totally rely on one source—this book, a web site, or the opinion of one expert to shape your future. You must carry out some of the ground school, so to speak, and the analysis yourself. We offer you plenty of resources to help along the way.

A necessary element to this discussion is also how much different the scope of our audience has become. Once, men and women in search of flying jobs almost naturally assumed the United States was the place they'd want to train, as well as the location they'd find work once they were ready. No longer. A story that also ran in the *WSJ* recently spoke to how the flavor of the pilot job hunt had changed over the past few years. Contract firms, such as PARC in Ireland or FlightCrews on the West Coast of the United States, have always hired contract pilots to crew transport category aircraft.

But today, we are seeing more pilots leave the environment of the U.S. for crewmember jobs in other countries. And, since 9/11, the necessary immigration paper workload to enter the United States, as either a pilot or a student who wants to learn to fly, has changed as well. Luckily, for people thinking of training in the U.S., the immigration work has eased, although it is still not as easy as it once was.

More than 85 percent of the pilots for Dubai-based Emirates Airlines (Figure 1-4) are expatriates. What that should mean to you is this: you'll need to be thinking that, someday, your choice of companies could well include one based outside the United States. Pilot citizens from other countries have been doing this for years. With a global economy upon us, it's now our turn.

Louis Smith is president of FLTops.com, one of the major airline pilot-employment information services. Smith also just retired as a DC-10 captain for a major carrier. I asked him to think about this industry over the past ten years and to make a few predictions. Here's a first chance for that second opinion.

Figure 1-4 Emirates Boeing 777. (Courtesy Nik French: www.jetphotos.net.)

"In 1996, the major airlines were in the third year of an eight-year expansion," Smith recalled when I interviewed him in late 2006. "The major airlines hired nearly 2,500 that year. All the pilots furloughed during the early '90s were back to work, except for US Airways. The pilot employment application process was still a paperwork ordeal and the Internet wasn't an integral part of the communication and data structure. Very few pilots left the U.S. to find flying jobs.

"After 9/11 the industry didn't just cycle down as in normal recessions, it became an avalanche resulting in over 10,000 passenger airline pilots losing their jobs. And it wasn't just the job losses. The passenger airlines were in desperate straits and an an editorial in *Aviation Week & Space Technology* during that time correctly surmised. 'Restructure or die' became the name of the game.

"Although the percentage of furloughs was not as great as the early 1980s recession, the length of the furloughs was much longer and was exacerbated by major airlines developing code sharing agreements with regional jet partners flying smaller aircraft. So, the pilot furloughed from the major airline might have found a job flying a much smaller jet, but with greatly diminished pay and benefits. In 2002, the major airlines hired only 508 pilots, the worst year for more than 40 years as a percentage. It's interesting to note, however, that the air freight and business/fractional jet operators were continuing to grow and make money, recruiting pilots as a result.

"In 2006, the major airlines will hire nearly 3,000 new pilots, but none of those jobs will be at the three largest airlines: American, Delta, and United. We think Delta will need to hire in the summer of 2007, but it may be a couple of years before United hires off the street. American Airlines might require more than two years to recall all its furloughed pilots, many of whom were ex-TWA pilots.

"Over the next 12 years, we expect the U.S. pilot job market will need an additional 120,000 pilots coming from outside the system. When we say outside the system, we mean the military and the aviation schools and universities. The actual amount of hiring will be much greater than 120,000 because of turnover within the system, e.g., pilots leaving Pinnacle for Continental, etc. The true demand will be 120,000 and the business/fractional jet sector will account for more than 40 percent of that."

Who Do You Know?

In the airline world, as well as in the corporate or charter marketplace, the story is much the same as you've most likely been hearing throughout your life. Getting hired is not as much a function of what you know, as much as who you know. The reason is simple: hiring new employees is a time-consuming, expensive task for any company.

People being interviewed want the job being offered, but for a variety of reasons. Some want a career, some want to pay the rent and put food on the table, and still others want a place to hang out until something more lucrative appears. It's the employers' job to decipher the information they receive as answers to the questions they ask and the tests they administer and synthesize everything into a decision—Is this the person I want to hire or not? Will this person work as well with other pilots and ground personnel as I hope? Will this pilot stay around only long enough to pick up a type rating, and then head for greener pastures?

Most employers have their work cut out for them, but so do pilot applicants. One of the most important questions you'll face in your job search is how to deliver the right information to an employer during an interview. That means understanding not only the marketplace, but also every fact you can possibly glean about the company you're hiring on with and giving the interviewer what they want. Sounds a bit mercenary, but it relates to that necessary aspect of marketing.

There are two important elements to networking and how it integrates into your job hunt. The first is understanding how you might fit with the company if you win the position. The second is understanding that the potential employer is also trying to hire you. So, you want a job and they want to hire a pilot. But they want to hire the right one. Anything you can do—from a personal visit to a phone call to dropping the name of someone the chief pilot knows—to get your foot in the door can reduce that potential employer's anxiety about hiring YOU!

That's the name of the game.

Greg Brown, pilot and aviation author, offered his perspective on the reasoning behind most networking efforts. "It all boils down to personal credibility. As an aircraft operator, you want the best and sharpest people working with you. That's why networking is so important. The more people who know and respect you in the industry, the faster you'll move up the career ladder. The other reason networking is so important is because a job may not be available today. Your mission is to be remembered when they do have an opening and need to fill it in a hurry. A specific amount of your time should be dedicated to getting to know people, not simply logging flight hours."

Throughout the remainder of this book, we talk about web sites, and information resources and techniques you can use to make you and your résumé stand out to a potential employer. We also look at how to learn more about the company you're trying to hire on with by talking directly to people who work there now or—sometimes even more valuable—talking to someone who recently left a company.

What if you don't know someone at the company you want to work for? Prior to 9/11, a pilot applicant could simply hang out at the local airline terminal and try talking to pilots. Those days are gone, unfortunately. Now, the work is a little harder, but it's not impossible. Pilots hang out at message boards, web sites, and web logs (blogs). Some of the free sites, such as PPRuNe.net, have been around for ages and are actually free. Often, simply listening in on the conversations through the message board can deliver a considerable amount of valuable information. If you're unsure of how to proceed on any of the boards, try beginning with a private e-mail to someone posting on a site. Tell them the truth. You're a new pilot in search of information . . . they call it *gouge*. We've all been in the job search mode, so most, but not all, pilots will be pretty good about sharing information once you meet them, even online.

Also important is the realization that the true definition of networking implies an interactive component. As you'll hear a number of my comrades say throughout this book, you cannot simply take—you must return something to the system—to those beneath you on the way up and as feedback to those above—to preserve the industry information flow.

A Note About Immigration

No discussion of the flight training necessary to become a pilot would be complete without a look at the changes to that aspect industry since 9/11. Not only did a considerable number of flight schools shutter their doors after the grounding of flight training airplanes here in the states following the terrorist attacks, but the U.S. government also ordered tough new rules that made it more difficult for any foreign national to gain access to the country for many types of education, most importantly, flight training.

Despite the fact that foreign-national students have continued to attend American universities since 2001 to study a variety of academic curriculums, starting flight training is still cumbersome at times for non-U.S. citizens. But, it is not impossible by any means. Most large flight schools will employ someone who specializes in helping students work their way through the American system of government and gain approval for a flight training admission. Many flight-training industry business owners do not agree with the rules the U.S. government has set down for new foreign-national students. Unfortunately, as with any country, these are the current policies, and we all must make the best effort we can to muddle through.

General details for exchange students are available on the U.S. State Department web site at http://exchanges.state.gov/education/ jexchanges/private/trainee.htm. Specifically, the State Department adds these details about flight training. The U.S. Transportation Security

Administration has also set up a page to specifically answer questions for flight training students who are not U.S. citizens. That address is https://www.flightschoolcandidates.gov:443/.

In addition to complying with the regulations for training programs, sponsors applying for designation of a flight training program must also meet the following requirements.

The program must be a Federal Aviation Administration–certified pilot school according to Title 14, Code of Federal Regulations, Part 141.

The program must be accredited as a flight training program by an accrediting agency listed in the current edition of "Nationally Recognized Accrediting Agencies and Associations" (U.S. Department of Education), or accredited as a flight training program by a member of the Council on Post-Secondary Accreditation. If the sponsor is not accredited at the time of application, it must have formally commenced the accreditation process. Programs pending accreditation may be designated for 12 months; continued designation is predicated on the successful completion of the accreditation process.

The maximum program duration for Airline Transport Pilot (ATP) certification is 24 months. The maximum program duration of all other flight training programs is 18 months total. Requests for extension must be submitted to the Department of State for approval.

To meet the evaluation requirement for training programs, sponsors and/or third parties conducting the flight training may use the same training records required by the Federal Aviation Administration to be maintained pursuant to 14 CFR 141.101.

The Possibilities

I can't imagine a better profession than flying 35,000 feet above the Earth and looking out from a vantage point that few others have, finding yourself in one of the greatest theaters in the world, as you watch thousands of miles of wondrously changing clouds and landscape pass beneath. It's a treat that's tough to surpass. In addition to enjoying the beauty and majesty of the world beneath you, I've spent the better part of my life making money at some form of flying . . . and so can you.

Where do you see yourself next year or, perhaps, in five years? Can you imagine flying airplanes for a living? Thousands of people can see themselves involved in the profession, but only a few will become pilots. Why? Because many people still believe learning to fly is complicated. Certainly transforming yourself from a fledgling student into a commercial and, eventually, into an ATP involves a considerable amount of work, considerable sacrifice, and not a small sum of cash. By taking advantage of a wealth of resources—both written and in the form of personal relationships—and by developing a solid plan, you can do it. The work is worth the reward. And, in case you're thinking you've already read some of these concepts before, like a belief in goals, you have. And, it won't be the last time either.

A funny story helps illustrate this. There was an old preacher somewhere in the South, so the story goes. Each Sunday, he stood in front of his congregation and preached the same sermon. Over and over, he spoke about values and the need to focus to be a successful member of the world community. One day, a small boy of 12 walked up to him outside the church. "Sir," the young man asked quizzically, "Do you realize you gave the same sermon today that you did last week and the week before?" "I do," the preacher replied with a smile. "And I'm going to keep on preaching the same thing over and over until every last person in the place finally gets the point."

Flying airplanes for a living is a proud profession that dates back nearly a hundred years to an era when flying meant a serious risk to a pilot's life and limb, not to mention the passengers' lives. Although pilots were always viewed as somewhat crazy, they were also looked on as supermen and superwomen of sorts: brave individuals, adventurers who defied the odds and cheated death every time they took to the skies.

In the days of the barnstormers, aircraft engines were simple but unreliable, making engine failures commonplace. The fuselage and wings of the first commercial aircraft were nothing more than pieces of exquisitely cut spruce, nailed and glued together to form the frame, and, later, covered with fabric and many layers of paint for strength.

The first commercial aircraft carried names like Ryan, Douglas, Curtiss, and Boeing (Figures 1-5 and 1-6). They ranged from simple, single-engine aircraft, often carrying only a single pilot and a few pouches of mail, to more sophisticated machines that would prove capable of flying six to eight passengers with a little room left over for baggage. As aircraft grew more sophisticated, the title of commercial pilot began to take on the meaning of someone with, if not yet a respectable job, at least a regular salary. Notice, though, no one said anything about longevity because, in the '20s and '30s, being a pilot was still considered a risky living. Over the years, aircraft reliability improved as did the size of aircraft and their capabilities.

Even today, walk into a social situation and people's faces turn impressively toward the man or woman who tells the crowd they haul people around the sky in shiny silver airplanes. If you haven't seen that, I can only think of two reasons. One, you don't hang out with the right people yet. And, two, you haven't been watching the faces of people at parties very well. You might think I'm bringing up some rather egocentric topics here, and I must admit I am, but with good reason. Pilots tend to have outgoing, assertive personalities. This is a large part of what makes a pilot good at

Figure 1-5 Billy Mitchell helped build the military side of aviation.

Figure 1-6 Oshkosh B' Gosh early corporate aircraft (1920s).

their job, but it's certainly not a requirement. A flying career can bring years of pleasure, job satisfaction, adventure, and even a few financial rewards. If you're looking for a profession to turn people's heads, as well as your own, I might add, there's nothing better than flying.

What Is a Professional Pilot?

Many people still believe a professional pilot is an airline pilot, but there are many other categories of professional aviators. Corporate pilots shuttle the chief executive officers of Fortune 500 companies, as well as many smaller ones. All over the world, there are aircraft loaded to the gills with boxes and mail destined for every other point on Earth . . . and someone needs to fly those airplanes. Freighters run the gamut from Boeing 747s and Airbus aircraft flown by cargo carriers, such as Federal Express (FedEx) and United Parcel Service (UPS), to older commercial jets, such as the DC-8s you'll find flying their tails off for smaller freight carriers like DHL.

At the local general-aviation airport, you'll see another group of professional pilots—the flight instructors. Teaching is a noble profession in which these men and women turn the sometimes confusing chapters of your journey from private-to-professional pilot into knowledge designed to make you the safest aviator possible. Ever passed by a farmer's field and watched a small aircraft or helicopter swoop low over a soybean field and spout a cloud of mist from behind? These agriculture pilots earn their living every day saving fields of produce from destruction by pests and disease. If you want to talk about thrills, imagine flying 20 feet off the ground at 100 mph and making some wild-looking gyrations at the opposite ends of the field to maneuver the airplane back for the next pass!

There are also air ambulance pilots. They're on call 24 hours a day in case an emergency requires them to transport a severely ill or injured patient. In this kind of flying, the hours are never the same, and the destinations are seldom repeated. Ever watched banner-towing pilots fly low over a field to grab a long sign, trail it behind the airplane, and display an advertising message to thousands of people below? You don't cover much territory, usually, but the flying is a real challenge. I think I have a story or two later in the book about towing.

Still hungry for professional pilot jobs? How about a forest-fire tanker pilot? Obviously, they don't operate in as many locations as other forms of flight operations, but tanker flying can be a real adventure. For Hollywood's look at this side of the profession, try renting the DVD *Always*, starring Richard Dreyfuss and Holly Hunter.

Consider the heroes of the Part 135 Air Taxi, known as on-demand charter pilots. They could be called out to fly a quick load of freight from a parts plant in central Illinois to an auto maker in Detroit to protect an assembly line from a needless shutdown. On-demand could involve last-minute plans by a dignitary or the local appearance of a rock-and-roll group. Not long ago, I saw a crew load up a Learjet for a trip carrying only a single shoebox-size crate. It contained a human heart headed for an anxious recipient 900 miles away. And don't forget the pilots who report highway traffic from inside a cockpit each day.

Pilots are needed to fly corporate helicopters from heliports, as well as from the local airport. *Helicopters* are special-use flying machines that can be stopped virtually on a dime and descend vertically into a tight space to accomplish their mission. Besides their use as a corporate short-range shuttle, helicopters, by virtue of their agility, are one of the prime pieces of transportation used by Emergency Medical Services (EMS) for roadside evacuation of accident victims. In some locations, helicopters often compete with aircraft for aerial spraying and traffic-reporting jobs.

A Partial List of Pilot Jobs

- Airline pilot—major, national, regional
- Corporate pilot
- Freight pilot
- Fractional ownership pilot
- Charter pilot
- Certified flight instructor
- Traffic patrol pilot
- Military pilot
- Forestry pilot
- U.S. Customs Service pilot
- FBI pilot
- Contract pilot
- VLJ pilot
- Pipeline patrol pilot
- Banner tow pilot
- Law enforcement pilot
- Ferry pilot

- Package forwarder pilot
- Sight-seeing pilot
- FAA Airspace check pilot

I don't believe any of these terms define a professional pilot, however. They simply define a job. Being professional is a philosophy, a way of life, and it begins with the person, not the job, said Bill Traub, United's former vice president of flight standards earlier in the Introduction.

If you measure your self-worth and position in the world by the size of the aircraft you fly or the numbers on your paycheck, you're in the wrong career. I've known flight instructors who flew around in four-place Piper trainers with a shirt and tie who were more professional than some airline pilots who had become truly jaded by their aircraft type and their six-figure income. Make no mistake, a six-figure paycheck is something we'd all like. But, if all you pick up from flying is a paycheck every two weeks, you're missing some of the best this career and life itself have to offer. Fly because you love it. Fly because you can think of no greater work in the world that can possibly take you to the places a cockpit career might. But don't be surprised if, after you've earned all your ratings, companies don't fall all over themselves lining up to hire you. The best positions are about you *and* your qualifications.

Attitude: How's Yours?

If you're going to be successful at becoming a professional pilot, you need an attitude … the right one. You must mentally prepare yourself for the challenge— not only of learning to fly, but also for the even greater contest of finding the right job after you've learned to fly and gained the experience necessary for attaining your goal.

What's the attitude, the right stuff, so to speak, for getting ahead in aviation? Rest assured, it's about more than simply thinking positive. Consider for a moment; don't you know people who are very bright, people who know just what they should do to become successful, and yet never really succeed? No matter how bright they seem on the outside, they never seem to get the right break, or be in the right place at the right time, or sell themselves in the best way possible when the time is right. Positive thinking alone isn't going to make you a professional pilot. I believe the reason most people don't succeed at the flying game is not because they don't want to succeed, but because they're afraid of failure.

Fear of failure probably squelches more careers than just about anything else. You must believe you can accomplish your goal, but even more than that, you need a plan that includes a road map toward that goal, a path designed to help you leap past those points where the rejections arrive.

One young woman I interviewed for a story a few months back told me she had been asked to interview three different times by United Airlines before she was finally hired in 1992. Many pilots might have given up after being rejected once or wouldn't bother to interview at all because they're afraid of not being accepted. You can't fail if you don't apply . . . some say. Others believe they were rejected because the people at the airline or company specifically didn't like them. Don't give up; keep trying, no matter what. This is the only attitude for success.

Sure, there'll be times when you might not give a good interview, but ask yourself, and answer yourself truthfully. Were you really prepared? Did you possibly set yourself up a bit for failure? The woman we just mentioned who was hired at United; how did she cope with so much rejection from such a large organization? She knew her goal was to fly for United, and she wouldn't let anything turn her from that path . . . not anything. If they rejected her, she didn't take it personally. She took a deep breath and asked herself which way to turn around this obstacle in her path.

And, as long as I'm dumping reality on you, here's a bit more to chew on. I have been through many, many interviews in which the interviewer said, "If only you had a type rating in this airplane," or "Sorry, but we need someone with more experience in this machine." Don't take those comments personally, even though that's the way they are intended. What they really mean is "We don't think you'd be a good fit with our company." Do yourself a favor. Don't agonize over it. Just walk away. In the end, no matter what you think at that moment, there will be other days and other times. And a rejection today means just that ... today. That same company might call back in three months because something changed.

The Plan

Too often, pilots—young and old—believe they aren't a true pilot unless they're an airline pilot because that's where all the status, as well as all the money, is hiding. Personally, I don't agree. If you really believe all there is to aviation is sitting in the cockpit of a 747 on a long international route watching the islands and ice packs flow by beneath you, then I think you're headed into the wrong field. I believe what a retired Lockheed L-1011 pilot told me. "Do what you like, and the money will follow." In other words, if you chase only the money, you're setting yourself up for unhappiness in the long run. Avoid the "I want top dollar and I want it right now" disease.

The gurus who research employment, as well as personal goals, tell us the most important item you can accomplish in your plan to become a pilot is to write that goal on a piece of paper. Writing the goal down in some concrete fashion makes the goal take on a life of its own. This is the first step toward transforming your goal into reality.

Your plan is only a road map, a route to eventually arrive at your goal of becoming a professional pilot. But, like any route taken on the ground, your career plan shouldn't be so rigid that it's incapable of allowing you to take a detour where necessary. Some situations might force you to change your plan slightly, such as an unplanned opportunity to fly an aircraft in a state 1,000 miles away on a short term contract. Do you ignore the chance just because you didn't have it written down in a plan you produced a year ago? I hope not. Certainly, too, a time might come when a obstruction to your career will appear. Consider all your alternatives before you make any decisions. If part of your plan was to spend the summer months dusting crops or towing banners, and the job you'd hoped for disappears because the company went out of business or they hired someone else, what are you going to do? You must have some alternative course to steer your plan out of the muck and on to high ground.

Having a plan that only talks about goals, with no way of dealing with the hurdles life is going to toss at you along the way, is useless. The people who make it are the ones who understand that, while they can't control life, while they're unable to predict what kinds of roadblocks will be tossed in their paths, they can control the way they react to events. You're in control of your own life. If such a thing exists as a positive attitude, I believe it means being realistic about what part of your career you do have control over.

One of the major portions you do control is the level of education you obtain along the way, education additional to those much-needed flying credentials. If the industry troubles of the early 21st century have reminded us of anything, it's that a flying job can evaporate in the wink of an eye with little input from you. With a general college degree under your belt, something that was designed only to check off the "College Degree" box on an application could leave you in a lurch. Most successful people understand the need for an education that not only allows them to win a flying job, but allows them to earn a living using some other skills when times are tough.

Currently, I read seven aviation magazines each month, as well as a few business publications. The *WSJ*, both the print and the online versions, are also a necessity to stay on top of the news. One of the aviation magazines is published on a weekly basis, so that's like reading 11 magazines a month. Often, you'll notice the same story more than once and consider reading it again a waste of time, but no two writers tell the same story the same way or include all the same facts and perspectives. Each writer provides a different insight into the industry that might be useful at some point.

I recently read a story in one of the trade magazines about a company producing new aircraft in Wisconsin, and I wondered whether there might be a place for me, even in a small operation such as that. I had extensive experience with the aircraft they were producing, and I thought it was worth a chance to contact the firm and offer my services as a pilot. Unfortunately, this particular opportunity didn't turn out favorably. But it might have. My motto is this: I have absolutely nothing to lose by asking. The worst that could happen is the company would say no, and no one ever died from being told no.

The plan you produce for yourself should be realistic. This means asking a variety of questions and developing answers for each to build the shell of your plan. If you expect to fly right-seat on a 757 after college in the U.S., your chances are slim—not impossible, but slim. Despite the increased numbers of flying positions available, the search for a flying job is still extremely competitive. Don't expect instant success. How long does your plan include for you to win your ratings and when do you expect to begin? Another extremely important issue is how to pay for your education. More on money a little later.

Picking up your license is not enough either. You need to keep logging time for experience and to reach the magic goals various companies require to apply. What kinds of flying jobs are you willing to take on to reach each step of your goal? What won't you do? When will you complete your instrument rating? What's your organizational plan for sending out résumés? How will you locate the addresses of companies that hire pilots?

Would you consider giving up a few years of your freedom to join the U.S. Air Force or Navy to learn to fly?

The list will grow as you spend more time on your plan, but much like a business plan, a career map is likely never finished. Think of it more as a work in progress. The trick is knowing when to stick resolutely to your plan and when a detour makes sense. If I can provide any advice on this front, it's not to jump too quickly at anything, no matter how good it looks. Think how this decision will affect the rest of your plan. If you're not sure, ask a close friend to sit down and talk with you about an opportunity when it presents itself. *Brainstorming*—simply thinking about the possibilities without restrictions—as well as *networking*—talking to everyone you know and meet about your job search—can reap incredible benefits in terms of information and, perhaps, even a job.

Now's the time to ask where you see yourself in one year, five years, or even ten years. Don't hold back. Let your mind run free. Do you see yourself in the left seat of a corporate G-V flying transatlantic? Maybe it's the left seat of a 777 for American or United Airlines. Perhaps you enjoy flying, but enjoy being home more often. A regional airline or local charter job, or possibly even some kind of utility flying, is what might be best suited for you.

If you're a woman or a member of a minority group, I have good news for you, too. Besides the federal legislation that mandates that a company not discriminate on the basis of sex or race, most of the major airlines and corporations actively seek candidates from minority groups. Unfortunately, as one vice president of flight operations at a major airlines noted, "We don't even receive enough applications from either of these two groups to hire very many as pilots."

If you're a woman, a place you can certainly begin your search for information about a career in flying could be Women in Aviation, mentioned again in Chapter 4, or the members of other minority groups, all you need to know is this: once you complete your flight training, many of these companies will be waiting to interview you. Remember, no one is promising anyone a job.

Decisions, Decisions: Where Are You Headed?

Before you make the decision about what kind of flying you're interested in, think about your ultimate flying goal in terms of your interests and the type of flying that will keep you happy and productive. I've said before that solid information is the key to an informed decision. This need to research should be ongoing, not only for your job search, but also for your flying career.

To arrive at the right career decision, you need to know where you're headed. A simple goal might be, "I want a flying job." I challenge you that an objective like this is too broad. Dig a little deeper for some additional insight. What kind of flying job? "Big airplanes!" Still too simple. Keep digging. "Fly big airplanes on international routes." That's better. At least now your potential flying job is beginning to take shape. Will it be airlines, corporate, freight charter, or what exactly? Think about those people involvement issues again.

Will you agree, then, that any job short of flying large aircraft on international routes will be only a stopping point? If the answer is yes, you should realize that you'll keep a job with Air Tran only for as long as it takes to get hired by American, United, Delta, and so forth. Flying a King Air from Biloxi, Mississippi, then, is also not going to be your final career move, is it? What I'm after here is to make you think through your objectives. Ask yourself why you want to fly large aircraft. This might sound crazy, but why do you want that? Does it make you feel you've arrived? What do you want to prove and to whom? This kind of analysis can help you decide how you may need to—or want to—move from one job to another in that quest for Utopia (Figure 1-7).

Figure 1-7 Imagine flying one of these U.S. Air Force KC-10s lined up for takeoff.

Nancy Molitor, Ph.D., a clinical psychologist in Chicago, believes this: "For a pilot to choose the right direction for their career, it is important they ask themselves quality of life kinds of questions. How much of a sense of control do they want over their work environment, such as their schedules? An airline job is going to offer them fairly insignificant involvement with the passengers in the back of the airplane, while a corporate or charter flying position will bring the cockpit crew into direct interaction with passengers. They need to clearly understand which operation they like more. They should consider the capacity for change a particular position offers. A unionized airline job is fairly regimented. If you have an advanced degree and like to make decisions, other than basic flying, you may find airline flying somewhat disappointing, no matter how large the aircraft you fly or grand the paycheck."

To make the right decision, you should delve deeper into what makes you tick. Ask yourself seriously, "What kind of environment do you feel most comfortable in?" says Molitor. "You'll go further and be a lot happier if you know before you begin. Spend the time to learn what kind of working environment other pilots have. What kind of schedules do they work, or do they even have schedules? What are the opportunities for upgrade at the company? How will your job responsibilities change after you do upgrade? When you investigate all the aspects of a particular type of flying, you might find yourself changing your mind about where you're headed."

Other issues to wrestle with include how much you value money and why. Try to answer honestly: do you want a job that pays a ton of money, just so you can tell the world "I've arrived"? Are you willing to do any kind of flying to get that cash? You may say yes now, but what about a year after you've been flying long, monotonous international routes with 12 to 14 hour duty days and only one landing. Would you be happier at a smaller carrier like Southwest, or in a corporate position, both of which pay substantial salaries, but offer you more actual flying time, and more landings and takeoffs?

Many pilots tend to be rather tactical people, dealing with life and career problems much the same way they deal with the problems flying tosses at them—as they occur. This may work for some, but in the end, you may find yourself only flying an airplane for a living. Why settle for this when, with a little planning and forethought, you can design a life that offers as many of the things you value in life, while avoiding as many pitfalls as possible?

Let's take a look at the characteristics of some of the flying jobs we spoke of earlier. Each has positive attributes, as well as a few drawbacks. And, yes, even the major airlines have some disadvantages. Let's say you get on with the right airline, one of the big ones (only about six or eight of them exist). You successfully climb the internal ladder to larger and larger aircraft, and then you eventually move from the first officer to the captain's seat. You could well find yourself commanding a salary in excess of $200,000 per year, while flying as little as 10 or 12 days a month. Who wouldn't want this, right? Remember, those little dollar signs have a way of swaying just about anyone, so don't think I'm crazy when I say some pilots will look elsewhere for a career.

Some Flying Realities

The Major Airlines

We might as well begin here because everyone thinks they want to be a major airline pilot anyway. A major airline job often pays the highest salaries around for driving some of the largest airplanes in the world. Despite the changes in salary since 9/11, these jobs still pay pretty substantially.

But what's the real cost of a major airline job like this? First, only a small percentage of Northwest, Delta, or American pilots—or any pilots for that matter—will make $200,000 a year. In late 1998, the average airline pilot salary was about $130,000 per year. Not bad, but it's not $200,000.

Another factor to consider is how much time you'll spend away from home. If you're single, you might not care. If you're a newlywed, or you have small children, being gone for long periods of time may cause some heartache for both you and your family, especially if you're going to be a two-income family. Are you willing to hear about your daughter's school play while you're sitting in a hotel room in Tokyo?

One Northwest pilot I know—a Boeing 747 captain—called me recently—just before she left on a 12-day trip. Now I'll grant you, she probably only flies one of these trips per month, but that is a pretty long stretch to be away from home. If you have a family, perhaps you should consider jobs that require only a few nights away at a time. Airline flying also puts you on a schedule where you'll know which cities you'll see in a given month. While it is not uncommon for some pilots to bid "Flying Line 121" at their company, for example, because they'll know their overnight locations well in advance, there is seldom any variety to the destinations from what is scheduled.

Will you live where you're based? If you and your wife and your families are from Columbus, Ohio, taking a job with United that bases you in Los Angeles could cause some sorrow. Your family may be upset if they have to move, leaving the rest of their family behind. You could choose to commute to and from L.A., but be ready for your commuting time to eat into your personal time, just as it does to any other person trying to get to work these days. If you have a trip that begins out of L.A. at 9 A.M. on a Monday morning, you'll need to be in L.A. by Sunday afternoon at least. Where will you stay when you're operating out of your base, such as just before a trip? A *crash pad*—a place where four or five pilots often live together—may be a cheap answer, but your privacy will be almost nonexistent.

While we're discussing commuting, consider the current state of airline travel. Most airplanes are often full as load factors have surpassed 80 percent. As a nonrevenue passenger trying to get to work, you'll get a seat only if an empty one is available. You might be able to ride in the cockpit jump seat, but if a more senior pilot comes by, they will get priority over you. You may need to be at the Columbus Airport early Sunday morning to allow for a safety margin of missed flights to be in position for work in L.A. On the other end, if you finish a trip at 11 P.M., will you be able to get home to CMH that night? Perhaps. But realize that, with the flying time and the time zone changes, it will be tomorrow morning by the time you walk through the door, most likely exhausted. And, your thoughts begin almost immediately to when you leave for work again. Many pilots do it, but most dislike commuting.

All Airlines Are Not Equal

Although dozens of airlines fly around the U.S., only 15 of them make up about 75 percent of the large aircraft flying jobs. Why not consider a smaller airline with a base in your hometown? When you finish work, you would have a short drive to sleep in your own bed at night. Other carriers flying large aircraft might represent a much easier commute than L.A. If you work for Skybus, you might be based in Columbus. In bad weather or if all the connecting flights were full, you could drive to your base if you had to. Airlines tend not to be patient with commuting pilots who don't make it to work because they got bumped (airline lingo here) off a flight. But, of course, you'll never make the kind of money at Skybus that you'll make at United or American.

Time can cure many things, however. If you're 29 when you hire on with a major airline, you can put up with the problems for quite a few years before the base you want opens up and still have plenty of productive flying years ahead of you. An older pilot, though, may have fewer years until the magic age 60 (or perhaps age 65) bell rings, limiting some of their options.

The Nationals

A *national carrier*—based on the strictest definition—is one with total revenues of more than $100 million and less than $1 billion. These are airlines you might, at first, have considered to be regionals, such as Air Wisconsin, Spirit, and Frontier. The list also includes companies such as Kitty Hawk Air Cargo, Comair, Hawaiian, American Eagle, and Polar.

Essentially, the pay here is less and, often, the schedules are not as good, although scheduling seems to be relative to the pilot. On many airlines, there are no duty or trip rigs that offer pilots any pay guarantees for non-flying time spent at work. So, you might ask, why consider them? Because some offer a wide variety of flying experiences you might not see at larger airlines, such as the opportunity to fly more approaches in a day, if you're concerned about sitting in cruise for too long. The pilot groups are smaller—United has about 10,000 pilots right now—so the camaraderie at a national might be much better.

A note here, too, about the concept of guaranteed flight hours that you'll often hear tossed about. An 85-hour month may seem fairly easy, but what you need to understand is how much duty time pilots put in to fly those 85 hours. This applies to any airline you talk to. From my experience, duty time is normally double the flight hours and, in some cases, nearly triple that—making for some pretty long workweeks. This can also make commuting to and from work rather difficult and sometimes fruitless when you have only two or three days off in a row. Do your research. One regional I looked at recently was in desperate need of both captains and first officers to fly some rather large turboprop aircraft. But the major airline they feed had only agreed to renew their code-sharing marketing agreement—and allow them to continue flying as an express carrier—for two years. After that, many jobs might be up for grabs.

Why not find pilots who fly for some of these companies before you make any decisions? A great place to begin might be a board on the Internet, such as www.airlinepilotforums.com.

The Regionals

Flying for the *regionals*—formerly called *commuters*—is an adventure in itself. While most pilots point to money and benefits as *the* reason not to fly for a regional, the schedules often tend to be somewhat more grueling than major airline flying as well. Regional pay has fluctuated quite a bit in the past few years and it will never reach the level of the majors or nationals. A 30- to 50-seat airplane simply doesn't have enough seats to generate the revenue necessary to pay hefty salaries.

But, the upside for many regionals is just that—their location. Some pilots with families are willing to forgo the big airplane, hefty salary magnet because they've come to realize other things are more important to them. Regionals include PSA, Allegiant, Northwest Airlink (Figure 1-8), Midwest Airlines, Pinnacle and US Airways Express. Note, too, that the range of aircraft these companies fly is often vast.

Corporate

Corporate flying is as different from flying for the airlines as night and day. The relationship between passengers on corporate aircraft and their pilots

Figure 1-8 Northwest Airlines Airlink RJ. (Courtesy Dale Boeru: www.jetphotos.net.)

is also much different from the airlines. On board a Part 121 airliner, the cockpit crew will normally have the door open until they're ready to start the engines, when the door is shut. The passengers never see the crew again until after landing. Depending on the captain, the people in the back may hear an occasional announcement about what is happening, but often little more.

On a corporate flight, the crew does everything. They check the weather and file the flight plans, something an airliner crew has a dispatcher to do. The corporate crew also makes sure the aircraft has all the supplies, such as food and drinks, which passengers may require. The crew is responsible for making certain the aircraft is clean and that the lavatory is serviced. When the passengers arrive, the crew normally loads the baggage.

When a corporate crew starts the engines, they often leave the door open—that is, if they even have a door, which many corporate aircraft do not. Not uncommon during the flight is for the chairman or the CEO of a Fortune 500 company to walk up to the cockpit to chat about the flight, the state of the company, or to ask the crew about their families. In general, the crew of a corporate aircraft is much more involved with the people in the back. When the weather ahead is bad or a delay is ensuing, a corporate pilot would probably not pick up the PA. They'd more than likely walk back and talk directly to the boss. To make this work, you need to like being involved with people and you also need to be good at talking with them.

Corporate flying is not like airline flying in another sense—the regulations. The airlines, even the regionals, fly under Part 121. Most corporate crews fly under Part 91, yes, the same regulations you flew under when you rented a Cessna 150 to go tooling around the skies. The benefit is that corporate crews have no duty and rest restrictions as do airline pilots, although some view this as a drawback. Airline pilots can fly only a specified number of hours per day, per week, and per month, depending on which segment of the regulations their company operates under. A corporate crew could fly 12 hours a day if the company wanted them to.

Now for reality. Most corporations—at least the good ones—may be governed by Part 91, but they fly their crews to the same Part 121 standards the airlines use. No CEO of a billion-dollar company wants to put his life— or the lives of his senior officers—at risk by putting them on a $30 million aircraft with a crew who is likely to fall asleep just after takeoff. But the variations on corporate schedules can be vast. Some crews fly 50 hours a month, while others see only 15 or 20. Often, trips leave early in the morning, fly an hour-and-a-half away, and then oblige the crew to sit around for eight or nine hours before retracing their steps. But a corporate crew seldom visits the same city twice in a week.

Corporate flying often demands some collateral effort from pilots as well, such as helping with the scheduling, training, or keeping navigational charts in order. Each day and each destination, however, is new to a corporate pilot. Another good part of that kind of schedule is this: the time you have at the destination is usually your own to read, write, snooze, or call home. Salaries on the corporate side can also be substantial with captains of large jets making well in excess of $100,000 annually, with all the benefits any other corporate executive might enjoy.

A good corporation today is also more interested in your education status than an airline. A corporation views a pilot as a business asset, not simply as an airplane driver, which is essentially what you can become as an airline pilot. In a good corporate position, you have a future with the possibility of advancement into a managerial position within the flight department, if you hold the requisite education and skills.

Many corporate pilots report, too, that in times of financial difficulty, flying around with the man or woman who makes all the decisions about their future can be quite beneficial. Any information you hear comes right from your boss's mouth. If you enjoy interacting with people and if you want a voice in your future on a regular basis, a corporate job might be worth looking into.

Charter Flying

Also known as Part 135 operations (see additional story in Chapter 7) this type of flying uses a vast array of aircraft from propeller-driven Piper Navajos to a Gulfstream 5, or a Canadair Challenger, and everything in between. Charter companies, based all over the U.S., act as a sort of hybrid, on-demand corporate operation.

The major difference between a charter department and a corporate operator is—again—the regulations under which they operate. A Part 135 charter pilot operates with flight time and duty-time restrictions, as well as numerous safety and training considerations—much like Part 121. But the regulations are not quite as tough in a Part 135 situation. Pilots can fly longer days and fly with much less training than a Part 121 carrier. This doesn't mean they are less safe, they are just a little less regulated than the scheduled airlines, which brings us to the next major defining point in a charter company.

Part 135 regulations are mainly focused on unscheduled operations. This means not only will flight crews often not know for certain where they are headed, but they often won't know when they are leaving or when they are coming back. Some of this also occurs in a corporate flight department, but senior executives normally plan long trips ahead of schedule, simply because their time is so valuable.

A Part 135 pilot usually carries a pager and a cell phone, which can make them rise at 1 A.M. with a call to head for the airport ASAP. Most charter pilots keep a bag in their cars at all times. This can create somewhat of a chaotic life, but many love the variety and the fast-paced level of action. Schedules can be difficult for charter pilots. They are often on call a great deal of every month, but they seldom fly all the days they're on call. The regulations say a company only needs to give a charter pilot 13 days off in a quarter or about one day a week. Some run it close. Depending on the type of aircraft you fly, you might find yourself leaving home at odd hours of the day and night, but you might also find yourself sleeping in your own bed quite a bit.

Having flown for a Part 135 charter company, I can tell you that you'll quickly get to a point where almost no destination—or time of day—will surprise you. Salaries on the charter side can vary widely, too. A charter company makes money from their flying as their primary function, while a corporate flight department uses the airplanes as a travel element to their main business. On the corporate side, a Citation 5 captain might make $70,000 per year. On the charter side—where there is a slim profit margin to contend with—a captain might only bring in $50,000 to $60,000 per year. Again, though, charter flying offers lots of hands-on action with nearly every facet of flying and, often, into and out of short, remote places the airlines don't serve.

The Fractionals

Sandwiched somewhere between the corporate and the charter sectors is a quickly growing segment of business flying called the fractionals. These include companies with names such as NetJets and Flight Options. Fractionals sprang up out of nowhere a few years ago to serve an emerging marketplace—companies that did not own a business aircraft and could not justify the entire price of their own machine. *Fractionals* picked up their names, in fact, because they sell partial ownership in a business airplane. A company can purchase a 1/4 share in a Citation, a Gulfstream, or a Hawker, even down to as little as 1/16th of the aircraft. What fractionals have done for business aviation—besides offer aircraft to new companies—is to create an incredible demand for aircraft and pilots to fill the needs of their customers.

The largest of the fractionals—*NetJets*— is run by Executive Jet Aviation, based in Columbus, Ohio, and owned by billionaire financier Warren Buffett. Buffet is so bullish on business aviation—he also owns training icon FlightSafety International—that he placed orders for more than $3 billion worth of new corporate aircraft at the 1998 National Business Aircraft Show (NBAA) in Las Vegas. His order included 10 Gulfstream 550s, with options for 12 more, and 14 G450s for Executive Jet itself and 50 new

Cessna Sovereigns—an aircraft designed to compete with the Learjet 60 and the Hawker 800XP—and options on 50 more.

NetJets employs nearly 2,600 pilots to crew some 550 aircraft. American Airlines, by comparison, employs nearly 12,000 pilots for about 700 aircraft. The good part about flying with NetJets is you'll be flying—a lot. NetJet crews normally fly a schedule of between four and six days on and three to five days off. A NetJet pilot can expect to log somewhere between 60 and 80 hours per month. Pay is pretty good, starting at about $40,000 per year. Senior captains can make in excess of $100,000. NetJets plans to hire 250 new pilots in 2007. Commuting is minimal at NetJets, requiring pilots to live near the five gateway cities the carrier uses, including L.A., West Palm Beach, Columbus, Dallas, and Teterboro.

Freight Companies

If you're simply not a people person and you want to make certain the captain never asks you to go back in the cabin to deal with an irate passenger, flying freight might be your cup of tea. Freight companies vary widely, from a local company flying newspapers in a turboprop to UPS or FedEx, flying cargo all over the world in large air-carrier aircraft. During the worst of times in the early twenty-first century, FedEx and UPS have evolved as some of the most stable flying around, not to mention some of the best paid.

Traditionally, freight is flown at night, although some of the large carriers are increasing their presence during daylight hours. If you like flying at night when the air is smooth and the Air Traffic Control (ATC) system is not experiencing a coronary, freight could be for you. Then, too, there exists a rather closely knit brotherhood of night flyers, which many pilots say they will never give up. Often, among the smaller carriers, considerable on-demand cargo flying occurs at all times of the day and night. Cargo companies operate under both Part 135 and Part 121 regulations.

The pay and days off at the top end of the freight spectrum, where FedEx, UPS, and Airborne (ABX) reside, are top notch. At last report, a second-year captain starts at about $135,000 per year at UPS, and about $113,000 at ABX. At the other end of the spectrum, small freight carriers pay varying rates, and some can be pretty minimal.

In Appendix D, you'll see how one carrier, Air Net, explains its pay system.

Flight Instruction

If ever there were a group of unsung heroes, it is the teachers. Flight instruction (Figure 1-9)—as you'll read in Scott Spangler's editorial in

Figure 1-9 FlightSafety student and instructor. (Courtesy FlightSafety International.)

Chapter 4—has always been a low-paid position that almost everyone (yes, I used to think this way, too) thought was simply a way to build time until you could line up a real job. Nothing could be further from the truth, however, because when I look back on my instructing days, I realize that much of the pilot I am today, as well as the person I am today, is based on the experiences I had while teaching people to fly.

The best pilots—in my opinion—are those who rose from the flight instructor ranks. No matter what kind of a job you eventually land, you'll find that you'll be a flight instructor to some degree again. I think pilots who never flew a PA-28 from the right seat trying to teach someone how to land, will operate at a distinct disadvantage in a crew concept. Many of the worst corporate and airline pilots I've flown with avoided the flight-instructing route, often because they believed such work was beneath them. The true advantage of flight instructing—in addition to the time a Certified Flight Instructor (CFI) can log—is the training it offers in learning how to better interact with people. This is much the same kind of training they need later to relate to the people in the back seats as corporate pilots or, perhaps, as they train a new first officer for the airline.

No doubt, flight instructors are not yet looked on as the professionals they should be, but as the pool of well-qualified pilots continues to shrink, more and more flight schools may find themselves running ads like this one from Western Michigan University (see the following). Unless we find enough instructors to train new pilots, that pipeline to the airlines, corporate, and charter is going to begin to dry up. This represents a serious dilemma within the industry.

Flight instructors often work five- to six-day weeks and are required—at some schools—to be available at all hours during those days. Instructors need to be blessed with patience. If you have none, find another way to build time and save everyone else a great deal of grief. But, if you truly enjoy flying—a lot—and you have a talent for teaching other people about flying and a desire to share the joy you feel about aviation, then an instructing position is something you should consider.

Part of the problem with flight instructing is that often, the people who run flight schools are rather naïve about what kind of person will make a good flight instructor and they often end up hiring anyone who applies. Why? Because some flight schools are run by some pretty poor businesspeople. I hate to say it, but pilots do not traditionally make good managers. A new CFI looking for work would do well to consider some of the same questions that student pilots are thinking about as they choose a flight school. Why would you want to work in a school in which you could never imagine yourself training? Avoiding these schools is the only way to help clean up this part of the industry and make instructing a more satisfying job. Remember, changing the role of perception of flight instructors does begin with you!

IASCO, in partnership with Japan Airlines (JAL), began flight training at the Napa Flight Crew Training Center in 1971. Today, IASCO employees manage and staff the facility under contract with JAL. IASCO currently has 100 employees on staff at our Napa location.

The Napa Flight Crew Training Center's mission is to produce high-quality, entry-level airline pilots for Japan Airlines. Over the past 35 years, IASCO has successfully trained over 2,400 JAL pilots, flying in excess of 500,000 accident-free flight hours on a Monday through Friday training schedule. Our training fleet currently consists of 22 A36 Bonanzas and 11 B58 Barons maintained under an FAA-accepted Phase Inspection program written by IASCO for our specific maintenance requirements.

We have a proven commitment to safety, 35 years of flight-training experience, 500,000-plus injury-free training hours, and we have been an FAA Diamond Award Winner since 1995.

Certified Flight Instructor Position Minimum Requirements:
FAA Commercial Pilot Certificate—Single Engine, Multiengine (ME), and Instrument, FAA Certified Flight Instructor Certificate—Single Engine,

Multiengine, and Instrument. Must have logged a minimum of 1,000 hours of total flight time, 500 hours of multiengine, ME dual given time requirements may be waived, depending on the instructor's background and experience.

- Part 91 operation—flight training for Japan Airlines ab initio program.
- Pay scale currently starts at $40K and ends at $70K in ten years (reviewed periodically).
- Vacation, sick leave, 12 paid holidays, medical/dental/life/AD&D, 401(k), and severance plan.
- Full-time flight instruction position (Monday through Friday).
- This position does not build flight hours quickly.

Summing It Up

By now, you should have enough variations on flying bouncing around in your head to make you think carefully before you decide which direction to head. But, remember, no decision is cut in stone. The real crime would be to remain in some segment of flying that turns out not to be a good fit simply because you've already put in a few years.

The decision is yours.

The First Steps Toward a Career as a Professional Pilot

Before you even apply anywhere, you need to pick up a host of pilot certificates and ratings. While many are in the list, you only need a few to get rolling. You certainly need a private pilot's license to begin. Follow that up with a commercial license. Your instrument rating is normally a part of the commercial license. A multiengine rating will also be necessary in most cases because it enables you to pilot an aircraft with more than one engine. The flight instructor rating, while not required, certainly opens a wide range of extra employment opportunities.

All pilot certificates and ratings are won in basically the same manner, through a written exam, an oral exam, and a practical skills test. The basic requirements for the following ratings can be found by reading through a copy of the Federal Air Regulations Parts 61 and 91. Those books are available online at Amazon.com. You won't need to memorize the requirements, but you'd best become familiar with them because they are bound to be a part of your life from this day forward.

Certificates:
- Private
- Commercial
- Airline Transport Pilot

Ratings:
- Instrument
- Multiengine
- Flight Instructor
- Additional Aircraft Type Rating (for example, Boeing 737, Gulfstream 550)

International Flight Certificates

Foreign national pilots arriving to train in the U.S. can expect to work to FAA rules, but may not necessarily be expected to learn only the FAA standards. Pilots heading back to the European Community, as well as to portions of the Middle East and Asia, might well learn to fly in the United States, but be required to fly new check rides when they leave the U.S. That's because the European Community trains to much different standards than America does.

As one source told me when I was compiling material for this book, the traffic density and the geography with so many nation states operating in such close proximity makes the Joint Aviation Authority (JAA), soon to be superseded by the European Aviation Safety Administration (EASA), think that U.S. training may not be enough to qualify a pilot to fly in the region.

If you spend enough time reading through the resources in this book and researching job possibilities on the Internet, you'll run into jobs in other parts of the world that require a JAA ATP, for example. This means that an American citizen, or a pilot trained in the United States, will need to convert their FAA certificate to a JAA version when they return to their home country. That could be time-consuming and expensive. Be certain you understand the certification requirements of the country you're headed for before you begin training, so you will not have any ugly surprises in the future. An exception to some of the JAA requirements might be for expatriate pilots with FAA certificates flying in the Middle East and some parts of Asia.

Multicrew Pilot License

Areas of the world planning for rapid aviation growth have found themselves without the numbers of pilots they need to sustain the growth their local economies desire. The first beta tests for the new Multicrew Pilot License (MPL) are taking place in America through coordinated efforts by the Airline Academy of Australia and Alteon Training, a Boeing subsidiary.

The MPL can be applied anywhere, but is not expected—at least right now—to be used in the United States. It should do well in India and China, where airline growth has created strong demand for first officers. The goal for MPL is to develop a rigorous ab initio training program to take zero-time pilots to first officers' status quickly and at a much lower cost than other programs.

A story in the November 6, 2006, issue of *Aviation Week & Space Technology* about MPL explains that "Heavy reliance on simulators has provoked an outcry by some pilots concerned that simulators are not sufficient for beginning pilots to master the feel of operating an aircraft. The overall [MPL] concept conceives of training that's specific to each airline's criteria." Another concern about the multicrew pilot license was voiced in the *Weekly of Business Aviation*, which came from the International Federation of Air Line Pilots' Association (IFALPA). In a statement quoted in an October 30, 2006, article, IFALPA said the union had "concerns that an accelerated implementation of the license, without certain protections, could erode safety," and the MPL could be incorrectly used "in response to cost or time pressures . . . to address the current pilot shortage." The article also said that "under MPL standards, training will involve a 'multicrew environment' from the start. It requires the same number of flight hours as current rules, but a large portion of those hours can be flown in a flight simulator. An MPL pilot cannot fly a single-pilot aircraft without meeting additional requirements."

Another fundamental reason the International Civil Aviation Organization (ICAO) developed the MPL is this: few countries around the world have the benefit of the profound general-aviation fleet available to residents in the United States. Without that access to small aircraft to regularly hone their skills, pilots from other regions of the globe are at a competitive disadvantage in their search for cockpit jobs. The MPL gives them a direction and a means to make their dreams a reality, while also providing a ready source of cockpit crews trained to the standards of the airline industry outside the United States.

Medical Certificates

Before you can qualify for any pilot certificates, you must also possess the appropriate class of FAA medical certificate to verify the state of your health. In general, the medicals are categorized this way. The least stringent medical is the *third-class medical*, valid for 24 to 36 months after it's issued and necessary to qualify for a private pilot certificate. Next, is the *second-class medical*, valid for 12 months and necessary for a commercial certificate, which is normally needed for the flight instructor rating. The most stringent of all is the *first-class medical*.

Necessary for the issuance of an ATP certificate, the first-class medical is only valid for six months. Because an ATP certificate is necessary to serve as captain on nearly any aircraft, I'd find out early on in my training where I stood in meeting the first-class medical requirements. One of the nice things about the FAA medicals is they often serve more than one function. If, for instance, you were to obtain a first-class medical certificate and you were not exercising the privileges of your ATP certificate, that medical could be good for 12 months from the date issued and also be perfectly acceptable as a second-class medical. Hold on to your first-class medical for two, and in some cases, three years, and you'll still be able to enjoy the privileges of a private pilot.

Requirement by Certificate

With essentially only minor exceptions, a pilot must pass a physical exam and carry a first-class medical certificate when exercising the privileges of an airline transport pilot certificate, a second-class medical certificate when exercising the privileges of a commercial pilot certificate, or at least a third-class medical certificate when exercising the privileges of a private pilot certificate.

Duration of a Medical Certificate

A first-class medical certificate expires at the end of the last day of the sixth month after the month of the date of examination shown on the certificate for operations requiring an airline transport pilot certificate or the 12th month after the month of the date of examination requiring a commercial pilot certificate.

A second-class medical certificate expires at the end of the last day of the 12th month after the month of the date of examination shown on the certificate for operations requiring a commercial pilot certificate.

A third-class medical certificate for operations requiring a recreational pilot certificate, a private pilot certificate, a flight instructor certificate (when acting as pilot in command or a required pilot flight crewmember in operations other than a glider or a balloon), or a student pilot certificate issued expires at the end of the 36th month after the month of the date of the examination shown on the certificate if the person has not reached their 40th birthday on or before the date of examination; or the 24th month after the month of the date of the examination shown on the certificate if the person has reached their 40th birthday on or before the date of the examination.

Logbooks

In the section we'll begin to delve into the technology available not only to help find a job, but also to better manage the information you need to track during your search. More on technology for the job search appears in Chapter 8. One of the most important records you'll ever maintain is your personal pilot logbook. Guard your logbooks preciously. They are the only true records of the flight time and hard work you've put in toward reaching your goal of becoming a professional pilot. Here are some common definitions you'll run across when logging flight time.

- **Solo**—The only time you'll most likely log solo flight time is when you're a student pilot, but the regulation says you may log solo flight time any time you're the sole occupant of the aircraft.

- **Pilot in Command (PIC)**—You can log *PIC* when you're the sole manipulator of the controls for an aircraft for which you're rated or when acting as a pilot on an aircraft for which more than one pilot is required under the type certification of the aircraft or the regulations under which the flight is being conducted. An example could be when acting as a safety pilot for someone wearing a hood during instrument practice (See Appendix E for more information about PIC/SIC.)

- **Actual Instrument**—You may only log as *actual instrument* time that flight time during which you control the aircraft by total reference to the flight instruments under actual weather conditions. Simulated instrument flight, or hooded time, under simulated weather conditions must be logged separately. A rough estimate in the eyes of many interviewers is that approximately 10 percent of your total logged time should be as actual instrument.

- **Flight Instructor**—A certified *flight instructor* may log all time during which they act as a flight instructor as PIC. If you fly with an instrument student in actual instrument weather, the CFI may log the time as actual instrument time and PIC.

- **Second in Command (SIC)**—Pilots may log time as second in command (SIC) when they act as SIC of an aircraft for which more than one pilot is required under the type certification of the aircraft or the regulations under which the flight is conducted.

A Practical Look at Your Logbook: It's More Than Simply Numbers

Reprinted courtesy *Career Pilot* magazine

One of the great advantages of aviation—or one of its more serious drawbacks, depending on your point of view—is that much of this industry is cut and dry. Because aviation is also one of the most highly unionized of industries, a procedure to follow typically exists for nearly any situation. How many people can we carry with this fuel load? Check the aircraft's Pilot Operating Handbook. What happens when the weather goes below minimums at your destination after you're airborne? Check the operations manual. There simply is not a great deal of room for interpretation, until you come to a subject like pilot logbooks.

As Denis Caravella, safety programs manager, FAA Chicago DuPage FSDO, said, "There are not many guidelines for logging time other than FAR 61.51 and the Flight Inspector's Handbook." The only advice WestAir pilot Dan Dornseif received was "be honest." Part 61.51 states that the only flight time you must log is the "aeronautical training and experience used to meet the requirements for a certificate or rating, or the recent flight experience requirements of this part must be shown in a reliable record. The logging of other flight time is not required." One unwritten rule of thumb is that your logbook is used to verify the flight time you have documented is what you have actually flown. Any questions an interviewer might have about your experience should be easy to substantiate by reference to the remarks or the endorsement section of your book.

Some pilots believe logbooks are only logbooks. Actually, however, your logbook could be your competitive edge during an interview, especially because different companies look for different things in your logbook, such as exactly how your time should be totaled for their review. Bob Fiedler, supervisor for flight operations at United Airlines, said, "We have a different definition of dual time, for instance. We consider any dual received as student time." But, also, contrary to most pilots' opinions, most companies do not care whether the logbook you bring them was printed with a pen each night or with one swift keystroke of your computer. Rod Jones, assistant chief pilot in Phoenix for Southwest Airlines, said, "While the computerized logbook shows much more attention to detail and much more effort, either version is fine for us."

Most commercially produced pilot logbooks are divided between the small plastic-covered ones you might use for private flying and the larger, more expensive versions necessary to chart your career as a professional. Fiedler recalled a time when an applicant "brought in a box of small logbooks and loose pages tied together with paper clips and rubber bands." Another interviewer remembered "an applicant who brought in a box of fuel receipts three inches high and said that was how he tracked his time." If you intend on chasing a career as a professional pilot, spend the bucks for a professional flight logbook.

How important is neatness? That depends. United's Fiedler said, "To us, neatness really isn't that critical. However, when many people refer to neatness, they are referring to a book that is easy to read. Put yourself in the shoes of the person who must read your logbook—the interviewer for the airline or corporation that has a position you want. Your job is to make that interviewer's work as easy as possible. You'll accomplish this by making certain that there are a minimum of corrections and whiteouts. A correction on almost every page makes me wonder about that applicant's attention to detail."

Also, be certain the columns total up correctly. "We ask for a fairly detailed matrix of flight time on our application. That application information should be verifiable with the logbook," said Chuck Hanesbuth, director of Pilot Standards at Northwest Airlines. If you have a computerized logbook, offer a summary sheet that reflects the format of the application. Some pilots bring in their logbooks up-to-date for the interview, while others have no entries for previous months. The latter tells an interviewer that the applicant didn't do much preparation for the interview and is probably a procrastinator.

What to Log?

This is a question with many answers, depending on who is looking at your record. Most airlines and companies want to see at least the basics, including the date of the flight, point of origin and destination, the type of aircraft and tail number, the duration of the flight, the conditions—instrument flight rules (IFR) or visual flight rules (VFR)—and whatever remarks you believe would be of interest in recalling something about the flight later for the FAA, the company, or a potential employer.

Remarks might mention many items, like unusual conditions during the flight, Clear Air Turbulence (CAT), severe thunderstorms, missed approaches, and, perhaps, the altitude at which you broke out of the clouds on the approach. Five approaches to 200 and 1/2 display a different skill level than five approaches when the ceiling is 900 feet. But don't turn the remarks into a soap opera of your flying life either. "Thank heavens they made the remarks section rather small," said pilot Randy Ottinger. "The comments in your logbook really should mature as you do as a pilot."

Another logbook issue that causes considerable confusion is logging PIC time. "We see many errors in logging Pilot in Command time," the FAA's Caravella said. The problem often appears over the difference between when you are acting as PIC according to FAR 91.3 and when you can log flying time as PIC under Part 61. PIC, as defined in FAR Part 91.3, refers to the person who holds the ultimate responsibility for the flight. There can only be one. Period. But to log PIC time, according to many sources, the pilot need only be acting as PIC-rated and the sole manipulator of the controls. A pilot who is type rated in a Citation, for example, could be flying from the right seat. During the time, he is actually moving the controls and can log that time as PIC, even though he is not the PIC under the 91.3 definition. Essentially, the difference is who is moving the control wheel at any given time and who has the overall responsibility for the flight at all times. They might be the same person, but then again they might not.

And what exactly do the recruiters look for when they open your logbook? Sometimes, you never know. During some airline interviews, like those at Southwest and United, for example, the logbooks are often collected from all the applicants early in the interview process. The applicants never see them again until after the interviews are complete. This can have some serious drawbacks and could force you to pay close attention to the details of your logbook before you hand it over, because you will not have an opportunity to explain any inconsistencies or errors in your work.

The best advice again is to make certain your work is easily decipherable by most anyone. Try offering your logbook up to a few other pilots for their input before the interview. They may well catch errors you have passed over a dozen times. This effort can pay large dividends, because a pattern of errors says a great deal about your attention to detail here and, potentially, in the cockpit later. Also, if you plan to buy interview preparation from a company specific to this industry, ask if they have a logbook evaluation as part of their service.

One saving grace for pilots who are mathematically challenged is the newest in computer logbooks. One Southwest Airline's interviewer, Rod Jones, said, "Five years ago I probably saw one computer logbook in a hundred applicants. Today, that number is about one in ten." The reason is simple: flexibility. The electronic pilot logbook is, essentially, an electronic spreadsheet that can easily tabulate columns of numbers accurately and save you big dollars on correction fluid. If you've ever had to fill out an FAA Form 3710 for a check ride or insurance paperwork, an electronic logbook's capability to easily order the seemingly random sets of criteria the FAA and some airlines want to see can make it worth the price. With an electronic logbook, you can ask for the total amount of IFR PIC time you flew in a Beech 1900 during the November 1999 to January 2002 period, which included stops at MDW, DTW, MEM, and STL. The combinations are all selectable by the pilot before they run a report that takes just seconds to complete. Another potentially helpful aspect of the computerized logbook is you can define any of the additional fields you want, such as PIC time just for the Boeing 737.

The Drawbacks

A computerized logbook operation has a few shortcomings. First, computers can only do what you tell them. Before they can begin any computations, they need the raw flight data of your flight time, up to the moment of the request. You can only do that in one way—type it in. While that effort may seem colossal at first, balance the work against the speed and flexibility you gain over your flight history for the remainder of the 20 or 30 years you might continue to fly.

That convenience may not be enough for some pilots. Former WestAir pilot Dornseif said, "I thought about a computer logbook for a while, but I put it out of my mind. I've seen computers crash and do weird things. I don't want that to happen to my logbook." But, David Lewis, a Mesa pilot, said, "I like the flexibility and reliability of my electronic logbook. I use Aerolog. I know the math is always correct, and I can compile information for any application in just a few seconds."

To gain the flexibility of an electronic logbook, some pilots reported having tried a number of different routes to quickly enter the data into the system. One is a highly questionable single-entry-per-thousand-hours version. This technique puts one entry per year in your logbook for all the flying you've accomplished until this moment, for example, "flew 652 hours in 1998." Think about the eyebrows that sort of entry would raise, though, if you read Part 61.51 (b) literally—"Each pilot shall enter . . . information for each flight or lesson logged." Another version goes the single entry one better. It takes the annual total and at least breaks down the amount by aircraft, "of the 652 hours in 1998, 386 were flown in the Boeing 727." That's better, but it won't suffice in the minds of airline recruiters. Imagine using these techniques with those airlines that don't offer you the opportunity to talk about your logbook. Would you want this to be the only impression of you that a company walks away with?

The best—and the most accurate—method of making the switch from paper to computerized logbook is to type in each entry, one by one, verbatim from the paper logbook. You might consider paying someone else to type in your data, with you running through it later to verify the entries for accuracy. When it is all complete, simply run a summary report. If it does not agree with your paper logbook, you then have an option of believing the computer and moving on from that point or trying to track the error. If the time difference is not great, you might want to consider simply taking the new total as gospel.

Finally, no discussion of logbooks would be complete without some mention of backups—whether you use a computer or a paper logbook. Dornseif remembers the time his regular logbook was stolen from the flight bag in his car. Think about the implications of losing the only record of your flight time available anywhere. For people who pride themselves on planning ahead for all the possibilities of weather, fuel, or mechanical problems, this strategy certainly has a familiar ring to it.

If you have a paper logbook, consider what Ottinger does: "I make a fresh copy of each full page of my logbook when I've completed it." He then stores the photocopy version of his logbook at the home of his parents for safekeeping. If he ever did need to reconstruct a new logbook, the task would be relatively simple. If you are using a computer logbook, the backup process is normally as simple as inserting a blank disk into a drive on your machine and pressing ENTER. Making two backup copies of your data is a good idea as well. Make certain you store one of them at a location away from where your computer normally stands, such as a safety deposit box at a bank.

Jones summed up the logbook issue best: "Pilots should have some pride of ownership in their logbooks. Some treat them as just another document, while others I've interviewed show their log as a list of the accomplishments that got them to that interview. Their logbook says 'this is who I am.'"

This completes the essential requirements that will take you from rank beginner to airline transport pilot. This quest of yours requires time, money, and effort on your part if you're to be successful. In upcoming chapters, I discuss all these items to make certain you reach the rank of professional aviator with minimal fuss and maximum fun. If you aren't having some fun along the way toward becoming a professional pilot, you probably should consider another line of work.

Every group always has naysayers, people who believe you'll never be able to achieve your goal. "There are too many pilots now," they say. "The economy is too tough." "You're a woman" or "No one will hire a minority." I found a story that I wish I had written. It's the story of a pilot who had a dream she never lost.

The following is printed with permission from UND Aerospace, Grand Forks, ND, and Christy DeJoy, J. Patrick Moore, LaMaster Farmer, Minneapolis, MN.

Nine-year-old Jean Haley was to write an essay explaining what she wanted to be when she grew up. The assignment was simple; Jean knew she would be an airline pilot. But those dreams were nearly shattered when she saw the *F* at the top of her essay. Her third-grade teacher reasoned, "This is a fairy tale, not an essay." The year was 1959, and the reality was that there was no such thing as a female airline pilot.

Fast forward to the year 1993, and Jean (Haley) Harper is a United Airlines captain.

The 34 years in-between have been filled with hard work, bravery, and commitment. Harper credits much of her success to her father. When young Jean brought her F-graded essay home, Frank Haley, a crop duster, said, "You can do anything you want."

She wanted to fly, so seven years later she began flying lessons. "I wanted to prove to myself I could do it. I figured I could prove the naysayers wrong by doing the things they said I couldn't do."

Harper determined that she needed to meet three criteria to become an airline pilot. First, she needed the appropriate licenses. So, at 20 years old, Harper took out a bank loan and spent six months getting her private, commercial, and instrument ratings. Second, Harper needed flying time—at least 1,500 hours. She started logging hours hauling skydivers and studied to become a flight instructor. And, third, Harper needed a four-year degree. She enrolled in the aviation program at the University of North Dakota (UND) in Grand Forks on the advice of Mike Sacrey, a UND Aerospace graduate and the FAA pilot who checked her out as a flight instructor.

When Harper enrolled at UND in 1971, there were still no women airline pilots. "It was a scary feeling thinking I could spend the years and the money trying to attain it and still not be taken seriously. But not taking the chance of becoming an airline pilot was more scary," she said.

While female pilots were still a novelty at UND, Harper, nonetheless, became the school's first female flight instructor. "It was a first, but John Odegard was enthusiastic that I was there and was happy to offer me the position," she said.

During the next four years, Harper juggled her studies and a variety of jobs, including flight instructor, charter pilot, cloud seeder, crop duster, glider tower, and night airmail pilot. "I was spread thin," Harper admits, but she was doing well enough to earn several scholarships and awards.

The best gift Harper received was from a UND student pilot named Tracy Van Den Berg, who told her in 1973 that Frontier Airlines had hired pilot Emily Howell. "I thought, 'Oh, my God. It happened. The last obstacle has fallen.' I finally had an honest-to-goodness role model." A few months later, American Airlines hired Bonnie Tiburzi.

After graduation in 1975, Harper began writing to airlines asking for pilot application forms. She was getting little response until she started signing her name "J.E. Haley" instead of "Jean Haley."

Because of an industry downcycle, the airlines were hiring few pilots during this time. For nearly three years, Harper flew whenever she could—moonlighting as a flight instructor, and flying with small cargo and commuter carriers.

At Meridian Air Cargo, Harper met a pilot named Vic Harper. She used to switch shifts with his copilot, so she could fly with him. Three years later, they became matrimonial copilots.

Meanwhile, airline pilot hiring increased. Harper interviewed with Delta, Allegheny, and United Airlines, and she landed a job in Denver with United as a 727 flight engineer. She became United's third woman pilot, hired just weeks after the first two women. During that same time, Vic accepted a job with Frontier, also in Denver.

The person who most strongly supported Harper's dream of becoming an airline pilot never saw her in her United uniform. Harper's father died in an airplane accident not long after she graduated from UND.

In her 15 years with United, Harper has flown Boeing 727s, 737s, 757s, and 767s. In November 1992, Harper, then 43, flew her first line trip as a 737 captain—with her favorite first officer, Vic Harper, who had since joined United.

"Becoming captain was such an accomplishment for me. I'm so proud." She said her pride in her qualifications rubs off on other people, easing the minds of the few who might not be completely comfortable with a woman captain.

"The vast majority of comments I've received about being a captain have been very positive. Among the most enthusiastic congratulations I have received have come from the 'old timers'—the senior captains at United," Harper said. "I think my attitude has a lot to do with it."

In addition to her career achievements, Harper said she's also met her goals of a rewarding personal life. The Harpers juggle two airline careers and two children—Annie, age 7, and Sam, age 3-1/2. In 1985, Harper became United's first pregnant pilot. United didn't even have a maternity leave policy for several years after that.

"I remember having serious concerns whether I could have this career and have a family, too. It bothered me that I might have to choose. No one would ever tell a man he couldn't be married, have children, and have a career," Harper said. She has been able to have both career and family. "This is it. I have no more unfulfilled wants." Perhaps, Harper's third-grade teacher was right. She is living a fairy-tale life (Figure 1-10).

Figure 1-10 Are you ready to begin? (Courtesy Daniel Havlik: www.jetphotos.com.)

2
Flight Training

Before you think about flying for the airlines, a corporation, or the U.S. military, there's a simpler goal to accomplish: learning to fly (Figure 2-1). "Simple" is a relative term and I'm not trying to treat learning to fly in a matter-of-fact way. How and where you learn to fly is important to cultivating your professional flying career. Without the right flight education to teach the appropriate attitudes, skills, and habits, your career could take an early nosedive.

Although few people will mention this, it must be said. Your career as a pilot depends on your past and current life habits, to some extent. Every flight school today requires a background check, not only due to new regulations from the Transportation Security Administration and the FAA. If you have DUI convictions, or drug offenses classified as felonies, you are going to have a difficult time getting hired by some companies, such as airlines. I didn't say it would be impossible, but the road will not be easy. There is a continuing requirement to be drug and alcohol-free once you begin flying, as well.

Learning to fly involves more than running out to the airport for a few lessons once a week, taking a test, and picking up your license. During the months or years it might take you to gain the experience necessary to leap to the next stage of training for that commercial, instrument, or multi-engine rating, you'll be working on a regular basis with an assortment of flight instructors. You might be tempted to believe that the best progress through your flight training will occur when you find a "good instructor" and stay with that person all the way through to your first flying job. While that certainly is an option, it's not one I would recommend.

Most students view their first instructor as quite an aviation authority figure during those early months of learning to fly. It's this first instructor who will, I hope, sit patiently with as you learn how the various parts are put together to form a real flying airplane. It's this instructor who will teach you first about lift and weather and federal regulations. But, along with those first bits and pieces of practical knowledge will come bits and pieces of an attitude. Student pilots tend to emulate what they see.

Figure 2-1 This Diamond simulator incorporates the latest flight-training technology. (Photo courtesy Frasca International Inc.)

If your first instructor is extremely careful about preflighting the aircraft before they fly or is constantly picking up the microphone while in flight either to use the air traffic control system or to check on a threatening weather situation, you, too, will learn this is the way a well-managed cockpit is operated.

If, on the other hand, that first instructor is someone who flies carelessly, a person to whom just getting the job done is sufficient, an instructor who explains once and says, "Go back and read the rest on your own," and never spends the time to discover what you understand, the instructor is doing you a great disservice. If that same instructor does not use a checklist, if that instructor has little regard for precise airspeed control in their flying, you'll come to believe this is the standard method for piloting an aircraft.

And, too, it's not just what the instructor teaches you that's important. What this teacher leaves out of your training could be vital to your career as well as your safety. One of the strengths of flying with more than a single instructor is the regular exposure a student receives to another pilot's scrutiny.

When the major aircraft manufacturers were designing their own learn-to-fly curriculums in the 1960s, they included a number of *phase checks* or mini-reviews. In my position as an assistant chief flight instructor in a Part

141 school, I was required to fly many of these checks. They were designed to give me a look at not only how well the students were learning, but also how well instructors were teaching.

A Practical Side to Phase Checks

I remember a young female commercial student who arrived for a phase check one sunny Saturday afternoon. After a good chat about flying and regulations before the flight, the woman seemed well-versed in the subjects needed to pass both her commercial flight and oral exam, which was the next scheduled step.

After we departed the airport for the practice area, I noticed she was having difficulty maintaining a specific airspeed or rate of climb. Initially, I put some of this off to nerves and flying with a new instructor, but within 10 minutes, I realized it was a real problem. I asked her to try some basic flight maneuvers where I noticed she was not using a primary tool, the elevator trim control.

When used correctly, the *trim control* relieves pressure from the primary control wheel to allow an aircraft to be flown basically hands-off, which significantly reduces the pilot's workload. Performed incorrectly, a lack of proper trim in an airplane forces the pilot to work much harder than necessary and can actually become a distraction when other operations require attention.

I was amazed that, somehow, this student had almost reached her commercial flight test with no understanding of this most basic concept. After I spent a few minutes showing the student the trim wheel and what it could do for her, she was surprised. I saw her a month or so later, after she had successfully passed her commercial test, and she told me that learning about trim made life easier for her during the check ride. Without the opportunity for another instructor to check her progress, this woman would have missed some important training. If you were a history major in college, you'd never think of taking every single course with the same teacher, so why give your flying career such a handicap? Good or bad, the best, well-rounded flying education is with more than one instructor.

Emerging Industry Training Standards

Aircraft technology has leaped ahead in the past ten years to a point where single-engine aircraft now emerge from the factory floor equipped with electronic components, which have transformed them from mere trainers to capable business aircraft. A company in South Carolina—SATS Air—has in fact started an aircraft charter company using only Cirrus Design SR-22s in the operation with a single pilot on board.

New glass-cockpit technology has replaced the older round dial gauges common to earlier generations of aircraft with a few simple flat-panel computer screens. These screens generate electronic representations of aircraft instruments, much the way the newest glass panel systems once did for the Boeing 757–767 and Airbus airliners 25 years ago.

Most of the new glass panel aircraft, called Technically Advanced Aircraft (TAA), include systems once found only in airliners such as multi-axis autopilots, real-time weather displayed over the GPS navigation routes, and traffic avoidance systems. *Glass cockpit technology* is now considered the norm for most new aircraft whether they are single-engine trainers or the newest of the very light jets. The Cessna 172, for example, uses essentially the same Garmin 1000 (Figure 2-2) avionics system the company offers as standard equipment in its new Citation Mustang, making the transition from one aircraft to another much easier.

But, along with the proliferation of these new technologies has come the dilemma of how to sufficiently train a new generation of pilots capable of safely operating these machines, especially when they are flying as the only

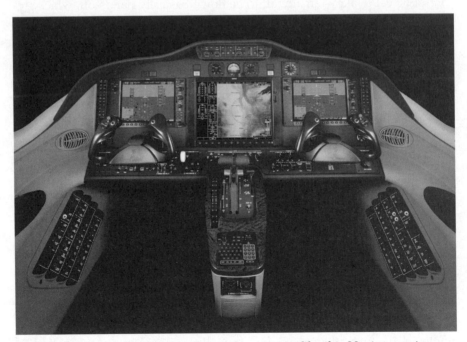

Figure 2-2 Glass cockpits make the move up to a jet like this Mustang easier. (Courtesy Cessna Aircraft Company.)

pilot. Flying the airplanes themselves still requires pretty much the same stick and rudder skills as it did 25 years ago. In the past, new pilots focused on attaining a specific skill level by demonstrating their proficiency at holding airspeed within + or − five knots, for example or remaining within a particular distance from an on-course indication.

Flying a new glass cockpit airplane, while accurately programming the navigation and communications system, has added a new dimension to flight education that requires new skills from cockpit crewmembers. The aviation industry and academia answered the call for this new generation of pilots with FAA/Industry Training Standards (FITS) designed to develop pilots who are not simply skilled at maneuvering an aircraft, but are also well-versed in the decision-making needed to keep many pilots— so to speak—in the air at the same time.

FAA has not taken a regulatory approach with FITS but, rather, focused on a practical acceptance, believing that additional regulations do not always increase the quality of the pilot. The industry and the FAA have begun to transform flight education in the U.S. to one that is more scenario-based by using a highly structured script of real-world experiences to address flight training in an operational environment.

For many years, the airlines and much of business aviation has focused on "training the way you fly" and "flying the way you train," as an answer to avoiding highly skilled pilots who might not be ready for flying in complex airspace environments. The new scenario-based training offers pilots an opportunity to practice for situations that require sound aeronautical decision-making. The FITS training also requires that a pilot's education directly relate to the specific aircraft they fly. For example, an SR-22 pilot's training would be different from that needed to safely pilot an Eclipse. Ideally, all flight training should include some degree of scenario-based training, which helps develop decision-making, risk management, and single-pilot resource management skills.

Choosing a Flight Instructor and Flight School

If receiving a well-rounded, professional flight education is your goal, another important concern is "who" will teach you. Good pilots always use a *checklist*, a list that prompts them to be certain they haven't forgotten some important item, such as the fuel pumps or the landing gear. Good consumers use checklists of sorts, too, as they search for the best deal on a washer and dryer, an automobile, or a flight school. Let's look at some of the items you should consider as you make the decision about which flight school will receive your training dollars.

Flight Schools Are Not All Created Equal

JoAnn Watzke was stuck in a rut and after six years as an art therapist, she knew she wanted out. She was already hooked on flying after 30 hours of lessons in a Grumman trainer. While JoAnn knew her future was in aviation, she also knew she needed a pocketful of licenses and ratings. But deciding where she should train for those items is the same problem facings thousands of other aviation professionals.

There's no shortage of companies willing and capable of putting a private pilot certificate or better in your wallet, from the local fixed-base operator to large schools, such as American Flyers and Delta Connection Academy. But what questions should a new pilot ask in the early stages of an aviation career to make certain they get pointed in the right direction? What answers should they listen for? Some considerations certainly include quality and quantity of training, accessibility, and cost.

Lynetta Sowder, director of admissions at Southwind Aviation in Brownsville, Texas, said, "Go and look at the school. Tell them you want to look at the housing. Tell them you want to go out and fly with an instructor for an hour. Look at the people who run the school, too. Are they pilots who really know something about aviation?"

Everything Flyable's director of operations, David Parsons, said, "People shouldn't be afraid to ask questions about their personal financial responsibility when it comes to insurance. Be sure and ask to see the aircraft's logbooks, too, so you can determine if the aircraft is legal to be flown."

But, before you make any solid decisions about where to train, you need a career plan. Michael Sullivan, former president of Comair Aviation Academy (now Delta Connection Academy) said, "If your objective is to become a professional pilot, you'll want to search for a professional atmosphere to learn in." Exactly where are you headed professionally? Is your flight training aimed at eventually placing you into the cockpit of a B-737 for a major airline or do you see yourself flying a corporate helicopter?

A school such as Phoenix East Aviation offers a course to take a student from zero time through the ATP and a type rating in a B-737 if the student wants because, as Fred DeWitt, the school's vice president, said, "We don't expect a guy with 200 hours to get a type rating, but they should start getting familiar with advanced turboprops and jets as soon as they can. After all, a doctor doesn't wait until he gets a job to start learning about advanced medicine."

If your goal is fixed-wing flying you might think the easiest method of paring down the list is skipping over schools that offer rotorcraft ratings, but that might not be the best plan of action. Some schools, such as Everything Flyable, offer both. If your goal is to fly for an airline as soon as possible, you'll find that a few flight training schools, such as Delta Connection Academy or FlightSafety International, offer the opportunity to put you into those Big Iron cockpits faster than many others.

AOPA's *Flight Training* magazine just released its 2007 College Directory (see Appendix A) listing colleges and universities in the United States that provide associate's, bachelor's, or master's degrees in aviation fields. You can access AOPA Flight Training's college database online (http://ft.aopa.org/colleges). If you're interested in learning to fly or to upgrade your pilot certificate and you already have a degree, or if college isn't the route for you right now, check out AOPA's online flight school directory (http://ft.aopa.org/schools).

The final advice is about goals: write them down. Putting goals on paper makes them take on a life of their own as they form the nucleus of your plan for becoming a professional pilot. Goals can be as simple as the choice of a climate to train, or as complex as how to finance that B-737 type rating or undertake the never-ending search for a good scholarship program.

Keep accurate records about each school you consider to make an efficient search. Don't trust your memory! Other questions to ask include: Will you train near home or spend the extra money for room and board to get "better" training elsewhere? Is a big school in the North worth the time you may waste on canceled flights when the weather is bad? Don't choose a southern California training facility, though, just because you like to surf in your free time. A school somewhere else may take you to your goal much faster.

Will your training be a full-time effort or a part-time struggle for ratings sandwiched between your current job? Parsons said, "Some of the local students make a pretty poor showing for lessons. If we had 25 local students all scheduled, only 15 to 18 would show up. Local residents (who train here) just have too many distractions."

Most pilots hired today hold a college degree. Recent Air Inc. figures show 87 percent of all surveyed pilots hired at global or major airlines held a four-year degree. That number is 62 percent at the regional airlines. Several large universities, such as the University of North Dakota, University of Illinois, Embry-Riddle Aeronautical University, Purdue University, Auburn, and Parks College, run large flight-school operations besides their regular degree courses (Figure 2-3).

Before you invest in a school that provides only flight training, consider a college curriculum that could grant your degree as you earn your wings. A good place to begin is with the University Aviation Association in Opelika, Ala. (www.uaa.auburn.edu) an organization that represents all facets of collegiate aviation education.

How much time are you willing to spend to earn your ratings? Some schools believe a zero-time, total immersion student can make the jump from first lesson to flight instructor in about 6 months, while other, smaller schools think the time is closer to 12 months. Phoenix East's DeWitt said, "All the way to the ATP is running about 24 months here." It may seem cheaper at first to stay home with Mom and Dad, and train when you have

Figure 2-3 Auburn University's Professor Emmett Johnson teaches both aviation and business to undergrads.

the money, but that method could turn your quest for a flying job into a Herculean effort, both in time and cash, stretching the training out for years. Can you afford to learn that way? Are you willing to put your career on hold for the extra years it may take for you to earn your ratings on a part-time basis?

How will you finance this venture? Most schools can direct you to a financial institution to apply for a loan, but few offer financing directly. "I have a deal with the bank," said Ed Pavelka, owner of Ed's Flying Service in Alamogordo, N.M. "They don't offer flight training and I don't offer financing." How much money will you need? The prices range from a low of $25,000 to a high of nearly $65,000 to go from zero time through the multiengine flight instructor certificate. Phoenix East quoted an approximate price of $42,000, but this took a zero-time pilot through the ATP rating.

Students need to be aware, not only of price, but also of the value they receive for their training dollar. An inexpensive school could turn out to be less than what you need, while the most costly could offer too many choices. Your career plan needs a budget to keep your eyes from making the decisions your checkbook should.

Will you train at a FAR Part 141 school or under Part 61 regulations? Part 141 is structured, with all paperwork constantly scrutinized by the FAA, while pilots can train in a less-organized, though not necessarily less-professional, atmosphere under Part 61. Many fine, well-structured flight programs are operating under Part 61, merely because a school doesn't want to deal with the hassle of paperwork and government "meddling" that comes with FAA approval under Part 141. *Part 141* dictates how a school may advertise its services, the requirements for the school's chief pilot's job, and how the school must keep records. The curriculum offers little flexibility over how a school handles training. Instead, Part 141 provides a set plan for what subjects the school covers and when.

A benefit of the Part 141 school over training under Part 61 is the total time needed to earn a certificate. An FAA-approved school (Part 141) can set you up for a private pilot-license flight test after only 35 total hours. A school under Part 61 requires you to have 40 hours. A Part 141 student at the commercial-instrument rating level could complete the course with only 190 hours compared to 250 under Part 61. But check with the school for realistic statistics. The FAA says most private pilots, for example, need about 72 hours before they see their license, no matter where they train.

A Part 61 operation could involve a single flight instructor with a single airplane or, as mentioned earlier, a large school that doesn't want to hassle with the paperwork for Part 141. Technically, a school under Part 61 needs neither a chief pilot nor a standardized, orderly curriculum to operate. Sullivan said, "It's sometimes advantageous for a customer to be under Part 61, particularly if they are bringing in previously logged time." Part 141 generally forces you to start at the beginning.

One of the main reasons some students choose Part 61 training, however, is price. Generally, a Part 141 school costs more, possibly because its facilities and fleet are larger and, therefore, carry higher overhead costs.

When you pin down your flight school possibilities to just two or three, you have no better way to pass final judgment than to visit them in person because, only then, can you decide if a school is right for you. You can often complete onsite inspections in a day if you will train locally. But, if one school is in Florida and another in California, you'll need cost-effective transportation for these visits. Don't think of skipping these site visits, however. There simply is too much important, although subjective, information to gain from them.

When Watzke visited her first California flight school, she found the place "too sloppy for my taste, with books and papers scattered everywhere. I decided the rest of the school must be like that, too." Eventually, she chose Comair Aviation Academy (now Delta Connection Academy) after reading some magazine reports about it. Although she did not visit the Sanford, Fla., school personally before choosing it, and even though she is pleased with her career now, Watzke recommended, "A new student needs to interview the school. Talk to other students, current and graduated. Try to meet more than just those students the school sets you up with, too."

It's good business to make an appointment. When you arrive at the school, find the chief flight instructor for an indoctrination chat to gain a good, gut level feeling for the company. How long did the chief keep you waiting? Did they seem interested when you spoke or were you quickly pushed off on an instructor to answer your questions? What did the school's lobby look like? Were the personnel friendly? Were you provided with added information quickly or did people pause when you talked about issues like refunds? Take a look at the aircraft, too. Are they recent models? How many are IFR equipped? What condition are the paint and interior in? A poorly kept outside could mean the school also skipped something under the cowling.

Check out the classrooms. Are they well lit and roomy? Can you imagine yourself spending several hours a day there? If you can't, find another school. Is there a place to eat nearby? Is there transportation to and from the school if you don't have a car? Does the school have simulators? How old are they? Does the school teach ground school only one-on-one or does it use a computer- or video-based system as a supplement?

Look at the flight schedule. Are airplanes and instructors kept busy or do you see large gaps in the day? Too many large gaps could indicate a problem picking up or keeping customers. Find out why. Ask to meet some of the instructors. Sullivan said, "Look at the interaction between the instructor and the student." How old are the instructors? What types of experience do they have? Do they seem friendly? Ask what the school's policy is on changing instructors if you don't click with the one they assign to you. Remember, your relationship with your instructor can be one of the most important of your career. You need to be nosy.

Some Value Decisions

You need only a few more items to round out the checklist before you make a final decision. Ask about each school's pass/fail ratio and how the school determined the figures. Some schools base the ratio on the percentage of students who pass the check rides the first time, while others exclude from the figures people who drop out of the program halfway. Ask about references. Will the school allow you to speak to some current students and some graduates? Sullivan said, "One of our best advertisements is our former students." A good school should be happy to provide this information.

Check the schools remaining on your list with the local Better Business Bureau (BBB). While the BBB won't offer specifics about any incident, it will tell you if any complaints were lodged against a school on your list. A check of a state's attorney general's office and the local FAA Flight Standards District Office may provide information about a school. FAA-designated examiners can also provide insight on the success of students from certain schools, and even certain instructors, when taking the flight exams.

Discuss money. Is there a cost to the student before training begins? Does the school want all the money up front or only part of it? Will the school allow you to pay as you go each month or after each lesson? Might a few more dollars on account deliver a lower hourly rate on its aircraft? What is the school's refund policy? Ask to see these items in writing before you sign anything. "Students can pay in full, up front, but I don't recommend it. I put a large sum of money up for a school once and four months later they went bankrupt," Parsons said.

Many schools require a contract or agreement before the training begins. A well-written agreement protects both parties if one of them does not live up to their word. If the fine print seems too confusing, consult an attorney before you sign. Parsons recommended students "pick up a copy of the school's rules and regulations, so they'll know what to expect."

The big question now is how to make the final decision. As you look at each school's name on your list, think back to your personal visits. Which of the schools left you with only a so-so feeling? Pull those schools out of the competition first. You might call the presidents of the remaining schools to tell them you are deciding between their school and one or two others. Ask if there are any final items they want to mention about their school to help finalize the decision. You never know what an owner may offer to entice you to train with that school, but it always is worth asking. Usually, though, the better facility will stand out in some way.

Watzke was not ready for one item in her selection of a professional flight school. "They work you very hard and some students don't expect that. I thought this was harder than graduate school, but the training was very good." After following through on her decisions, Watzke went on to fly as a Comair EMB-120 Brasilia First Officer. Those decisions, for her, resulted in the job that got her out of her rut.

Questions to Ask Before You Choose a Flight School

- Did the school allow you to fly with an instructor for an hour?
- Did it have any housing? Did they show it to you?
- What about insurance? What are the terms?
- Did you see the aircraft's logbooks?
- Will the training curriculum direct you to the job you want?

- If affiliated with an airline, does that mean a guaranteed job with that airline, or just an interview?
- Is there an affordable place to live at the school or near it?
- Does it offer a four-year degree in conjunction with the flight training?
- Is it a FAR Part 141 school or does it operate under Part 61?
- Does it offer or recommend someone for financing?
- What are its prices?
- How many hours will a person need before getting the licenses and ratings they seek?
- Is everything neat and clean in the main offices?
- Is there a decent facility to house the school?
- Do the instructors seem interested when spoken to?
- Do the personnel answer questions freely?
- Are the aircraft recent models?
- Are the classrooms well lit and roomy?
- What is the pass/fail ratio and how did the school come to those figures?
- Were you allowed to speak to current students, as well as graduates?
- Are there any complaints lodged against the school with the Better Business Bureau?
- What condition are the aircraft in?
- Does the school have simulators?
- How old are the simulators?
- How long has the school been in business at its present location?
- How much will the training cost be up front?
- Do you pay all the money up front or just half?
- Can you pay each month or at the end of each lesson?
- What is the school's refund policy?
- Does the school keep its instructors busy?
- What is its policy about changing instructors if you don't click with the one assigned to you?
- What is in the school's training contract or agreement?

More on Part 61 versus Part 141

The two sets of regulations that apply to flight school programs, FAR Part 61 and Part 141, outline the actual certification process for you to become a pilot. They also can be confusing to differentiate.

Part 141 is the regulation that tells a formal flight school—called an approved school—how it must operate, such as the requirements necessary for a chief flight instructor, how the school may advertise, and in what form the student records should be kept. Another advantage of the Part 141 school, as mentioned in the previous article, is that total times required before the flight test are less than if you were to pick up your license from a Part 61 operator.

Is a Part 141 flight school better than the freelance flight instructor whose ad you might see in the local paper on the weekend? In all fairness, it can be pretty tough to claim one is definitely better than the other. While the curriculum of the Part 141 school is certainly organized differently, it doesn't necessarily mean a Part 61 operator is disorganized. While being able to pick up a private pilot license in 35 hours instead of 40 is a benefit at the Part 141 school, the statistics say that few pilots complete training in 35 hours. The average private pilot picks up their license in about 70 hours total time. At the commercial, instrument-rating end of things, Part 141 training might make a bit more sense, depending on how immersed in the training you become. Realize, too, that while total times are less, the price per hour at the Part 141 school is usually higher.

If you're searching for flight schools that have been in business for a long time, you'll find that, for the most part, they'll be Part 141 schools. At a Part 141 school, you'll find yourself engaged in regular phase checks with other school instructors as a system of checks and balances on your education. You'll also have some assurances that the chief flight instructor needed a set amount of flight time under their belt before being put in charge of the school. A Part 141 flight school must even meet certain requirements for its pilot briefing rooms, as well as the ground-training rooms in which you'll be spending a lot of time. At a Part 141 school, the FAA pops in on a regular basis to be certain the quality of the students is also up to par. If you were engaged in any training to be paid for by the new Veterans G.I. Bill, the Veterans Administration would require your training be conducted at a Part 141 school.

So why would anyone *not* want to attend an approved school?

One of the first reasons could be that, in your location of the country, a formal Part 141 school doesn't exist. Another consideration is price, and that's often the most compelling reason people choose a freelance instructor. The reason the freelance price is usually less is simple business economics. A freelance instructor might only own one aircraft and only rent a tiny office for ground training. Of course, while all the amenities of the Part 141 school might not exist, the benefit comes at the end of the lesson with the lower overall bill. Recently, I saw an ad for a freelance instructor while I was traveling in a major East Coast city, so I called the teacher. She used an older model Cessna 172 that she didn't own. Another friend owned the aircraft and leased it to the instructor with the insurance. The instructor charged $60 per hour for the airplane and $20 per hour for her time. At the same time, a nationally known flight school charged $78 for a C-172, plus $25 an hour for the instructor. The bottom line is this: the facilities are nice and the Part 141 schools are more regulated, but you certainly will pay for them.

Freelance flight instructors might also be a little tougher to locate than a Part 141 school. Try searching the classified ads in your local newspaper, and you'll probably see, "Private flight instruction. Your airplane or mine, $20 per hour, call John. . . ." Make the phone call, just as you did at the Part 141 school; make the appointment to visit the aircraft, and then make your decision. In both cases, I'd ask for the names and phone numbers of a few recent graduates as references. While the privacy laws in some cases could make the operator a bit wary of giving out this information, any legitimate school shouldn't have trouble putting you in touch with people who have used their services. If they refuse, I'd walk out.

While prices might be less for the freelance instructor, you should consider some potential problems. If the instructor is only equipped with one aircraft, it might be pretty tough to book that airplane for some of the long cross-country flights you need to complete your license. What if the aircraft has a mechanical breakdown somewhere along the way? A large school will simply switch you to another airplane, while the freelancer might need to switch you to another day entirely while the aircraft is repaired.

Consider the difference. I've taken instruction in both kinds of operations, and I've also worked in both as a flight instructor. I once found that a simple cup of coffee drunk standing near the wing of a Cessna 150, listening to a freelance instructor tell me some of his adventures and discussing what I wanted out of my private license, was enough to make me work with him. Later on, for me at least, when I started working on my instructor ratings, I found that a formal Part 141 school fulfilled my needs because I thought it was better organized.

Finally, reserve the right before you even walk through the door of the flight school, that if things just don't look, feel, or sound right, you'll leave. You probably have more than one flight school near your home, so make sure you like the one where you'll possibly be spending thousands of dollars.

Let's take a quick check of some of the bare-bones costs you might expect to train for a stand-alone commercial pilot certificate in the United States. Outside the U.S., the prices are significantly higher.

Commercial Pilot

- 240 hours Cessna 152 (aircraft rental only) @ $70 per hour = $16,800
- 50 hours dual-flight instruction (instructor fees) @ $35 per hour = $1,750
- 10 hours Cessna 172 RG @ $140 per hour = $1,400
- 50 hours ground instruction @ $30 per hour = $1,500
- FAA check ride: $400 (approx.)
- Commercial Pilot Certificate cost from zero time = $21,850

Instrument Rating

A commercial pilot certificate holds little value without an instrument rating attached. If you picked that up from a freelance instructor, the totals might look like the figures below. I'm assuming most of the rating is conducted in the aircraft. An airplane has too many distractions—noise, bumps, and so forth—to make it worthwhile to do all the training while you're airborne. A good simulator offers you a place to learn and practice the procedures you'll refine in the airplane. My estimate includes some simulator time. Although the price may be less than a Part 141 school, I would search hard for an instructor who at least had access to a simulator, even a PC-based training device, which you read about in Chapter 8.

- 50 hours in Cessna 172 @ $90 per hour = $4,500
- 25 hours of ground instruction @ $35 per hour = $875
- 25 hours of simulator instruction @ $100 per hour = 2,500
- FAA check ride = $400 (approx.)
- Instrument Rating Total = $8,275

While the commercial certificate under Part 61 requires 250 hours, that does not guarantee you'll be able to learn all the maneuvers proficiently enough to pass the exam in that time. Most pilots take a little longer. Look at the total Part 61 price of $21,850. This seems like a great deal of money, and it is. Another benefit, though, in working with a Part 61 flight school is this: the owner might be willing to negotiate the rate down for you from the $21,850 total, if they know you'll give them this much business. This would not normally be something you could plan on accomplishing at the larger Part 141 operators. However, some of the large Part 141 schools might offer advantages that a smaller Part 61 operator might be incapable of, such as better classroom or simulator facilities. **Note:** The price of the written knowledge exams is not included in the previous costs because those costs can vary significantly from locale to locale.

The Knowledge Exams

All pilot certificates and ratings also require the applicant to pass an FAA written exam, now called a knowledge exam, prior to the day you take the final flight test for each rating. One rating—the multiengine—is an exception and does not require a written knowledge test. These written exams have always been a requirement and should be looked at that way, as a requirement. I personally question their value in deciding what kind of a pilot you'll become. I know of some pretty spectacular pilots who freeze up in a written test situation. Personally, I've never been good at them.

The writtens can be passed in a number of ways. You can buy any of a number of excellent books on each rating and study them on your own. You could sit down with an instructor for enough ground instruction to pass the test. You could also take a weekend ground school and spend ten hours on Saturday and Sunday having the needed material crammed into your brain, so you'll be ready for a Monday morning written exam. Another method, becoming more and more common today, is to buy a set of ground school video tapes or DVDs from any of a number of companies, such as King Schools or Jeppesen. The ads for their tapes can be found in many of the aviation magazines. The major benefits to this system are you not only work on the ground school at your own pace, but you can also spend as much time on a subject as you need to make it sink in.

Here's a sample of the videos available from the King Schools to help you prepare for the FAA Knowledge exams. I've also used the King Computerized Exam Reviews to help me study questions I once took on my commercial, instrument, and ATP writtens. It's amazing how many times questions from these exams reappear during the interview process.

Written Exam Video/DVD Courses

- Private Pilot
- Instrument
- Commercial/Instructor
- ATP-135
- ATP-121

- Flight Engineer
- Instrument Instructor

Computerized Exam Reviews

- Private
- Commercial
- Instrument
- Flight Instructor—Airplane
- Flight Instructor—Instrument
- ATP-135
- ATP-121
- Flight Engineer

   Call King at 1-800-854-1001.

Ab Initio Training

Another program to consider is the ab initio system. Basically, the *ab initio system* takes a pilot nonstop from absolutely zero flight time up through a right-seat job in a regional airliner or, at the least, all the professional pilot certificates, including the various flight instructor ratings. Although the expense for a flight training program such as this is considerably higher since it includes the specifics of a particular airline, the benefits are usually a flat price for all the ratings, as well as a specified time period for completion. In Europe, this training concept is used by major airlines such as Lufthansa. In Asia, Japan Air Lines (JAL) also uses ab initio training. What the airline eventually ends up with when a student pilot completes the training (which often takes years) is a pilot who is totally immersed in this particular airline's methods of operation. The program also provides the airline with a certain amount of employee loyalty.

   One young Italian pilot I recently met came to the United States as part of his country's ab initio training because the price to train a pilot in the United States is considerably less than comparable training in Europe. At the completion of his training (with a total of about 700 hours, much of it logged here in the United States), this Italian pilot returned to his homeland, where, with a final few months of training, he picked up a type rating in a Hawker-800, and then began flying in the right seat of a DC-9.

While some controversy about the qualifications of these pilots surrounds ab initio training from an operational status, there's no doubt in my mind that being totally involved with your career, be it medicine or airplanes, is the way to go. From an operational standpoint, many pilots don't believe a relatively low-time pilot belongs in the right seat of a high-performance turbo prop or pure jet aircraft. They believe only actual experience—total logged hours—can indicate a pilot's ability to cope with a difficult situation aloft. We've only to look at the military operations of the United States for another view of this controversy. In the U.S., pilots with only a few hundred hours of total logged time are out flying high-performance supersonic fighter aircraft by themselves. Obviously, then, low-time pilots can become productive members of a cockpit crew.

The only stumbling block with ab initio training is the price, but even that is relative because, again, value, not simply cost, is the most important aspect of training. Definite differences exist in the curriculum of any school, as well as differences in price. You'll have to be a good shopper and spend the time to check out everything one school offers compared to another. Location can also be important, as can the kind of housing you'll have while you're there. When you pare your decision down to only two or three schools, I'd take the time to visit the campus and speak to some of the people who run the school, as well as to some of the students. Price and paperwork are certainly going to give you some direction, but again, only an onsite visit can make the final decision for you. Let's look at two of the most well-known flight schools in America and how they approach pilots who begin from the zero experience level.

Commercial Flight Schools

Delta Connection Academy

Large commercial flight schools can provide a wide variety of flight education services. Sanford, Florida–based *Delta Connection Academy* is one of a number of the largest schools that specializes in just about everything a new pilot needs to send them in the right direction toward a cockpit job. Delta Connection Academy, a Part 141 school, provides an accelerated professional pilot-training program from absolutely zero time through the certified flight instructor ratings and a Jet Transition Training Course. With programs including training in a full flight simulator at Delta Air Lines and flying aircraft with the latest in glass cockpit technology, Delta Connection Academy truly offers the complete training package for a future airline pilot. The training is airline standards–based, which is reflected in academics, simulator, and flight instruction programs.

Headquartered in Florida, where the weather is great most of the year, Delta Connection Academy operates 117 aircraft. The Academy president is Gary Beck, a former chief pilot and senior vice president at Delta Airlines. Rich Morris is the Academy's director of flight standards and chief pilot. Delta Connection Academy flies the Cirrus SR-20 G2 aircraft (Figure 2-4) and the Piper PA-44 Seminole. This change to their fleet occurred in December 2006. Prior to this date, they operated the C-152, C-172, C-172RGs, and Piper Seminoles. These aircraft are operated by a highly trained core of 225 flight instructors across the company. The school also has 11 Aviation Training Devices (ATDs) and seven Level 6 Flight Training Devices (FTDs) manufactured by Aerosim Technologies, which the school integrates throughout all phases of the training.

In training circles, simulators are categorized by letters and numbers according to their capability to accurately reproduce a specific aircraft's flight characteristics. Today, a pilot can attend an aircraft-specific ground school, take a complete final check ride in the same Level D simulator they trained in, and win their type rating without ever having set foot in the actual aircraft. The Level 6 FTD used at Delta Connection is almost as powerful as its Level D simulator in reproducing the feel of a regional jet. A Level 6 FTD can also be used for up to 80 percent of the training required

Figure 2-4 Delta Connection Academy will begin training new flight students in the Cirrus in 2007. (Courtesy Delta Connection Academy.)

for an instrument rating. The Academy plans to integrate the use of a Level 6 device for training that soon will begin in a Cirrus aircraft that will reduce the total time needed for a pilot to pass a combined private/instrument rating check ride.

The standard of training at all schools might obviously vary some around the Part 141 regulations, but how those variances take their form is the crucial point. Delta Connection Academy, owned by Delta Air Lines, has chosen to market itself as an airline training school. Right from the start, academy students are taught with manuals written just like those provided to Delta's new hires. The Academy's Training Doctrine and Training Course Outlines are based on the central concept of standardization, as the term is understood and applied in commercial airline operations. This idea is reflected in training aircraft standards manuals and checklists, in construction and content of all courseware, and in the selection, initial training, and transition training of flight instructors and the details of operations.

What's the point of all this airline consistency? Why should a prospective professional even consider a place like Delta Connection Academy for their training? Most people need only to look at the chance provided to some of Delta Connection Academy's graduates of the school's Jet Transition "Bridge Course" for their answer. *Bridge Course*, built around the CRJ 200, CRJ 700, and the EMB 145, is primarily designed for pilots with some experience who are looking for that first step up into the right seat of an airliner. Delta Connection Academy maintains relationships with a number of regional carriers.

Rich Morris added, "Any pilot application we send through to any of the carriers we work with, such as Skywest, ASA, Mesa, American Eagle, and Express Jet, are handed right to the Human Resources recruiter. That's part of our bridge program agreement with students. If they complete the course, we guarantee them an interview." Currently, minimum qualifications are set at a commercial certificate with multiengine land and instrument ratings, 1,000 hours total time, and 100 hours multiengine time. Morris said, in some situations, regionals have begun talking to pilots with as little as 600 hours total time. In March 2007, American Eagle reduced their requirements to 500 hours total time.

While Delta Connection Academy is not the only flight school around that will train you and, later, assist with the interview and hiring process at a regional, they're unique because they're owned by one of the largest airlines in the world. This means they provide the most up-to-date training and airline preparation available.

Morris discussed an issue we also talk about later in this chapter—whether a pilot needs a college degree to be successful. Although many of the major airlines have changed their requirements just slightly by now using the word "preferred," rather than "required," Morris believes the four-year degree will again soon become the standard. "We have partnered with a number of institutions, like Jacksonville University, to help our pilots obtain their degree. Many of our students work on their degrees through online programs while they attend here and complete those degrees after they graduate."

The goal is to reach the magic numbers of 1,000 total time and 100 multi-engine. When each student arrives at that point, they're guaranteed an interview with a selection of airlines as a potential First Officer with Comair Airlines, ASA, Chautauqua, Mesa, Mesaba, American Eagle, SkyWest, Pinnacle, Trans States, and many others, which is definitely one of the benefits of attending a school owned by an airline. Delta Connection Academy Vice President John O'Brien reports, "In 2004, 2005, and 2006, 100 percent of the students who completed our training program have been hired by one of the airlines with whom we have guaranteed interviews. We give the airline a proven product in our pilots. They know the standards we train them to." You can find out more at www.deltaconnectionacademy.com or 800-U-CAN-FLY.

And, because one of the regular questions asked at the Delta Connection Academy is about the cost, Morris said in late 2006 that a zero-time pilot, who has never before seen the inside of an airplane, could expect to spend $72,000 from day one to the first airline job. Delta Connection Academy also offers their graduates a number of methods to build their time if they decide to remain with the company as a flight instructor, as well as upset flight training as a standard requirement for all pilots.

FlightSafety Academy

Based in Vero Beach, Florida, where the weather is great most of the year, FlightSafety Academy shares the name of the world's most well-known training company: FlightSafety International owned by billionaire Warren Buffet. In fact, Dick Skovgaard, the academy's director and center manager, says the number one way people find their way to his school "is through word of mouth." FlightSafety International trains 60,000 to 65,000 pilots each year, in the full-motion simulators at their airline and business aviation learning centers around the globe.

FlightSafety uses three different highly visible programs to direct candidates to any of three well-defined paths to a professional pilot career: the Advanced Airline track, the Business Jet Direct track, or the Instructor track.

The *Advanced Airline track* is designed for pilots with some previous flight time and a commercial license. Graduates are guaranteed an interview with a participating airline once they complete the course, which is comprised of four phases. This not only brings the pilot's skill levels to the required minimums, but it offers 25 hours of multiengine time in the Piper Seminole and 44 hours of time in an ERJ full-motion simulator, 22 from the left seat and 22 from the right.

The *Business Jet Direct track* is designed for a pilot whose ultimate goal is to fly a corporate aircraft (Figure 2-5). Because FSI has dozens of simulators in almost every part of the United States, they've developed a plan to take their own instructor graduates and place them as right-seat simulator partners for regular FlightSafety Initial and Recurrent Training customers. Once assigned to a Learning Center, the instructor goes through the same ground school as FSI customers and is, eventually, SIC checked in a variety of airplanes, from a Hawker 400A or a Gulfstream 550 to a Challenger 605. Not only does this arrangement offer young pilots considerable opportunity to experience the aircraft, but they are also presented with many opportunities to meet FSI customers who operate these aircraft. Many have been hired away to a corporate flight department (Figure 2-6).

The third method to build time and experience for FSI graduates is to remain on the property as an instructor (the *Instructor track*) in one of the company's 88 aircraft: 62 are single engine, 26 multiengine, and two are Zlin 242 aerobatic aircraft. The academy campus also uses four Level-D ERJ-145 and one Saab 2000 turboprop simulators. Once students complete the instructor course, they can stay on at FSI to teach. If they commit to give the company back 800 hours of dual-given instruction, FSI will reimburse the pilot for the cost of their instrument and multiengine instructor add on ratings. The instructor also has the opportunity to pick those up at an employee-discounted rate to begin with.

Figure 2-5 FlightSafety Academy offers students a number of professional options. (Courtesy FlightSafety International.)

Figure 2-6 FlightSafety Vero Beach campus. (Courtesy FlightSafety International.)

Like every other kind of academic institution, Skovgaard says, "the school you attend does make a difference." While the quality of the training received is very important, he adds, "Students can't forget that employers are looking at the entire person during an interview. They want integrity, good working skills, and people who are well-groomed."

Skovgaard notes, "There is definitely a perception about an upcoming pilot shortage. That's why I believe the regionals are looking so far ahead to their own needs. The FlightSafety Academy also works with students from other parts of the world. We work with Swiss, RWL Flying College in Germany, and PTC College in Ireland. Those companies start these employees on some of the European Joint Aviation Authority (JAA) training they need and we continue the work. The work here is under FAA rules, but geared to JAA guidelines." When asked about bringing new students to Vero Beach from overseas, Skovgaard says there "is more the perception of difficulty than there really is. We have a person on staff that has expertise in filling out all the forms an international student requires."

At the FlightSafety Academy, Skovgaard says the zero time to commercial, multiengine instrument course costs $53,000 as of January 2007. The CFI track adds about $11,000 more, for a total of $64,000, while the Direct Airline Track brings the final total to about $75,000. Instructors who remain with FlightSafety after graduation are paid on a per hours basis, as well as with employee discounts and reimbursement plans for additional ratings (http://www.flightsafetyacademy.com/).

Training Standardization

One of the toughest parts of making it to the big time for many light aircraft pilots is the transition from flying small, single-engine airplanes the way their flight instructors taught them, to piloting the fast, sophisticated aircraft that most corporations or airlines use. While some people are better pilots than others, it is the training, more often than not, that separates the successful pilots from the unsuccessful. Why?

When you begin flying a large turboprop or a jet, the qualification process involves performing the same kinds of air work you did in a C-172—stalls, steep turns, GPS approaches, and even emergencies. Many pilots fail simply because they have no idea of how to perform the maneuvers. And many companies will not take the time to explain them all to you. They expect you to know what to do.

What follows, then, is an example of an airline training manual that explains precisely how this company expected its pilots to fly the company's aircraft—both in revenue service and during training. Learning the maneuvers procedure-by-procedure made learning the actual handling characteristics much easier and considerably less stressful.

The other strength of a standardization manual emerges during emergencies. Committing many of these procedures to memory offers both pilots a little extra time to think through problems when they pop up . . . and they will.

In this updated edition of *Professional Pilot Career Guide*, I seriously thought about editing the following piece out because few companies fly the EMB-120 any longer, having moved on to the Regional Jets and even larger machines. I decided against removing it though, because the process is the critical element. Certainly other companies may use slightly different speeds of criteria for their operations because the aircraft are larger or smaller, but all professionally operated companies will include some version of this kind of manual, hence, my decision to keep it.

EMB-120 Brasilia, Flight Crew Standardization Manual

This manual and training curriculum has been implemented to develop high standards of training and coordination among all flight crew members (Figure 2-7).

Figure 2-7 Embraer's EMB-120 Brasilia—the first of the fast regional turboprops.

Flight crews are required to observe these procedures for the continued safety and comfort of our passengers in daily line operations.

Each crewmember will be trained exactly to the same performance standards to ensure total conformity among all crewmembers.

The following material is a guide to the procedures, maneuvers, and methods involved in line flying, flight training, and flight checks. Specific maneuvers and standards are indicated for particular types of flights, as dictated either by company policy or by Federal Aviation Regulations.

The description of procedures, maneuvers, and recommended techniques in this section compliment the material contained in the Normal, Abnormal, and Emergency section of the Aircraft Flight Manual (AFM). Pilots should be completely familiar with, and operate in accordance with, material in this section. The acronym for Flying pilot is FP and for Non-Flying pilot, it is NFP.

Flight maneuvers are to be flown to the following criteria for minimum proficiency:

- Ability to maintain altitude (1 or –) 100 feet
- Ability to maintain airspeed within (1 or –) 10 KIAS
- 0 to 15 KIAS engine failure climb
- 0 to 15 KIAS on approaches
- Smoothness and coordination of turns

- Constant degree bank
- Recovery of turns within 5 degrees of predetermined heading
- Precision approach
- Localizer and glide slope within one quarter scale
- Non-precision approach
- Localizer or VOR within half scale
- RMI or bearing pointer 1 or –5 degrees
- Tolerance on MDA 150 to –0

The Captain is always responsible for the aircraft, completion of checklists, and cockpit procedures, regardless of who is flying the aircraft. Duties pertaining to aircraft operation may be delegated to the First Officer, but final authority is vested in the Captain.

Checklists

The checklist will be consulted at all times during ground and flight operation. On the ground, the First Officer will read the checklist using either the Challenge and Response (C&R) method or a participatory method. In flight, the NFP will accomplish the checklist.

Maintaining good crew coordination by means of continuous communication is vitally important. At all times, the Captain and the First Officer should be fully aware of what is happening in the cockpit. This can greatly enhance the primary objective of our daily operation—SAFETY.

Aircraft Preflight

The First Officer normally assumes the responsibility of accomplishing the aircraft preflight. Both the cockpit checklist and the external checklist will be used to full extent. The First Officer will check that the maintenance release has been signed on the Aircraft Flight Log (AFL). The logbook will then be signed by the First Officer on completion of his preflight. Any problems found will be reported to both maintenance and the Captain. If the aircraft is deemed airworthy by the Captain, they will then sign the AFL and the logbook will be placed in the aircraft. All other paperwork should be checked, including: airworthiness certificate, registration, radio license, P.O.H., M.E.L., 135 manual, weight and balance information, and all aeronautical charts and plates.

Visually check the exterior of the airplane for its condition, security of access panels, and for signs of damage. Check for any liquids on the ground that may indicate engine, hydraulic, or water system failures. Check all areas visible from the ground. Emphasis should be placed on go/no-go items, such as nose wheel steering (NWS) over steering pin, oxygen and fire bottle discharge discs, defuel switches safety-wired, and so forth. Ensure removal of landing gear pins.

Cockpit Preparation

The first flight of the day, the Captain and First Officer will complete the turn around pre-start checklist by the Challenge and Response method. Thereafter, the Captain or First Officer (at the Captain's discretion) will perform the checklist alone. It is still acceptable, however, to complete the checklist by Challenge and Response, if desired.

Engine Start Procedures

Each pilot will start their respective engine. Under no circumstances (online) will a start be initiated without ground marshaller approval. Normally, the #2 (right) engine will be started first, followed by the #1 (left) engine. Always visually clear the propeller area prior to start initiation, and ensure all doors are closed prior to starting the #1 engine.

Taxi

Captains should practice taxi with and without nose wheel steering. NWS failure can be simulated by the IP depressing the NWS disengage button.

Ensure the aircraft is not moved until removal of attitude flags and presentation of attitude is displayed on the electronic attitude indicator (EADI).

Set power levers to Ground Idle (GND IDLE) and release the parking brake. Next, smoothly advance power to allow the aircraft to roll forward, and then reduce power to GND IDLE.

Control taxi speed with the power levers in the beta range.

Rudder pedal steering is used while taxiing where small directional changes are required. After completing a turn and before stopping, center the nose wheel to relieve the tire loading.

Scan instrument panels and observe instruments for normal indications, including the magnetic compass.

Both pilots should verbally clear the propeller and wing area, at which time the Captain will advance the power sufficiently to start the aircraft moving, and then retard the power to ground idle for the initial turn away from the ramp area. Condition levers are set to minimum rpm. Limit Propeller Speed (Np) to 65 percent. Brake smoothly whenever necessary, and then release brakes to allow cooling.

Turns should be limited to the total steering limit of 57 degrees. Rudder pedal steering is available to 7 degrees nose wheel deflection. Locked wheel turns should be avoided as they cause excessive loads on the gear.

Before Takeoff

After leaving the gate, the Captain will call for the pre-takeoff checklist. The checklist will be read by the First Officer, and the Captain will respond to all items pertaining to him, that is, altimeter setting, flight instruments, V-Speed review, and so forth. Preferably, the crew will complete the checklist prior to reaching the end of the runway.

Notes

A. Taxi speed will be slow (Jog). This not only enhances safety on the ground, but it also helps save brakes and tires, and enhances passenger comfort.
B. During periods of lengthy delays, single engine taxi is permitted to conserve fuel. However, both the starting engines and after start checklists will be used any time an engine is shut down and restarted.
C. Crew briefing: The following items will be covered by the flying pilot:

   1. Type of takeoff (normal VFR or IFR, crosswind, and so forth).
   2. Procedure to be followed in event of engine failure prior to V1, after V1, and after takeoff.
   3. Procedures to be followed for any emergency with respect to weather, gross weight, mechanical failure, airport facilities, runway length and condition, and so forth.

The Captain will always review the departure clearance, regardless of who is flying.

V1. Takeoff: General

Takeoffs may be broken down into normal, engine out, crosswind, day or night, rejected, instrument (lower than standard), or any combination thereof. Flying pilot is noted as FP, while the non-flying pilot is noted as NFP.

Normal Takeoff

The Captain will align the aircraft with the center line of the runway. The NFP will hold the ailerons to compensate for the crosswind conditions. The flying pilot will smoothly apply takeoff power to both engines and maintain directional control with the rudders. Before starting the takeoff roll, both pilots will verify the takeoff clearance.

The FP will:

- Advance the power levers to 75 percent torque
- Command set power at 75 percent torque

The NFP pilot will:

- Set the final takeoff power by 60 KIAS and call power set
- Monitor the flight & engine instruments
- Call out any malfunctions or abnormalities
- Call 80 knots (FP cross checks his airspeed indicator)
- Call V1
- Call Rotate
- Call Positive Rate of Climb (with positive indication on VSI and altimeter)
- Additional callouts as briefed

The use of nose wheel tiller steering after the aircraft is aligned is not recommended unless rudder pedal steering is inadequate. Approaching V1, forward control pressure can be reduced to just maintaining nose wheel contact.

At Vr smoothly rotate (approximately 2 degrees per second) to a positive angle of attack (10 degrees), and then make pitch adjustments to attain V2 110 KIAS or 10 degrees nose up maximum.

As soon as the aircraft is definitely airborne and climbing, the gear is retracted. Runway or noise abatement heading should be complied with initially.

Climb should be made at a speed of V2 110 KIAS or higher as necessary, limiting the deck angle to 10 degrees or less.

Accelerating through V2 120 KIAS and out of 400 feet AGL call "Flaps Up" Establish a climb at 150–180 KIAS.

If a close-in turn is required, flaps will remain at 15 degrees until the aircraft is established on the assigned heading, and then proceed at 400 feet. At 1,000 feet, AGL, the NFP will call "1,000 feet." The FP will call "Set Climb Power, After Takeoff Checklist," at which time the NFP will comply.

Note: The after takeoff checklist will be completed in its entirety with the exception of calling company operations. This should be accomplished only after climbing through a safe altitude and away from any terminal area. During operations in a terminal area, both pilots should be monitoring ATC and devoting full attention to the safe operation of the aircraft.

Use of Aircraft Lights

The anti-collision lights are to be on at all times when engines are running. The strobe lights are to be used at all times in flight unless they cause distraction (that is, in clouds). Taxi and/or landing lights will be used at the Captain's discretion, and below 10,000 feet, unless cruise altitude is lower, especially operating in terminal areas. Use of wing ice-inspection light(s) are at the Captain's discretion. *Logo lights* are also a valuable anti-collision tool and should be displayed at all times in terminal areas at night. Light bulbs are easily and readily replaceable. In effect, if you feel you need lights, by all means use them! Aircraft lights are the most effective anti-collision tools outside of your own eyes. Flight crews are expected to practice "See and Avoid" at all times.

Power Settings

Takeoff power should be maintained to 1,000 feet AGL, at which time the FP will reduce torque to 84 percent (or T6 not above 720 degrees) and the NP will reduce prop rpm, as required.

Instrument Takeoff

The following procedures will be followed when executing an instrument or lower than standard takeoff.

No rolling takeoff will be approved. Align aircraft with center line and position heading bug on lubber line heading of the Electronic Heading and Situation Indicator (EHSI).

After positioning aircraft on runway, also check EADI for level-reference mark, make initial power runup, and stabilize.

Continue takeoff using normal procedures: Initiate T/O roll, NFP call out 60 Kts.—airspeed alive, monitor power settings, and call V1, Vr.

At Vr, rotate to 110 degrees pitchup EADI (reference mark 110 degrees). At 400 feet AGL minimum, lower nose to 7 degrees EADI and accelerate to 150 kts, while ensuring a positive rate of climb and constant heading.

NFP will call out deviations from target pitch, airspeed, and attitudes.

Through the first and second segment, climb and maintain maximum takeoff power. At 1,000 feet, power reduction may be made and normal procedures followed.

Emergencies During Takeoff

Takeoff is one of the most critical phases of flight. Therefore, crews will abide by the following procedures.

Any abnormal indication observed by either pilot will be immediately called and identified, that is, "oil pressure right engine," and so forth.

Either pilot has authority to call for a rejected takeoff up to V1 by calling "ABORT! ABORT!"

The actions are to retard power levers to ground idle and apply positive forward pressure to control column. Using brake and reverse power (Beta), the FP will bring the aircraft to a stop. The Captain may, at any time, command control of the aircraft by calling "I have it." If the First Officer is flying, he will immediately relinquish control on command.

If the takeoff is rejected because of a fire warning or other such critical circumstances, the aircraft will be brought to a complete stop, parking brake set, and condition levers placed at minimum rpm prior to further action.

If the takeoff is rejected due to a power failure of either engine, extreme caution must be exercised in the use of reverse on the operating engine due to the yawing tendency that will be created. Maximum braking will be necessary, and again, caution must be exercised to avoid skidding, especially when runways are wet or snow covered.

Crosswind Takeoff

The maximum crosswind component, including gusts, is 25 knots. (If braking is less than good, reference appropriate crosswind limits.) The techniques required for this kind of takeoff are not much different than those used during the normal takeoff. The upwind wing will have a tendency to rise and aileron deflection should be applied in the direction of the crosswind to keep the wings level. This deflection will not materially affect the takeoff performance.

Some forward yoke pressure should be applied and maintained to ensure positive nose wheel contact with the runway. (Avoid excessive pinning of nose gear.) Rudder deflection will maintain directional control with nose steering initially and, as speed increases, aerodynamically. Aileron input should be decreased as speed increases. The primary objective of aileron input during crosswind takeoffs is to *keep the wings level*.

Engine Failure at or Above V1

If an engine failure occurs after V1, the takeoff will be continued. Use the rudder to maintain directional control and aileron to counteract the roll.

At Vr, rotate the aircraft initially to approximately 10 degrees pitch and adjust aircraft pitch to maintain V2 speed.

Retract gear when definitely airborne and climbing.

Check Auto feather complete.

If the engine has not feathered or is on fire, the memory items of the Engine Fire Checklist will be performed at this time. Continue to climb to 400 feet at V2. Level off and accelerate to V2 120, retract the flaps, set maximum continuous power, and establish a climb.

The Engine Fire Checklist should then be accomplished if it was previously started or the Precautionary Engine Shutdown Checklist will be accomplished.

Note: Retracting the flaps causes the aircraft to sink. Compensate for this tendency to prevent altitude loss.

Note: The FP will devote his full attention to the flying of the aircraft. The NFP pilot will monitor the progress of the flight, retract the gear, check for fire and feather the propeller of the inoperative engine, and make the appropriate callouts.

Caution: If an engine failure occurs, proper control inputs must be applied and maintained. All performance parameters are met even with the use of reduced power takeoff. If ground contact appears imminent, however, power on the operating engine may be increased, even to the physical limit of power lever movement.

Consideration may be given to returning for landing on a single engine, proceeding to an alternate airport, or attempting an engine relight.

Area Departure

The flight crew shall adhere to ATC clearances and use available navigation facilities and equipment, according to established procedures.

Climb

Climb power is normally set after the aircraft accelerates through V2 120 KIAS and flaps are retracted. Initial climb power setting of 720 degrees T6 may be used; reference the climb-power setting charts as soon as practical. Maintain climb speed of 150–180 KIAS, unless conditions dictate otherwise.

Single Engine Climb

Single engine climbs above acceleration height are made at maximum continuous power, until such time that the aircraft arrives at a safe altitude. At that time, power may be reduced to max inflight torque until the aircraft accelerates to desired airspeed. A safe altitude is determined by considering the airport traffic pattern, altitude assigned for radar vectoring, minimum sector altitude, MOCA, MEA, and so forth.

Turns

The pilot should be familiar with flight control inputs for coordinated turns in different configurations. This includes climbing and descending turns. The necessity for smooth and minimum control pressure use must be emphasized, especially at higher speeds and in turbulence.

Cruise

Care must be exercised in cruise not to exceed Vmo. Engine operation should be adjusted to maintain maximum or optimal cruise speed. Should a power split occur with power levers evenly matched, the power should be set according to equal torque values.

(A slight power split of up to 1/2 knob width is not uncommon in engine installations and, is, from the company standpoint, acceptable.)

At all times, vigilance must be kept by both pilots to "See and Avoid." Remember, ultimately in or out of radar, this responsibility is yours! Per FAR 135.100, any operation except normal cruise below 10,000 feet MSL is considered the *Critical Phase of Flight*.

Engine Failure Enroute

The single most-important aspect of an engine failure enroute is to maintain a safe airspeed. Power should be increased on the operating engine, and engine fire memory items should be accomplished until the propeller on the failed engine has feathered. After the propeller has feathered, Engine Fire or Precautionary Engine Shutdown Checklist should be completed, as appropriate.

Caution: At higher altitudes aircraft may be above single engine absolute ceiling which means the aircraft will be unable to maintain altitude. In this case, altitude must be sacrificed to maintain a safe airspeed.

Descents

Enroute descents do not require the review of a checklist. However, the following procedures apply:

- FP—Initiates descent, not to exceed Vmo.
- NFP—Calls 1,000 ft. above assigned altitude, 500 ft. above assigned altitude, and target altitude. Will call deviation of plus or minus 100 ft. from target altitude.

Descent in Preparation for Approach   Approach Descent—Requires use of the Approach and Descent Checklist. The checklist will be called for by the FP within 30 nautical miles of the airport of landing and/or 15 minutes. In all cases, it must be completed prior to the final approach segment.

- FP—Initiates descent, and calls for approach and descent checklist.
- NFP—Performs the checklist in command and response format.

During ATIS copying, or company calls, the NFP will always verify that they are *off frequency*, meaning not monitoring or talking with ATC. Comm. 1 will be used for ATIS and company calls, Comm. 2 will be used for ATC as a company standard.

Crew Brief   Approach Procedure—Will review appropriate approach plate.

- Any abnormal procedures
- Altitude callouts
- V-speeds (covered in Landing Data)
- Missed approach procedures
- Items to be covered in an approach briefing (specifically)
- Type of approach (ILS, VOR, GPS, NDB, and so forth)
- Calls for intercepting course (Localizer Alive)
- Calls for course deviation (1 or – dot or outside of 5 degrees for NDB)
- Call intercepting glide slope ("One dot to intercept")
- Calls at FAF/Clocks started
- Altitude calls of 1,000 feet to DH or MDA, 500 feet to DH or MDA; each successive 100 feet to DH or MDA
- Runway in sight or missed approach
- FP—Acknowledges checklist complete
- NFP—Completes checklist and advises FP "Approach checks complete"

Landings   Maintain an approximate 3-degree glide slope on all approaches (VFR and IFR). (Always transition and cross-check from the electronic glide slope to VASI when available.)

- Use correct approach profile and avoid low and slow/high and fast.
- Remember—a good approach leads to a good landing.
- Land in the touchdown zone. Do not attempt the "First Turnoff."
- Never position power levers aft of flight idle until the aircraft is firmly on the ground (all three gears!). When you are too high and fast—a go around is the best choice. Never force a landing.
- Retard power levers to ground idle smoothly prior to applying brakes.
- Airspeed may vary in line operations for traffic requirements. Under no circumstances will airspeed limitations per Pilot's Operating Handbook (POH) be exceeded.
- Always tune navaids (localizer) to appropriate frequency for landing runway.
- Electronic glide slope will be followed to touchdown zone in both IFR and VFR. If no localizer or glide slope is available, VASI will be followed (red over white).
- If a flap setting other than Flaps 25 is used, it must be covered in the crew briefing and a new Vref calculated.
- Radar Altitude is a good reminder of landing clearance. When cleared to land, set radar alt. to zero feet.

Parking and Engine Shutdown   The Captain will maneuver the aircraft into the gate following the ground marshaller's hand signals. (A marshaller must be present to park the aircraft.)

- Turning off the taxi lights upon arriving in the gate area is a signal to the marshaller that he has assumed control to direct parking.
- Upon coming to a stop at the gate, the Captain will set the parking brake and call for the "Shutdown Checklist." This checklist will be performed in the C&R method, with the First Officer reading the action and the Captain responding.
- Ensure that the flight controls are locked to prevent gust or jet blast damage.
- Where a GPU or APU is unavailable, it is necessary to conserve battery power as much as practical. Crews should exercise good judgment regarding the use of electrical power only when batteries are in use.

- If the aircraft is left unattended, all doors must be closed and locked, switches in the off position, and, if applicable, covers, plugs, and gust locks installed (overnight).

Engine Failures or Precautionary Shutdowns   Each emergency is unique, therefore, the circumstances prevailing at the occurrence cannot be predicted. Thus, standard procedures cannot be precise and well defined because priorities will vary according to circumstances.

The following notes are intended as a guideline to priorities in handling an engine failure.

- Fly the aircraft.
- Alter power and configuration as necessary to achieve required performance.
- Identify and confirm the failed or affected engine.
- In the event of a failure, the Dead foot, Dead engine rule to identify the inoperative powerplant applies.
- Verify the engine with engine parameters (by both pilots).
- Generally, torque is the best indicator of an engine loss.
- In any event, both pilots are instrumental in determining a failure, regardless of who is flying.
- Carry out shutdown/fire vital actions per AFM. Standard terminology will be for the FP to first state "right" or "left" regarding the switch or lever positioning, for example, Engine on fire Right (#2) Engine.

Crew Briefings   The first flight of each day will include a full briefing between pilots for both departures and arrivals. The briefing will be concise and specific for the circumstances of the operation. Avoid unnecessary details where both pilots already know the procedure. If either pilot does not fully understand the briefing, they will discuss the problem. The Captain will always review specific ATO clearances, but in all cases, the FP will deliver the crew briefing.

Between two crew members who have not previously flown together, a pre-departure briefing on Standard Operating Procedures (SOPs) should be given by the Captain prior to boarding the aircraft to ensure standardization of procedures.

When both pilots are fully conversant with the intended procedures under normal conditions (that is, takeoff, landings), the items pertinent to each procedure may be shortened to "Standard Briefing." A full briefing will still be covered the first flight of the day.

Use of Checklists   The appropriate checklist will be used at all times. During ground operations, the First Officer will read and complete the checklist. All items necessitating a response from the Captain will be C&R method. Line Ups and After Landings should be accomplished silently by the First Officer.

Airborne, the NFP will read and complete the checklist once called for. The checklist will not be started until the Captain calls for it during ground operations, or the FP while airborne.

The "After Starting," "Line Up, After Landing" and "Shutdown" checks may be accomplished by flow pattern method. However, on completion of the pattern, the checklist will be consulted to ascertain items completed. The pilot performing the checklist will always acknowledge "Checklist Complete" when finished with a checklist. Do not become "checklist complacent." A wrong response or action must be corrected immediately.

Pay-for-Training

Some training concepts last forever, such as standardization, while some, such as pay-for-training, come and go depending largely on the supply of pilots. When pilots are plentiful, companies look for opportunities to reduce their costs. One huge bite out of every company's budget is pilot training. Here's a story about a training concept that, while essentially nonexistent today, might return.

During every airline hiring boom, thousands and thousands of well-qualified pilots move closer toward what they believe is the perfect flying job—an airline cockpit. As these airmen move ahead to replace senior pilots retiring from the majors and nationals, as well as to support airline expansion, they leave behind thousands of pilot positions that must now be filled by qualified pilots who have not yet had the opportunity to make it to the majors.

For example, when a pilot moves to the majors from a regional, someone gets a chance to move up to Captain—perhaps someone who is currently a First Officer. This change, then, spells opportunity for the regional pilot. And, as this soon-to-be Captain is trained and allowed to fly the line, some junior First Officer will move up in seniority to fill the vacancy of the new Captain. This also means the airline will probably hire a new class of First Officers, as well.

While all of this moving around spells success for pilots, it spells red ink for the company. The red ink comes from the extra dollars that must be spent to train new pilots and make them ready to assume command.

How much it costs to train a new pilot varies significantly by the type of aircraft and the size of the company in question.

To retrain a pilot to fly a Piper Navajo will probably cost no more than a few hours of flight training and some ground school. Let's assume that for a small charter company, bringing someone on payroll and training them to fly single pilot in that Navajo costs the company $3,000.

To train a new Captain on a Gulfstream 550 to replace a pilot who left for the airlines could set a company back $35,000—$40,000 once travel time to FlightSafety or Simuflite, hotels, and payroll are figured in—a pretty substantial sum. At the regional level, training a new class of Captains might fall somewhere in between these figures.

Some of the overall cost can be absorbed at various levels because most airlines have a training department that will normally include ground and flight instructors whose job it is to make these new Captains ready for the line to replace those who have left or are expected to. No matter how you slice it, though, training costs can be a significant part of any aviation company's budget. And, remember the small company that was only going to spend a few thousand to ready a new Captain? While that amount may seem small compared to the others we've looked at, that small sum might represent a sizable percentage more of that small operator's training budget than for the G-550 operator.

As small airlines and corporate and charter operators watched pilots leave their employ—some less than a year after they began—a few decided to take action. While most realized they could not stem the tide of pilots leaving the company for better pay and benefits, as well as the lure of larger aircraft, they did understand the need to control their costs and demand a better return on investment (ROI) than they had been willing to settle for in years past.

One solution for some companies came to be known as pay-for-training. Now, let's not to confuse this with the method you'll use to pay for your flight training, because the two situations are vastly different. *Pay-for-training* has also been called "Buy Your Job" by some, because only pilots with enough money to pay for the training were offered work, leaving many well-qualified pilots on the sidelines.

Essentially, an airline takes a look at your paperwork and decides to hire you. But, before they extend a regular job offer, they demand you successfully complete training in the aircraft that company flies—but you pay the bill. At times, this has also occurred in corporate and charter flying.

Some new-generation pilots who have graduated from college in the not-too-recent past may see this as an opportunity if they have the money to pay for their training. Indeed, some new pilots have found employment with regional carriers around the United States with unbelievably low total hours. One new hire at Chicago Express Airlines—the former commuter carrier for ATA that's based at Chicago's Midway Airport—had just under 300 total hours when he began training in the Jetstream 31 aircraft the company flies. He saw it as a chance to bypass the dues-paying that most pilots have done in the past. He had spent very little time as a flight instructor when he was interviewed with Chicago Express.

The company made a tentative offer of employment to this pilot with the understanding that it would be withdrawn if the pilot did not complete training successfully. A week later, this pilot was on his way to FlightSafety's St. Louis Training Center to begin ground school. Total cost to the pilot was about $9,000. This pilot was successful and began flying the line for Chicago Express just a few months later.

You may be reading this and scratching your head trying to figure out what the problem is then. Man or woman wants to move ahead, gets the training they need, and, well, moves ahead. They get a bill just like going on to grad school, right?

Well, not exactly.

One pay-for-training issue—and this may not seem like much of a factor to some newer pilots—is that traditionally in the airline industry, pilots have not paid for their own training. When you reached a point in your career when an airline gave you the thumbs up, they paid for the training.

Pilot qualification is also a factor in pay-for-training. During the recession of the early 1990s, thousands of well-qualified pilots were put out of work as carriers such as Eastern, Midway, and Pan Am ceased operations. While many of these pilots were qualified to fly aircraft that many of the regional airlines were flying, they could not be offered a position unless they were willing to write a check to pay for their training—often in an aircraft in which they were already qualified.

So why would an airline want to put a pilot in the right seat of an airplane when they have just four or five hours of time in it, when a pilot who has a few hundred or even a few thousand hours in that aircraft cannot get the same job? Quite simply, it's all economics. Paul Berliner, an airline Captain for a national carrier, says, "Pay-for-Training is also an ethical and a moral issue."

Many regional carriers began outsourcing their entire training departments to companies such as FlightSafety—this has not been much of an issue with the larger jet carriers normally, although a few did require pilots to pay for training. On the surface, they looked as if they were simply contracting out for a service that was too expensive for them to provide, but this was not the case. If you look through any of the magazines that offer flight training services, you'll see advertisements for dozens of companies that provide type ratings in large aircraft, such as the Boeing 737, for example, for a price.

The intrigue becomes more visible as you work some simple cost-analysis figures. If you can purchase a type rating in a 737 for $6,500, why should it cost nearly $10,000 to train you to right-seat standards in a Jetstream 31 (a 19-seat turboprop) where a newly trained Jetstream pilot leaves Flight Safety with no type rating to show for their efforts, although the company will type rate some pilots for an extra fee? Berliner comments, "A doctor takes their diploma with them after they complete their training. In a pay-for-training environment, the pilot walks away with nothing." If the pilot is later furloughed from the job they've paid for, they may get some help from Flight Safety to find a new position, but there is no guarantee.

So where is all the extra money going that pilots pay for their training? Certainly, Flight Safety is entitled to a fair profit for the work they perform for the airlines that contract with them. But what is a fair profit when you compare what a pilot receives versus what they spend? The solid profits of pay-for-training can only go two places—either as a bonus to Flight Safety or back into the airline's pocket. Pay-for-training, then, has transformed what was a cost center at some airlines to a profit center at others. That's tough for many pilots to stomach because these airlines don't make other professionals pay for their training to work there.

Berliner adds a note of caution, too. Pilots who see this as a quick way to climb the ladder should consider before they sign on the dotted line. "There is backlash from the pilot community at the majors (about pay-for-training). On every pilot-hiring committee I know of pilots who are vehemently opposed to pay-for-training. They don't believe it is fair for a pilot to bypass the traditional dues-paying method of gaining flight experience because they have money." Every applicant who gains their job through a pay-for-training environment must ask themselves whether they want to go into an interview at a major airline knowing they are blackballed to some extent. "Even a one-percent chance is not worth it," according to Berliner.

But the economy is changing and with it—perhaps—the way some companies view pay-for-training. While many pilots would like to believe the change has evolved through some sudden moral insight at the airlines that use it, the law of supply and demand has simply begun to run its course. As more and more pilots are being brought into the major and national carriers, the supply of well-qualified pilots willing to pay-for-training has dropped beneath the level that some carriers can use to still keep their operations running—despite low-time pilots who are willing to pay for training.

Significant changes to pay-for-training have occurred at USAir Express carrier, Commutair, and others. Even *Continental Express* (now *Express Jet*), the former regional airline for Continental Airlines, a company that tried to make pay-for-training palatable for some by agreeing to cover training costs for pilots with more than 2,000 hours total time, has now reduced—but not eliminated—the requirements to qualify for free training. Effective October 1, 1998, all Continental Express applicants with more than 1,200 hours total time and 250 hours multiengine no longer have to pay for their jobs.

So how can companies recoup their training costs from pilots who were not staying with a company long enough after training to make it all balance? What we are left with, as pay-for-training begins to make its exit, are training contracts. These *training contracts*—while not forcing a pilot to pay for the training out of their own pockets before they begin to work—do require a pilot to sign a legally binding document that asks for money in return if a pilot leaves employment before a specific number of months after completing training.

A Citation 3 Captain might be asked to agree to remain employed with a company for at least 12 to 18 months after that company pays for the type rating. Some contracts are worded in such a way that the pilot is denied any right to legal recourse and, normally, they contain no provision for getting out of the contract once it is signed, no matter what happens. Many would force a pilot to pay for their training in full if they left one month before the agreement was set to expire, with no proration clause available. Legally speaking, most training contracts have been found unenforceable in a court of law, and it costs more to prosecute the pilot than the company stands to gain if it wins.

But the policy of asking or forcing pilots, depending on your perspective, to buy their jobs still exists. Watch for it when a job opportunity looks a bit too good and avoid it at all costs. This can be dangerous to your career!

This Letter to the Editor about pay-for-training appeared in the November 9, 1998, *Aviation Week & Space Technology* magazine. The letter was called "Profits on the Pilots' Backs," and it read: "I'm glad my employer—Continental Airlines—has had its 14th quarter of record profits. In the meantime, the 1,287 pilots at Continental Express—who will soon be flying the first all-jet regional—are in their 18th month of working without a contract. The majority were required to pay a $9,280 'training fee' to get a job that started at $13.25 per hour. Any wonder how this company can be so profitable?" Name Withheld by Request.

The Four-Year Degree   But FAA ratings aren't the only items necessary to become a professional pilot today. While a pilot's position used to be reachable without the benefit of a college degree, that sheepskin is effectively a requirement in most segments of flying. Today, many airlines still list a college degree as encouraged, while other airlines and many corporate operators list the degree as a requirement. Either way, if you intend to compete with others in the aviation game, you're going to need a degree. Many fine state universities, such as Purdue, University of North Dakota, University of Illinois, Auburn, Embry-Riddle University, and others, combine a standard four-year degree with programs specifically designed to offer the student all the professional flight ratings they need to make their mark in the profession.

It's a tough call whether picking up your ratings on the side while you complete your college degree is less effective toward the goal of flying for a living than attending a four-year university that also happens to run a flight program. Chief pilots I interviewed varied considerably, from only caring about whether the pilot held the ratings plus the degree to asking what the degree was in. Many corporations are currently asking about the degree itself, so a pilot can become a more useful member of the corporation during the time they aren't flying. Some chief pilots did seem to hold a soft spot for pilots who attended the same school as themselves, at least long enough for the new applicant to get their foot in the door. After that, the applicant needed to compete on the same footing as everyone else.

Financing

Most careers offer the student not only the chance at a lifelong job they'll hopefully enjoy, but also a method of spending substantial amounts of cash in a relatively short period of time on the necessary training. Where do professional pilot students look for the financing they need to cope with the big bills that accompany this kind of training? The best place to search for money initially is to talk to the school where you intend to train. If you're planning to work with a local freelance instructor, the chances of picking up the financing you need are pretty slim. (Remember Ed … He doesn't finance and the bank doesn't teach flying.) If, however, you approach the people at some of the larger flight schools or universities, they should be able to steer you in the right direction.

While Delta Connection Academy is dedicated to the high-quality products and services it sells, it must stay in business to be able to offer those services—something that might be pretty tough if it had no way to help its customers pay their way. I doubt Sears or Wards would try to sell a washer and dryer without offering a way to finance the purchase.

Most local banks offer student loans normally backed by federal financing. Whether the rates are competitive will be a matter for you to decide. Because of the variables in financing (such as interest rates and repayment plans), shopping for a loan can become just as big a project as finding the school in the first place. Depending on the state of your finances, be ready for the fact that many financial institutions might not be willing to make you a loan without a cosigner to guarantee repayment. Regular state universities normally offer tuition financing through the campus financial aid office, so the best advice on financing is, first to locate the school you're interested in, and then talk to the school's financial aid counselor. Remember, too, that while the lowest monthly payments might at first look like the best way to finance your education, a slightly higher payment or a slightly shorter repayment period could shave thousands of dollars off the total amount of money to be repaid. Be sure to ask about these possibilities.

If you head to your local library's reference desk, you can find a number of good books available to locate financing for your educational venture, so don't just depend on your local financial or educational institution for help. One school we interviewed was willing to finance a pilot's flying, but at annual rates near 14.5 percent, considerably higher than a regular student loan. People took them up on this rate on a regular basis because of one reason . . . the financing process was relatively quick and easy to complete. Many students, the financial adviser admitted, never even asked about the interest rate, being more concerned about how long the repayment period would be and when the payments would begin.

Another option to consider is scholarships. The University Aviation Association (UAA), http://www.uaa.aero/, has put together an up-to-date scholarship directory available for purchase, in either book or CD form, at its web site. The price of the "Collegiate Aviation Scholarship Listing" is currently $19.95. UAA also offers a *Collegiate Aviation Directory* for $29.95. Call UAA at 334-844-2434.

The UAA's Executive Director Carolyn Williamson said, "The 2006 edition of the 'Collegiate Aviation Scholarship Listing' includes 777 scholarship awards worth over $1.2 million. While many are based upon need, some are based on achievement and proven ability to succeed, as well as other criteria."

Williamson offered students a few tips to consider when applying for financial aid. "The best strategy is to follow the application process to a *T*. If the guidelines say, submit three typed copies, do not submit two or four copies and do not submit a handwritten application. Students should also apply for as many as scholarships as they can. If a student meets the minimum criteria, they should apply. Many aviation scholarships go unawarded every year because there are no (qualified) applicants. If an essay is required, applicants should align their essay with the goals of the sponsoring organization. Be honest on the application. Write legibly! And keep the grades up."

Banks

For some strange reason, many pilot applicants do not think much about traditional academic financing when they think of flight education. Here's a short list of institutions that understand aviation and are ready to listen.

- PNC Bank Resource Loan Program—1-800-762-1001—
  www.pnconcampus.com

- Key Alternative Loan—1-800-539-5363—
  www.key.com/educate/alternative
- Wachovia Education Loan—1-800-338-2243—www.wachovia.com
- TERI Undergraduate—1-800-255-8374—www.teri.org
- Key Achiever Loan—1-800-359-5363
- SunTrust eCareer Educational Loan—1-888-816-2797
- Gate Loan—Bank of America—1-800-344-8382—
  www.bankofamerica.com/studentbanking
- SLM Financial/Sallie Mae—1-877-834-9851
- Citibank Flexible Education—1-866-293-8761

The G.I. Bill

If you're a veteran, you might be eligible for government assistance that could pay as much as 60 percent of the cost of adding additional flight ratings. As with any government program, there are requirements and possible red-tape delays, so if you think you might qualify, contact your local Department of Veterans Affairs (see Figure 2-8) office as soon as possible.

To begin G.I. Bill–qualified training, you must hold a private pilot certificate and meet the physical requirements for a commercial certificate. In addition to the school you select being FAA approved, that school must also meet the VA school requirements. Just being FAA approved doesn't necessarily mean the school is VA approved, either. Visit www.gibill.va.gov.

Internships

Internships can be a part of anyone's career path, but they are neither plentiful nor easy to win. Internships in most careers have been around for years as a method to show young people a possible career track, while at the same time, giving a potential employer a chance to see how the young person performs in the real world.

Most internships are arranged during the third or fourth year of college, as is the one we look at here: the pilot internship jointly arranged with United Airlines and participating colleges. The National Business Aviation Association is also organizing an internship program for students interested in the corporate-flying end of the profession.

At United Airlines, this program, officially called the Flight Officer College Relations Program, has been running since 1983. A student

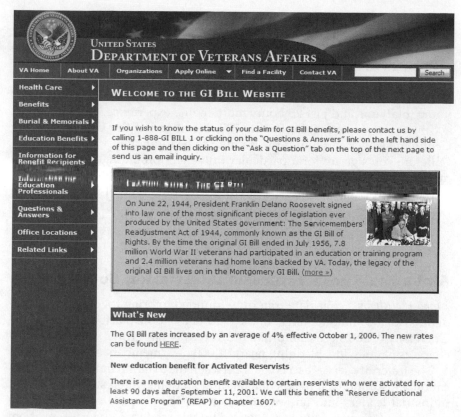

Figure 2-8 The Veterans Administration page for G.I. Bill recipients.

enrolled at one of the participating universities fills out an application. If, after interviews, the applicant is accepted, they take a full semester off school to work at one of United's facilities.

In the fall of 1998, United Airlines had 22 interns on board around their system. United's Instructor Training Manager and Intern Program's Director John Bauserman said, "We interviewed 60 students to come up with 22 interns. To apply, a student must be in their senior year of college, already hold a commercial pilot certificate with an instrument rating, and have a grade point average of about 3.0 on a four-point scale. In the past, about 250 interns of the 400 who have applied, have successfully joined United as pilots. One student I interviewed did his internship at Washington [D.C.'s] Dulles International Airport (IAD)."

Intern and Boeing 737 First Officer Arnie Quast began working in flight operations at IAD. "I had a chance to work closely with the pilots and get a good idea of what it was like to be a commercial airline pilot," Quast said.

But what made Quast's internship truly valuable was that, "I had the chance to jump seat on all the United aircraft that flew from Dulles and observe firsthand what crew coordination was all about. I also had the opportunity to fly many of the 27 United simulators in Denver."

After graduation, Quast continued to build his flight time as a flight instructor, while he waited for the call from United. With just over 1,000 hours total time and just 25 hours multiengine experience, Quast received a United class date for the Boeing 727. He attended the United Training Center in Denver and found the training to be top-notch.

Before you sign up for a particular university program, check for information on these intern programs. A phone call to some of the other airlines will reveal whether they're also involved in some type of internship program.

Alpha Eta Rho, a Professional Aviation Society

Alpha Eta Rho is a collegiate fraternity founded to bring together those students with a common interest in the field of aviation. Started in 1929 at the University of Southern California, the fraternity has grown to more than 70 chapters nationwide, pledging nearly 1,000 new members each year.

The fraternity serves as a contact between the aviation industry and educational institutions. It bands outstanding students, interested faculty, and industrial leaders into one organization for the purpose of studying the problems of everyday life as influenced by this modern industry—aviation.

Alpha Eta Rho serves to actively associate interested students of aviation with leaders and executives in the industry. This close association, strengthened through the bonds of an international aviation fraternity, establishes opportunities for all members to inspire interest and cooperation among those in the profession who are also members of Alpha Eta Rho. This fraternity continues to grow, and it serves as a lasting tribute to the farsighted understanding and vision of its founder, Professor Earl W. Hill.

Philosophy and Goals of Alpha Eta Rho:

- To further the cause of aviation in all its branches.

- To instill in the public mind a confidence in aviation.

- To promote contacts between the students of aviation and those engaged in the profession.

- To promote a closer affiliation between the students of aviation for the purpose of education and research.

The constitution of Alpha Eta Rho declares that the success of aeronautics depends on its unified development in all the countries of the world and on the cooperation of different phases of aviation with each other. Alpha Eta Rho affirms its character to be international and declares that eligibility for membership is not dependent on race, religion, nationality, or gender.

Visit them on the Web at http://www.alphaetarho.com/.

Let's take a minute and listen to how the entire process worked for one airline pilot.

Profile: Steve Mayer, First Officer, Boeing 757, Northwest Airlines

"Flying was always a part of my life," says Steve Mayer. "My dad is a former military pilot and now flies a 767 for Delta. Before that, he was a pilot for National and Pan Am."

Mayer was a keen observer of events around him, as well as a pilot who listened when a more experienced voice offered advice. "On my 20th birthday, I looked at my dad, a 727 Captain at the time. One of his friends, however, who had never gone through the military, was a 747 Captain. The 747 Captain was making a hundred grand a year more than my dad." Mayer decided to forgo the military route to an airline job.

"I really didn't start flying until I was 18 or 19 because I just didn't have the money to do it," Mayer recalls. "A pilot told me to get my commercial as soon as possible and I started when I was in college. I worked a lot every week while going to school and got my degree in political science."

After college, Mayer again listened to a more experienced pilot who told him the best way to an airline cockpit was with a ton of multiengine PIC time under his belt. "I decided to hop an airplane to the Virgin Islands because I heard they were always looking for pilots there. I arrived one night with my backpack and nowhere to sleep and just walked up the first hanger I saw. I asked if they needed any pilots and some guy said, 'Come back at 4 a.m. and you're hired.' I showed up and got the job, delivering newspapers, bank checks, fish—anything you could put in the Aztecs and Aero Commanders we had. I worked there for two years, logging nearly 2,800 hours, most of it PIC. I think I worked seven days a week actually."

Although the work was excruciating at times, Mayer had a plan to get himself through the long days. "I had a theory it was just a time builder. After I left this company, I got hired into the right seat of a

Jetstream 31 for Express Airlines 1 and stayed there two years, but never upgraded."

But, during his time as a regional pilot, Mayer says he "applied to every airline imaginable, here in the United States, and even internationally. I sent Northwest an application when they were not even hiring. I filled out the application in pencil, believe it or not. I just wanted to get into their database. Then out of the blue, Northwest called me. Later, during the interview with Northwest, they asked me about the penciled application. I just told them the truth. I also got the job."

Mayer's advice to applicants is this: "The airlines don't care what kind of a four-year degree you have, just as long as you have one. Avoid the fancy flight-school colleges. I have friends who are still paying off school loans and will be for some years to come. I thought maybe I was making a mistake by just trying to build PIC time any way I could, but I'll be a Captain here at Northwest when some of my friends are just being hired. But I also paid the money to get some simulator time before the interview at Northwest and it was worth it. I also bought a subscription to *Air Inc* to get a leg up on any information I could. Getting an interview at the airlines is like hitting the lottery, so you have to be prepared."

U.S. Military Flying

Since the winding down of the Cold War in Europe, the United States military entered what some experts called a slow-growth mode. This had become the same kind of downsizing that corporate America began experiencing in the early 1990s, with thousands less personnel being kept on active duty and many military bases around the world being shut down or reduced in size. What this change to the military means to you, if you're considering the military as a possible place to pick up your flight training, is best outlined in the following story from *Aviation Daily*: "Military Pilot Pool Will Shrink under DOT Cuts, Industry Told." It was penned before the invasion of Iraq.

The nation's airlines, which traditionally recruited 65 percent or more of their pilots from the military services, will face a shrinking supply later in the decade due to Defense Department budget and manpower cuts, and as the services entice more of their pilots to remain on active duty, Senator John McCain (R-AZ), ranking member of the aviation subcommittee, warned yesterday. In the near term, however, airlines will reap a bonanza as the military downsizes in the wake of the breakup of the Soviet Union, McCain yesterday told a committee looking into possible future shortages of pilots and maintenance technicians. Obviously, the conflicts in the Middle East have changed the direction [see Figure 2-9] of the U.S. military as this goes to press, so check with the recruiters mentioned later for the most up-to-date information.

James Busey, a former Department of Transportation (DOT) deputy secretary, told the panel, which is expected to issue a report in about a year on what measures can be taken to ameliorate possible shortages, that a shortfall in either category could have a profound impact not only on the industry, but also on the nation as a whole as far as its capability to compete on a global scale. "We're about to lose an important source of trained pilots from the military. The balance has now shifted and the supply will not be there," Busey said.

Because of the slowdown in the air transportation industry, it's now getting only about 45 percent of its pilots from the military, a figure that could get lower as the military pilot supply dwindles, said John Sheehan of Phaneuf Associates, the consultants to the panel that's chaired by Kenneth Tallman, President Emeritus of Embry-Riddle. Sheehan said air carriers will need about 2,400 pilots per year for the next ten years and regional airlines will need about 28,000 pilots over that period. The panel was originally to be a joint military-civil effort, but the downsizing military services aren't as concerned as they previously were. This will prove ominous for the industry later in the decade as a prime source of pilots disappears, both Sheehan and McCain said.

Figure 2-9 F-16 departs Shaw AFB, S.C.

If you plan to try to enter the U.S. military at some time in the future, you need to know that fewer flying jobs and tougher competition will exist for those remaining cockpit positions. Certainly, you'll need to talk to your Armed Forces recruiter when you're seriously ready to make this decision about your career because the volatile situation of the American military community could change at any time.

Let's look at the United States Air Force, for example. Because all Air Force pilots are officers, you need to apply and be accepted at Air Force Officer Training School (OTS) before anything else happens. The U.S. Air Force OTS is small and highly specialized. Candidates selected to attend OTS are college graduates and OTS prepares them for positions of responsibility, so they can lead the Air Force of tomorrow. The school's motto, "Always with Honor," reflects the ethical and professional standards expected of Air Force officers.

OTS is a fast-paced, three-month course. You complete it on-campus at Lackland Air Force Base in San Antonio, Texas, studying communication skills, leadership, management, military history, Air Force customs and courtesies, world affairs, and more. You take part in organized sports and physical conditioning to develop your confidence and teamwork abilities. The course is designed to aid your transition into the Air Force way of life.

To be eligible for OTS, you must be a United States citizen, 18 to 29 years of age, in good health, and able to pass a physical exam. You must be a graduate of an accredited college or university and have excellent moral character. You must also score well on the Air Force Officer Qualifying Test. To enter a technical or nontechnical career, you must be commissioned before age 30. Pilot or navigator training applicants must pass a flight physical and enter training before age 27½.

Pilot-training selectees without a private pilot's license attend the Flight Screening Program before going to OTS. This four-week course identifies your potential to complete undergraduate pilot training. Training time includes cockpit time in some single-engine aircraft to evaluate a potential candidate's talent at coping with the traditional flight environment during the Air Force's Flight Screening Program. The T-6 Texan (see Figure 2-10) is now the U.S. Air Force's primary turbine aircraft single-engine trainer. If you already hold a private pilot certificate when you apply for undergraduate pilot training, the screening exam is waived.

Training time includes cockpit time in a single-engine Cessna 172, called a T-41 by the Air Force. You need a solo flight to complete the Flight Screening Program. If you already have a private pilot license, the Flight Screening Program is waived. After OTS, you begin a one-year, intensive flight-training program. (See Chapter 7 for more details on Air Force training.)

Once accepted for OTS, you enlist in the Air Force in the rank of Staff Sergeant. The Air Force pays your way to OTS, as well as your flight train-

Figure 2-10 The T-6 Texan is now the primary trainer for the U.S. Air Force.

ing costs. While in training, you'll be paid nearly $500 every two weeks. You live on base and eat in the OTS dining hall. You pay only for personal items, such as laundry, postage, and telephone calls. A clothing allowance helps you defray the cost of uniforms. Upon graduation, you're commissioned a second lieutenant. OTS graduates who attend undergraduate pilot training incur an additional Air Force commitment of eight years after training is complete.

To learn whether you might be eligible to compete for a military flight training program, contact the information service of the branch of your choice at the numbers listed in Table 2-1.

Table 2-1 U.S. Military Information Numbers

Air Force	1-800-423-8723
Air Force Reserve	1-800-423-8723
Navy	www.navy.com/careers
Army	1-800-872-2769
Marines	1-800-627-4637
Coast Guard	1-800-424-8883
Air National Guard	1-800-864-6264

One U.S. Air Force Pilot's Perspective

I recently spoke to a young Air Force pilot, 1st Lt. Michael Fick, about some of his experiences during his Air Force flight training since he graduated from school. I think his answers about the training will tell you how viable this part of aviation is as a career field, although you certainly need to check into each different branch of the service for its particular requirements.

In 1983, the Air Force enlisted 1,590 new pilots. In 1993, that number had dropped to 700. The Air Force reports, of that current 700, 35 percent will be flying bombers and fighters, while the other 65 percent will serve out their tour of duty in airlift/transport aircraft.

Q: Why did you choose the USAF, Mike?
A: I had a great desire to continue my education after high school and very little money to do it with. I didn't want to be burdened for years with school loans, so I decided to apply to the U.S. Air Force Academy. My flight experience was limited prior to applying to the Academy, but my desire to fly was strong. I was turned down for the Academy my first year, but received a partial Falcon Foundation Scholarship to New Mexico Military Institute. This prep school provided me with the chance to prove my continued interest in the Academy. I graduated at the top of my class a year later and won an Academy slot.

Q: Do you plan on making the Air Force a career?
A: Right now I do. But the final decision depends on how well the Air Force treats me in the future, much like any other job.

Q: Would you tell me a little about Air Force flight training?
A: Sure. The training really began for me, though, at New Mexico Military Institute, where I learned about the attitude necessary to be a professional. That became even more firmly set at the Academy. My actual flight training began in the summer between my junior and senior year at the Academy, when I went through the required preliminary flight training in the T-41 (Cessna 172). Performance in the program, both G.P.A. and M.P.A. (Military Performance Average), allowed 125 out of 600 students to apply for the Euro NATO Joint Jet Pilot Training at Sheppard AFB, TX. Forty were selected. The program ran 13 months, where standard Undergraduate Pilot Training ran 12 months. A good majority of the training revolved around fighter tactics.

Q: What aircraft did you fly in training?
A: I flew the T-37 Tweety for the first six months and the T-38 Talon for the second six months.

Q: *What effects have the military cutbacks had on your career?*
A: Due to the cutbacks, there were very few fighter assignments when I graduated from training. Most were for tankers, transports, and instructor pilots. Some of the pilots were taken out of flying for a few years first, before being given their assignments. I had the C-130s, C135s, C-12 , or T-37 Instructor pilot jobs to choose from. I chose the C-12 (Beech King Air 200) assignment at Andrews AFB, MD, just across the river from Washington National Airport. This assignment should be followed by one in the C-141.

Q: *What did you enjoy most about training?*
A: I think the two and four ship formation flying provided the most serious adrenaline rush, as well as two-ship low level, where I flew at 420 knots about 500 AGL.

Q: *Have you thought about your flying once you return to civilian life?*
A: Most definitely. One of the driving factors because I couldn't get into fighters was the flying experience that will transfer to civilian life. I'll have turboprop time, in the C-12, as well as multiengine jet time in the C-141. And the C-12 time is in some of the busiest airspace in the country to boot.

About to Leave Military Service?

Because so much of this book is devoted to pilots sharing their experiences with other pilots, this message sent by my friend Derek Martin, an ex-U.S. Navy pilot and now a First Officer for Southwest Airlines, should be of immense value. Derek was passing on some advice to his old boss about what to expect when he left the Navy and began searching for a job as an airline pilot.

Hi Skipper,

Got your message—yeah, life is good at the show (the airlines). Unfortunately, I have good news and bad news. First, the bad news— if you're retiring in the fall, you're behind the power curve by about six months (minimum) if you want to have a job waiting for you. The time is now to flood the market with your resumes and/or applications, and I mean everybody. This includes carriers that you have no real intention of working for, for it's going to be the interview experience that you'll have under belt that's going to pay big dividends when United, Delta, or American come calling. They're probably not going to call for 6–12 months after you put your app in, so use that time by practicing the art in interviewing with others.

Currency—an issue that depends on the carrier. Most look for being recent in complex airplanes (that is, jets or turboprops), but flying regularly in a Cessna or Piper is the next best thing. Most carriers look not only at your total time, but also at your time in the last 6 to 12 months. Allowances might be given if your last tour was flying the proverbial desk. My advice is to join one of the flying clubs and go out and start flying ILS approaches with an instructor to get you going on flying in the civilian world. There is a difference.

Now the good news—your timing is still good. Don't even think twice about the age factor. In my class at Southwest, our oldest guy was 52 and there were four military retirees, plus two 15-yr retirees. American just started hiring and all the others are in full swing. It's not going to be the heyday of the mid-to-late 80s, but the numbers are still really good. It's all a matter of doing the work to position yourself for the opportunity when it comes calling.

From my limited perspective, the most difficult thing you're going to have to do is break yourself out of the cocoon of comfort you've spent your whole career building. You've still got a job to do in the Navy, but you're going to have to put yourself on a war footing that things are going to be tough and difficult, and that you're going to do whatever it takes to get you (and your family) to the promised land. The wife and kids are going to have to realize that some big sacrifices are going to have to be made as far as time on your part is concerned. The stress level is going to go up the closer you get to retirement and will not abate until you get hired. At this point, it will be replaced by training, IOE (Initial Operating Experience), and your off-probation, check-ride stresses. Include her (your wife) in the process, have her quiz you with flash cards, and let her run you through some mock interviews if she's good at that. This will go a long way in maintaining peace and harmony.

Where to start? Join Air Inc. or FLTops.com as Step one, and buy the full package ($160 to $200) or pay the monthly fee. The Air Inc. binder and the monthly newsletters are chock-full of good gouge, as well as resume boilerplates, and a bunch of other odds and ends.

Next, buy yourself a copy of the FAR/AIM and study, study, study. Parts 91 and 121, and the whole AIM is where to concentrate. Make flashcards and study in the car or wherever.

Study your aircraft flight manuals. They're not going to ask you to draw a schematic of the S-3s electrical system, but they will expect you to relate an aircraft system to them, usually in the form of an emergency that you experienced.

Learn how to read Jepps charts and plates. In the sim ride, if they offer you a choice between Jepps and DOD plates, take the Jepps.

Get on AOL or the Internet—it is one big gouge machine. In the airline forums, you will find a ton of information on what's going on, what the latest interview was like, and so forth.

Get a subscription to *Aviation Week & Space Technology* and save every article or blurb on any company you've got an app/resume in with. Prior to any interview, you can catch up on current events. Nothing will impress an interviewer more than when you know more about his company than he does.

Read the book *Hard Landing: The Epic Contest for Power and Profits That Plunged the Airlines into Chaos*, I believe by Thomas Petzinger, Jr. This book is required reading as far as I'm concerned, for anybody wanting to get to the show. It's a well-written history/business lesson on the workings of the airline industry. Once you know the environment, then the craziness of this industry will be a heck of a lot easier to understand.

Interview suit—dark navy, white shirt, black wingtip shoes, muted tie with maroon in it. It always amazed me how many guys showed up for interviews with their dream airline "out of uniform" and no haircut. The interview was somewhat predictable after that.

The art of the interview. All the gouge in the world doesn't mean anything if you're not ready for the interview. Practice and real experience will lower the stress level, but for most carriers (or companies for that matter), the delivery of the answer is just as important as the data itself. What most Captains are evaluating when they sit across the table from you is "could I spend 8–12 hours a day, 3–4 days at a stretch, over the course of a month with this guy?" Frame your answers in a situation-action-result format. Lose the acronyms unless you know your audience. Be yourself. They understand you're under a lot of pressure and they want to see how you handle it. If you don't know the answer to a technical question—say so and where you would go to find the answer. Irv Jascinski has a book on airline interviewing that's been around for quite awhile. The gouge and mindset that's needed are still valid.

A word on application fees. It's pricey, it's expensive, and it doesn't seem fair. Pay the fee and get on with your life. Last week, I had a Captain jump seat with us from the commuter I flew with between the Navy and Southwest. I asked him if he had an app in with American, and he replied no because he thought the $100 app fee was ridiculous. Well, I think the $100 app fee is stupid, too, but I'm not going to forego a career with a pretty stable airline flying some pretty good equipment making a pretty good salary over a $100 principle issue. It's nothing more than an investment that could pay some pretty good dividends. Look at it that way and the check will be a lot easier to write.

Flying the commuters. Don't you let anybody tell you that a military-trained pilot is better than a civilian-trained pilot. I flew for WestAir/United Express, and some of the best and most professional (and most fun) pilots I ever saw were Captains I flew with, most of whom were my age or younger. The pay stinks and the lifestyle might be bordering on horrible (at least for an FO), but the education and experience I gained were priceless, and I wouldn't trade it for the world. How good or professional a pilot is is completely up to the individual, no matter what their training or background might be.

Remember, every dollar you spend on your career transition: buying flight time, Air Inc., subscriptions, AOL, your suit, dry-cleaning that suit, and so forth is all tax-deductible under professional expenses.

Finally, what to do if it's Day 1 of retirement and the phone's not ringing? It's easy to say, "Don't worry," because you will and that's OK. I'm a firm believer in making your own luck. Every effort you make, no matter how small or insignificant, will eventually bear fruit. Some rewards happen right away, others will take time to manifest themselves, but if you want it—it will happen.

Hope this helps and good luck, Skipper!

No matter what route you use to gain your ratings, realize that everyone else started out pretty much the same way toward their goal of becoming a professional pilot: sweating it out on long, cross-country trips, building the flight time for a commercial certificate, or under the hood with a view-limiting device over their heads as they worked toward that instrument rating. There's no easy, quick way to a job in this industry. But all the work is worth it. You might not really believe that during all the tough months of training, but the first time you step into the cockpit of an airplane as a required crew member, you'll know all the work was worth it.

Unmanned Aerial Vehicles (UAVs)

Although not much information is available about Unmanned Aerial Vehicles (UAVs) (see Figure 2-11) yet, the production of them is on the upswing through the U.S. military. These may be pilot-less aircraft in the sense that the pilot is not on board but, somewhere, a pilot is sitting behind

Figure 2-11 U.S. Air Force Unmanned Aerial Vehicle (UAV).

a Microsoft Flight Simulator–kind of cockpit, controlling every move the aircraft makes. Because most UAVs are equipped with hi-resolution cameras, and some even with missile technology, watch for the military to begin recruiting pilots specifically to fly these machines. Currently, the military is pulling regular line pilots off duty to drive them, but the demand is not expected to be able to keep up (Figure 2-12).

Busting Minimums

When I wrote this story a few years ago, I almost could not believe the details I was reporting. Two intelligent pilots flying a large-cabin business jet busted minimums and crashed into a mountain while on approach to Aspen, Colorado.

Figure 2-12 Typical UAV ground cockpit.

Aspen Crash Prompts Approach Controversy

by Robert P. Mark
Reprinted by courtesy *Aviation International News*
www.ainonline.com

On March 29, 2001, a series of operational and instrument approach procedural errors led to the crash of N303GA, a Gulfstream III, just 2,400 ft short of the approach end of Aspen-Pitkin County Airport (ASE)'s Runway 15 while attempting to complete the VOR/DME C circling approach. Eighteen people, including three crewmembers, lost their lives in the accident. The NTSB report of the crash (AIN, July, page one) was riddled with lessons about what pilots flying under instrument flight rules should not do, ranging from examples of poor crew coordination to serious misunderstandings of basic instrument procedures, such as the fact that the GIII—certified as a Category D aircraft—was not even authorized to shoot the approach to Aspen under any conditions. Descending below the MDA without adequate reference to the airport environment was also noted.

This was hardly the kind of performance expected of professional pilots operating a sophisticated jet under Part 135 into a mountainous area at night. Adding to the irony of the accident was the pressure put on the flight crew by the customer to get the passengers to Aspen at all costs. In addition to listing a number of errors made by the crew of N303GA, the NTSB strongly recommended the FAA revise some ambiguous procedures that even the board found confusing and potentially unsafe.

But the confusion about issues related to the Aspen accident and the accompanying IFR procedures is not limited to only the pilots of the Gulfstream. What became abundantly clear on the NBAA's aviation manager Internet discussion board in the weeks following the release of the accident report and in response to a survey question posed by AIN more recently in just how many pilots joining the discussion might have also suffered a similar fate. Not because they were bad pilots, but simply because of the high-bewilderment factor generated by some instrument approach procedures today. Especially confusing are those involving nonprecision approaches, such as the one at Aspen. Some pointed an accusatory finger at the FAA's communication methods.

Not surprisingly, few of the pilots and experts AIN interviewed wanted to have their names attached to opinions about the agency that regulates their livelihoods. One pilot said bluntly, "The FAA is often a bunch of specialists who don't proofread the procedures they write very well. Few can understand what they have written and often do not understand how people will interpret this material. Realism often doesn't play a part in much of what the FAA does."

What's So Confusing?

Some pilots AIN interviewed since the Aspen accident expressed confusion about approach issues, such as category selection and minimums, as well as the nighttime restrictions at this mountain airport. The crew of N303GA departed late from LAX for the trip to Aspen with a minimal fuel load, which added pressure to the flight because the crew was aware the circling approach was not authorized at night. There were also snow showers in the Aspen area at the time of the accident, leading to numerous missed approaches by other aircraft prior to the crash.

The Aspen approach plate simply noted that circling was not authorized at night. Most private pilots can recite the regulation that says the beginning of official nighttime begins half an hour after sunset, but at Aspen there are practical circumstances that made this approach unusable long before sunset, much less the official beginning of night. Flying into mountainous terrain as the sun sets or rises can often mask the black hole a crew would be descending into if the airport was also obscured by weather. Because of the low-light levels at the time of the accident, the pilots of the GIII most likely would have been unable to see nearby unlighted terrain, even if they could have identified the runway early in the approach.

So why would an experienced crew not be aware of the confines of the approach as darkness fell? Why would they descend below the MDA for the approach more than once during the letdown? Why would they continue down past the missed approach point on what is already a demanding approach—both for them and the aircraft—in a mountainous area without adequate visual reference? And why would the tower even clear the aircraft for an approach nearly half an hour after official sunset, considering that the GIII is not authorized to make the Aspen approach under any conditions? Was the company dispatcher confused or simply uninformed when he sent the aircraft to Aspen? The NTSB never mentioned the category prohibition issue in its report. Was that because the board, too, believed the aircraft was operating under Category C?

How large a part will air-traffic controller training assume in the responsibility for this accident? A Kansas City aviation attorney who did not want his name used said, "Controllers should also not be issuing clearances for unauthorized procedures. The FAA is way behind the power curve on this." Pilot Robert Tod quickly recalled similarly perplexing issues during a recent approach to Santa Barbara: "The ATIS called for the ILS 7 Approach, but the glide slope was NOTAMed out of service with no further explanation."

Is the Aspen approach unsafe at night? Perhaps. The charter company in question—Airborne Charter—has operations specifications that prohibit landing at airports such as Aspen if the flight cannot be on the ground before sunset. An FAA flight check crew deemed the approach unusable at night, a week before the accident, which generated a NOTAM.

But the NTSB cited the FAA for sending an ambiguous NOTAM stating that circling on the Aspen VOR DME-C approach was not authorized at night. It did not state that the approach was not authorized at night but, rather, that circling was not authorized after sunset. Since the final approach is actually lined up fairly close to a straight-in with Runway 15 at Aspen, the NTSB believes the Gulfstream crew may have interpreted the NOTAM to mean they could still shoot the approach if they did not need to circle. The FAA, however, intended that no one should shoot the approach after dark. The NOTAM was updated the day after the crash with the less confusing, "Procedure not authorized at night."

Tod said, "Just saying the Aspen circling approach was not authorized gives the impression there may have been another option. But at Aspen, a pilot must be smart enough to recognize more than just what the book says about defining night. Although it may not be official nighttime, it might be just as dark. Safety begins with being able to correctly interpret known procedures, such as at Aspen, where the terrain affects maneuvering ability during approach and departure." The NTSB put it this way: "Nighttime restrictions [at Aspen] may not sufficiently mitigate potential hazards associated with flight operations into ASE and other airports with mountainous terrain during periods of darkness."

But one NBAA pilot wondered, "What exactly does it mean when a procedure is N/A at night? Is the procedure available as long as ATC clears you for the approach before official nighttime? Does it mean that the procedure must begin—cross the IAF—before then? Or does it mean that all elements of the approach must be completed before official night?" Another said, "There is plenty to be learned from this discussion, as well as a chance to reflect on our own abilities, procedures, and decision making." The cockpit voice recorder of N303GA recorded no conversation from the crew on the specifics of the Aspen approach.

As in most accidents, a combination of factors is responsible. But no one can ignore the fact that so many pilots seem to be unclear about circling approaches and the Aspen approach specifically. But how many other pilots might have suffered the same fate as the Gulfstream crew, not because they were distracted, but simply because of a lack of knowledge?

Under the best of conditions, the Aspen approach is also highly unstabilized, something that is counter to every aspect of flight training today. Measure the distance from the missed approach point to the threshold at Aspen, some 1.4 nm in which an aircraft must lose approximately 2,400 ft. At 123 kt, this equates to a descent of more than a 3,000 fpm, hardly stabilized in anyone's book.

A number of pilots said that in recurrent sessions, the major training organizations focus on aircraft systems and flying the approaches, with only a small amount of time spent on aircraft performance and no time spent exploring the U.S. standard for Terminal Instrument Procedures (TERPS). Pilots reaching this level of their professional lives are expected to know what makes an approach safe and how to determine the particulars for flying them. On type-rating check rides, however, it would be highly unusual for an examiner to test an applicant about protected airspace or anything more detailed than asking the particular category and airspeed at which a pilot might choose to fly a circling approach. Dave Stohr, director of training services for Air Routing International, said, "Pilots are often not well versed in these areas, because in training, we don't spend adequate time on subjects other than flying."

Robert Wright, FAA's manager of general aviation and commercial division in Washington, said basic instrument training and a crash like Aspen leads one to ask some pretty fundamental questions. "We are not preparing people to operate in the [ATC] system today. We may only be getting them ready to take and pass the knowledge test." Mac McWhinney, a San Diego flight instructor and aviation curriculum development specialist, said some flight-instruction programs address TERPS, a bit, but none, to his knowledge, offers an in-depth explanation of these building-block procedures. He added, "The FAA is not looking for the right things on the knowledge tests anyway. They are not thinking about what knowledge a pilot needs to be safer. All they are trying to do is make the tests hard."

The Cessna Pilot Center's instrument training program uses the book, *Cleared for Approach*, which offers some insights into how approaches are developed and should be flown. But, in general, the publications necessary to find the specifics of instrument-approach procedural answers are not easy to locate. An Internet search by AIN revealed a few web sites that offered links to TERPS-related information, such as www.terps.com, but only the U.S. Government Printing Office offered the TERPS publication for sale. FAR Part 97, Instrument Approach Procedures, is not normally included in the Jeppesen regulations available on a subscription basis, although that company offers an FAR/AIM book that does include this information. By contrast, international instrument procedures documents normally include ICAO's Procedure for Air Navigation Services Operations (PANSOPS) that is divided into two publications. One volume outlines how approaches are developed, while the other includes the nuts and bolts about how procedures should be safely flown. TERPS does not offer a similar "On The Job Training" publication.

Pilot Tod said, "The category issue also confused me. If you read about straight-in approaches, it is very easy to understand. But circling approaches are referenced to speed and don't talk about the categories at all." Tod recalled that his instrument training discussed the fact that categories existed, but not how they all fit into the instrument approach package.

Bob Johnson of Air Castle Corporation said, however, "I think the approach category system we use right now is fairly straightforward, but misunderstood."

And John Williams, a Houston-based Falcon 900 pilot, said, "I think the associated regulations are fine as they presently exist. I don't believe there is any ambiguity in our present state of [approach] design and certification. I'm not an aeronautical engineer, but I have no problem recognizing the difference between a certified value and the necessity of using values greater than the minimum certified when circumstances dictate."

"I was just at FlightSafety for recurrent," said one NBAA member. "We did an 'enrichment' course on this [Aspen] accident. The Category C versus Category D question came up and the instructor's response was that at the accident weight and the configuration they were in, they were actually legal to use Category C minimums."

Another experienced GIII pilot responded. "There is not any circumstance or weight that a Gulfstream GII or GIII can be Category C. Never. Circling with full flaps is not authorized in a G1159 aircraft either. The GIV and GV can be Category C because they are authorized to circle with full flaps, and due to their great wing, they fly slower. No matter what category the aircraft is in, if you fly faster than 141 knots, you are Category D."

Others speculated that because the flight was operated under FAR 135, it may have been over maximum landing weight for that runway to comply with the 60 percent landing runway rule, not to mention the fact that the runway was contaminated. Was the GIII crew confused or simply under that much pressure to land?

Most people believe a Hawker 800, for example, is certified under Category D for circling, but are often not sure why. FAR Part 97, Instrument Approach Procedures, says the aircraft category is determined by multiplying the aircraft's stalling speed in the landing configuration by 1.3, at maximum certified gross landing weight. So, indeed, the manufacturer determines the base category of the aircraft, because that weight cannot change. But circling approaches are based upon speed, which can change in the landing configuration on the basis of weight. Explaining that there are both certification and operational categories, the FAA's Tom Penland, air carrier branch manager in Washington, said, "Categories may be perceived as coming with the airplane, but if you go outside of those criteria, you have to make appropriate changes to ensure adequate clearances [from terrain]." He added, "But big airplanes don't circle. There's a reason they don't." The rule of thumb according to Penland is simply, "Fly faster and fly higher."

There is also a difference of opinion on which direction categories change. Most pilots believe categories can vary upwardly, such as a Category C aircraft falling into Category D for circling, but not the other way. One FAA spokesman said, "You can go up a category, but never down from the way the aircraft is certified."

Wright disagreed, saying, "If you can go up a category, you can also go down one," from the one in which an aircraft is certified. Some pilots believe, too, that if they can fly a Category D aircraft slow enough to remain in Category C's protected airspace, they may fly an approach under Category C.

Williams said, "An aircraft is certified in only one category, but may be operated using minimums associated with a higher category, but never a lower one." FAR Part 97 also seems to support this opinion, defining the approach category as downwardly unalterable: "Aircraft approach category means a grouping of aircraft based on a speed of 1.3 Vso (at maximum certificated landing weight). Vso and the maximum certificated landing weight are those values as established for the aircraft by the certificating authority of the country of registry."

A Kansas City attorney explained the overall dilemma. "It's up to the crew to reconcile the differences between the certification category and the operational category. There is no way for the FAA to give a final word without being unduly restrictive. You have to bring training, practice, standardization, briefing, and discipline all together to get the right answer."

If you agree that the operational category chosen for a circling maneuver is flexible, depending on the speed necessary to perform the maneuver as long as the aircraft remains within the protected airspace, there is also a question on whether the speed used to circle is indicated, ground, or true airspeed. Consider again the approach into Aspen. The flight data recorder showed the aircraft holding approximately 130 knots IAS on the way inbound from the final approach fix, Red Table VOR. However, at nearly 14,000 ft msl, the aircraft's TAS would have been considerably higher than 130 kt, as much as 13 kt in one expert's opinion, further reducing the safety margins on an already-tricky approach in IFR weather. Did the crew not care about this issue or were they simply not aware of the affects of altitude on their approach speed? FAA's Dave Cook, a helpful guy who designs instrument approaches in Oklahoma City, said, "I only develop protected [approach] airspace. We don't make our calculations based upon speed."

AIN asked some pilots what speed they reference in a circling maneuver. Two replied groundspeed, three said indicated, and one answered indicated, but accounting for true airspeed factors. When defining approach categories (below), the FAA references only the word "speed," although on straight-in, nonprecision approaches, groundspeed is the guideline. The *Airman's Information Manual* says that if a pilot is circling at a higher speed than the normal approach category, the higher approach speed should be used. One pilot said, "The regulation should state that determining your approach category should be based upon the actual fuel load versus the maximum fuel load and weight. That would be a much better way of determining what approach category an aircraft should be in. However, that is not what the regulation says."

Aircraft Approach Categories

Aircraft approach category means a grouping of aircraft based on a speed of 1.3 Vso (at maximum certificated landing weight). Vso and the maximum certificated landing weight are those values as established for the aircraft by the certificating authority of the country of registry. The categories are as follows:

(1) Category A: Speed less than 91 knots.

(2) Category B: Speed 91 knots or more, but less than 121 knots.

(3) Category C: Speed 121 knots or more, but less than 141 knots.

(4) Category D: Speed 141 knots or more, but less than 166 knots.

(5) Category E: Speed 166 knots or more.

If pilots don't learn the details, nuances, and blind alleys of various kinds of instrument procedures and their related categories at initial, recurrent, or when working on their instrument ratings, where do they learn? Some pilots claimed to have learned on the job or through their own reading. At issue, however, is that a pilot doesn't know if the person teaching understands instrument approaches any better than their teacher taught them. Stohr said, "Pilots should be learning this aircraft category and minimum information during their instrument rating."

Wright, the FAA manager of GA and commercial aviation, said he thinks it is time to start filling in the cracks in the system. He is at work on a white paper that addresses training issues in relation to real-world performance in the evolving ATC environment. "I'm looking at technology and airspace, and how this all may affect the current training system," he said. "It is time we stimulate some new thinking on the subject."

Apparently, many pilots share many different perspectives on instrument-approach procedures. Williams said, "Finding good guidance on any aviation-related question can be a challenge. There's not usually a single source you can go to in order to find answers. In most cases, simply asking the FAA's FSDO inspector won't be satisfactory either. These are pilots, who likely came from the same pool of confusion as other pilots in the industry."

Williams offered a few useful tips to break through the regulation and procedural clutter. "Never take what the FAA has interpreted and told you over the phone as error-free. Get second opinions and verify the sources. Professional pilots must be responsible enough to pursue clarification of the things they find confusing. Sometimes applying a little logic can go a long, long way."

3
Winning the Ratings: The Real Work Begins

Understanding the necessary ratings, as well as how to pay for them, may seem like your only goal right now. But don't forget a considerable amount of work still lay ahead to pass the necessary exams to win those ratings. And the pace and volume of material to be absorbed during training can often overwhelm people who are only mildly aware of what they've gotten themselves into. Choosing the right flight instructor is crucial, too. Without a strong one-on-one relationship that allows students to question or ask for and receive regular guidance, all the hard work will become more difficult than it needs to be. Let's take a look at the ratings and what makes them tick. This review begins with the assumption that the reader already possesses a private pilot certificate.

The Commercial Rating: When to Begin

In their quest toward a professional pilot career, most people begin with the commercial pilot certificate, perhaps because this rating offers a solid glimpse of things to come. It's usually the most time-consuming of the ratings and, hence, the one some pilots want to tackle first.

I enjoyed the commercial training more than just about any other. It seemed more like play to me than work, but then many professional pilots look at flying, any kind of flying, as anything but work. The commercial pilot certificate training takes you from the rank of novice, the private pilot, to the rank of someone who is really beginning to understand what makes an airplane fly in many different kinds of situations. Commercial pilot training can begin right after receipt of a private certificate, but I encourage you not to go that route.

Training for the private pilot certificate takes a great deal of time and energy. By the time most pilots receive their first set of wings, they've logged an average of 65 to 70 hours total time, with considerably more on the ground. This is when you should take a training break. Go fly for 25 or 35 hours around the local area and on a few cross-country trips. This is, of course, a little easier said than done for some pilots outside the U.S., where general aviation aircraft is not as plentiful. Before you bog yourself down with more flight training, classroom instruction, and written exams, have some fun using the skills you've spent many hard months learning. You need to experience the joy—the fun of flying—before you become embroiled in too much work.

Total cost is another advantage of putting some time in after you receive your private license. If you're working in a non-approved school again, you'll need 250 total flight hours before you can be recommended for the commercial exam. With a special exemption, an approved, Part 141 school only requires 190 hours. If you pick up your private at 65 hours and start right in on the commercial package, you'll end up spending dual instruction prices for time leading to the requirements you could have logged solo. If you start the commercial a bit later and have logged 25 hours on your own, that's 25 × $35 per hour (the instructor's fee), or nearly $875 you could save. That's a pretty hefty savings in anyone's book. After you have about 90 to 100 hours or so, sign up for the commercial course with the school of your choice.

The PTS

Before you're through with aviation, many years down the road I'll wager, you'll have encountered more acronyms than you care to even think about. Here's one that will become quite familiar on the road to becoming a professional pilot. Practical Test Standards (PTS) is the label given to a small book of guidelines, one for each certificate and rating the FAA offers (Figure 3-1).

These books describe exactly what subjects you need to be familiar with to pass the flight test for the particular certificate or rating you seek—in this example, the commercial. You must be able to perform and you will be asked to perform each task set forth in the PTS. That might sound pretty straightforward right now, but when people were picking up ratings a few years back, the booklet in use was the FAA's *Flight Test Guide*.

U.S. Department
of Transportation
**Federal Aviation
Administration**

FAA-S-8081-4D
Effective April 2004

INSTRUMENT RATING
Practical Test Standards

for

Airplane, Helicopter & Powered Lift

Figure 3-1 The PTS is the primary guiding document for flight training.

In that earlier guide, examiners could pick and choose the items they wanted to test. Unfortunately, as happens with human nature, some examiners picked the same maneuvers each time they gave an exam, and the word eventually got out. If you were weak on non-directional beacon (NDB) or VOR approaches, find Fred—he never asks you to fly one anyway. Herb never cared about Chandelles on a commercial, so if you had trouble with those, you'd do your best to locate Herb to give you the test. To say the least, this made the entire test procedure rather unfair and at times much too easy to predict. The really uncomfortable part was that people of various degrees of skill were sometimes awarded a certificate they might not normally have received

With the establishment of the PTS, much of the ambiguity of the testing procedure was eliminated because each applicant was told, long before the exam, that they would be tested on everything in the book. This made things a great deal easier, even if somewhat more complex, because if the students could perform all the maneuvers to the required standards, they were prepared for the flight test . . . period. The FAA says this about the PTS concept:

[Federal Air Regulations] FARs specify the areas in which knowledge and skill must be demonstrated by the applicant before the issuance of a rating. The FARs provide the flexibility to permit the FAA to publish practical test standards containing specific tasks in which pilot competency must be demonstrated. The FAA will add, delete, or revise tasks whenever it's determined that changes are needed in the interest of safety. Adherence to provisions of the regulations and the practical test standards is mandatory for the evaluation of pilot applicants.

An appropriately rated flight instructor is responsible for training the student to the acceptable standards as outlined in the objective of each task within the appropriate practical test standard. The flight instructor must certify the applicant is able to perform safely as a pilot and is competent to pass the required practical test for the rating sought.

The examiner who conducts the practical test is responsible for determining that the applicant meets the acceptable standards as outlined in the objective of each task within the appropriate practical test standard. This determination requires evaluation of both knowledge and skill because there's no formal division between the 'oral' and 'skill' portions of the practical test. It's intended that oral questioning be used at any time during the practical test to determine that the applicant shows adequate knowledge of the tasks and their related safety factors.

Some applicants also complained about the methods examiners used to determine just how qualified an applicant was. Primarily, their concerns focused around the examiner's use of distractions during the flight test. Some applicants believed it was unfair to toss in little questions or problems during a time when the applicant needed to concentrate on more important things. In the new PTS, the FAA addressed this issue:

Numerous studies indicate that many accidents have occurred when the pilot's attention has been distracted during various phases of flight. Many accidents have resulted from engine failure during take-offs and landings where safe flight was possible if the pilot had used correct control technique and divided attention properly.

Distractions that have been found to cause problems are: preoccupation with situations inside or outside the cockpit, maneuvering to avoid other traffic, or maneuvering to clear obstacles during takeoffs, climbs, approaches, or landings.

To strengthen this area of pilot training and evaluation, the examiner will provide realistic distractions throughout the flight portion of the practical test. Many distractions may be used to evaluate the applicant's ability to divide attention while maintaining safe flight. Some examples of distractions are:

- Simulating engine failure.
- Simulating radio tuning and communications.

- Identifying a field suitable for emergency landings.
- Identifying features or objects on the ground.
- Reading the outside air temperature gauge.
- Removing objects from the glove compartment or map case.
- Questioning by the examiner.

A look at the commercial PTS shows just what's required to be successful at the test. As you may know, a commercial pilot certificate allows you to carry passengers or cargo for compensation or hire. Another way of looking at the commercial is a rating that allows you to make money for flying. But, besides the ability to make money, the capture of a commercial certificate for your wallet means you've begun to look at aviation in a whole new way—as a potential profession. The maneuvers you'll be asked to perform, as well as the subjects you'll learn, are designed to deliver a significantly more intense level of flying knowledge which you'll be expected to demonstrate on the flight test. Though some of the maneuvers you'll fly on the commercial test might appear similar to those you learned for your private ticket, the tolerances are much tighter.

Commercial Subjects

Here's a look at some of the subjects on the Commercial Flight Test to which I'd like to offer some insight. As in the private pilot test, the commercial pilot will be asked to prove that the aircraft is airworthy by displaying and being able to discuss the various sorts of paperwork involved in making the aircraft legal to fly. This includes a possible discussion of the Minimum Equipment List (MEL) for your aircraft, not to be confused with the FAA-required minimum equipment for day and night VFR and IFR flight. You'll also be expected to prove you're legal to conduct the flight test, right down to being certain the information on the FAA Form 8710-1— the Application for Airman's Certificate—is correct.

Something that almost caught one of my students once was the location of the ELT in his new Bonanza. During basic questioning at the preflight, I asked the student where the ELT was located. When we looked in the small door in the rear of the aircraft where the arming switch should have been located, the student found nothing. He couldn't prove the ELT was there. It took a mechanic with a flashlight to finally locate the device for us. But better for me to ask than to have been caught on a flight test with this question. Don't just listen to your instructor. Take an active part and make certain you understand what's being taught. Learning to fly is no place to be passive.

Figure 3-2 Skywest Airlines CRJ in Delta Connection colors departs Atlanta. (Courtesy Chris Starnes: www.jetphotos.net.)

In every aircraft I've ever checked out in, I've always found the systems test to be the most difficult part of the oral exam given with the flight test. The key to passing this section is not only to understand that the landing gear, for instance, is hydraulically actuated, but also to understand the troubleshooting aspects of the system. What happens when the system falters? How do you lower the gear when the hydraulic pump fails? What's the backup if the alternate system fails? If you can't lock the left main, but the right main and the nose are down, should you land on two or bring them all up? If your landing gear is electric, take the time to learn more than simply the emergency gear extension. If the discussion turns to electrical, know what a voltage regulator is and where it's located on the alternator assembly. Realizing what kinds of electrical problems you must cope with is important for the oral exam, but so is being able to tell the examiner when your system has run out of options.

When it comes to weather, the examiner will be trying to establish your ability to read prognostic and radar charts, as well as sequence reports and forecasts, either on paper or by looking at the weather display you might pull up on your laptop computer. The examiner will be looking for your ability to discuss trends. I don't mean forecast the weather for the next 12 hours, but to be able to relate what you're reading to how that weather will affect the flight you're planning, in terms of routing, winds aloft, altitude selection, fuel required, and icing considerations. Sure, you might be able to tell the examiner that rotation about a low is counterclockwise, but on a cross-country trip, what does that mean regarding the weather and the changes you'll encounter? When will you encounter a headwind or tailwind, when will the pressure change, and which way will the altimeter move and why should you care?

Another area that often causes a great deal of stress is emergencies. This isn't really so strange because emergencies, by their nature, are anxiety-producing. A common emergency is total or partial engine failure. (Most instructors will talk about an engine failure right after rotation in a single engine aircraft, but it's not safe to simulate this event below 500 feet above the ground.) Know the memory items of your aircraft checklist and be able to recite them while you're being distracted with something else because that's most likely how the engine will die. It could quit in a steep spiral or during an eight around a pylon. Be able to tell the examiner what you should do if the door pops open right after rotation.

I distinctly remember the time I learned how to show the examiner a simulated in-flight fire recovery in a BE-55 Baron during VFR conditions. When he yelled "Fire," I chopped both throttles and almost rolled the airplane on its back as I headed down toward the ground. He liked that.

On cross-country planning, you get a bit of a break. Expect to spend a portion of the oral explaining how you determine fuel burn and appropriate cruise altitude, and whether NOTAMs are current for your route. But be ready to explain how to plan for the unusual. You'll spend half an hour looking up data for Fred's Airpark and the examiner will look it over and say, "Yes, but now I want to go to Hee Haw International."

Find all the airport information and explain what you've located. Before the flight test, be sure your regular instructor has quizzed you on every single symbol, color change, information box, and navaid symbol on the charts you'll use. It's not good enough to point and say "That's a VOR-TAC." You must understand what it means. So, here's the break. You don't have to fly a cross-country trip like you did for your private. You just have to be able to explain it.

During ground-operation questions, be ready to explain what different colors relate to the various grades of fuel. Realize that no color could be jet fuel in your tank instead of avgas. Be ready to explain just how frost might form and how you should remove it, as well as what you shouldn't try if the contaminate were ice (like beating it off with the handle of an ice scraper, which could dent the leading edge—guess how I learned this?). Do you taxi and pay attention only to taxiing, or do you try to write ground control instructions or ATIS while you're moving? Both of these would show some serious judgment problems. Before you take off, how well do you organize the charts you'll need? Do you put them somewhere easy to locate, but where they won't blow around on takeoff? Do you have a pen close by?

On your pre-takeoff checks, you'll be expected to demonstrate a professional attitude toward the checklist, a place where many applicants make mistakes. Too often, students read the item and look, but don't actually check to see if the proper operation was performed. For example, "altimeter . . . set." They look, but they don't change the setting in the Kohlsman window.

You'll be expected to handle the radio and ATC communications like a pro. That doesn't mean never missing a call or always understanding what air traffic is trying to say, but it does mean when you only catch part of the call, you take the proper action to fix the problem, such as asking ATC to "Say again!" This doesn't sound tough, but you'd be surprised at how many pilots simply sit there when they've missed something on the radio or look at the examiner and say, "Did you catch what they said?" Not a good idea.

If you live near Class B, C, or D airspace, absolutely be ready to explain when and how you'll request and enter that airspace. One student was recently caught when the examiner asked if the ATC facility repeating his call sign for entry into a Class C meant he could keep going toward the airport. The answer is yes. When asked if it worked the same way in a Class B, the student said sure. Wrong! Without teaching you the airspace designations, know what happens if one type of airspace overlaps with another. (I already know. You look it up.)

When it comes to in-flight maneuvers, eights around pylons for example, you either understand how to compensate for the wind or you don't. You either know where the reference point is or you don't. The most important part of the commercial preparation is this: there's no substitute for plain old practice. Too many students want to rush it as soon as they meet the requirements. Whether it's crosswind takeoffs and landings or eights on pylons, you must get out and practice if you intend to display professional grade competence on the exam. Personally, I'd be out practicing a few hours, three times per week before the flight test.

Finally, in this discussion of the PTS, there's one more simple concept, at least it seems simple to me. Know the tolerances the examiner expects you to perform to on all maneuvers. For a lazy eight, understand that when you're selecting your reference point you should never drop below 1,500 AGL as the maneuver progresses or that the altitude tolerance is plus or minus 100 feet at the 180-degree point, or that airspeed must be within 10 knots at both the 90- and 180-degree point. The commercial license examiner expects a professional looking and feeling flight. Another point about tolerances is that the test is not over if you get off altitude or speed a little bit. No one is perfect. The examiner will be watching, however, to see if you notice and correct the situation. Finally, if you don't understand what's expected of you at any point during the exam . . . ask! At times, you might not know exactly what to ask, but you will know if something is confusing—most likely you at this point. Say something about it.

There is no required sequence to the way in which you earn your ratings, as long as you begin with your private. You might follow a private with a commercial, and then the instrument, or do it the other way around. You might earn your multiengine rating in between both. Opportunity or availability of a particular aircraft type might also affect the training sequence. The order normally makes little difference, as long as you complete all that are necessary—the commercial, multiengine, and instrument rating being the minimum required for almost any flying position. And be sure to e-mail me during your training. Our feedback for new readers is significantly improved as we hear from all of you. That's rob@propilotbook.com.

Here's the perspective of a Regional Jet captain employed at Mesa Airlines.

Profile: Warren Cleveland, Regional Jet Captain, Mesa Airlines

"My dad used to fly remote-control airplanes when I was a kid," remembers Warren Cleveland. "That was my first exposure to flying. He always wanted to fly, but never had the opportunity before he passed away."

The younger Cleveland took the money from his 14th birthday and bought a flying lesson from a flight instructor in Baton Rouge, Louisiana. "I started lessons on a pretty limited basis. You could see years pass on one page of my logbook," Cleveland adds. He got a job washing airplanes at age 15 and had much of his salary credited to a flying account at the FBO he worked for. He soloed at age 17.

Looking back on this, Cleveland recalls that "I really didn't have a clear plan of how to get my ratings. I wished I had a mentor. I went to college because someone told me I'd need it to be a pilot. I even applied for Navy ROTC, but was turned down. I really hated college." But he did meet someone who had enrolled in a flight program at San Juan Community College in Farmington, NM. It was Mesa's Airline Pilot Development program.

Cleveland spent some time talking to the people there and felt at ease early on. "I felt they had a more organized view of how to help me get where I wanted to go. They told me that 100 percent of their grads were getting jobs. Embry-Riddle was just too expensive for me."

Cleveland enjoyed the New Mexico program from the beginning. "We began flying brand new 36 Bonanzas with HSIs and RMIs. It was sweet! I picked up all my ratings and even completed a turboprop transition course before I graduated. You got ten hours in the right seat of a Beech 1900 and ten more watching your partner fly. If you had a B average at graduation, you were guaranteed an interview with Mesa. I got hired and flew as a First Officer in the Brasilia for 18 months, but could not upgrade because I wasn't old enough (you must be age 23 to hold an ATP). But, eventually, I made Captain on the Beech 1900, and then the Brasilia. Now, I've just upgraded in the regional jet. As it turns out, I was the youngest RJ captain in the world when I upgraded."

And how does Cleveland like his new position flying a jet at FL410? "It's really hot stuff! Mesa's new management team, headed by Jonathan Ornstein, is really great, too. He communicates with the pilots. I think Mesa is on track to being one of the top regionals in the country. It's a great place to work."

Cleveland offers aspiring pilots a few insights. "In training and in the hiring process, attitude is everything. You'll get more opportunities with a good attitude than anything else. Show enthusiasm, make lots of contacts, and never, never, never burn your bridges behind you."

Now That You Have the Ratings, What's Next?

This seems like the grandest question of them all, but what's most important once you've picked up these various ratings? Keep flying. That might sound a bit simplistic, but do anything you can to keep logging and building your total time, as well as the variety of your experience. One way to accomplish this is to stick close to an airplane any way you can, whether it's by buying an airplane of your own or renting. It all counts toward your total time.

Let's talk about that coveted logbook for a moment. Everywhere you go in this industry, you'll hear people asking how much time you've logged. It seems, however, that the problem of logged hours is not quite as significant for people who have surpassed the point of 3,000 to 5,000 hours or so. Less than 1,000 hours pretty much puts you in the range of unproven more than anything else. When you're working with total hours in the hundreds, then, total time to a potential employer is significant, depending on the job you're trying to win. Since the pilot shortage really began to take hold in both the U.S. and in other parts of the world such as China and India, the total experience needed to win an interview with a regional airline has dropped significantly. American Eagle Airlines reduced their requirements to 500 hours total time in March 2007. That good news for job applicants means there will be jobs for pilots with 200-400 hours total time such as flying sight-seeing demo rides or, perhaps, towing gliders.

Another significant career hurdle might revolve around the amount of multiengine time you have under your belt. Multiengine time is always much tougher to come by because not only is it expensive, but it's also difficult to obtain since most insurance policies written on twin-engine aircraft pretty much prohibit low time pilots from flying them solo. What you might be left with is trying to find a charter operator (Part 135) that will allow you to fly Second in Command (SIC) on one of their airplanes or, perhaps, a small company that will allow the same kind of SIC time in their twin. I won't pull any punches with you here. This isn't going to be easy. I remember the struggle: begging, borrowing, and running around the airport at all hours of the day and night, trying to pick up that multiengine time.

I started out flying right seat on a light twin for a small Chicago manufacturing firm and, eventually, ended up in the left seat. The airplane was an early Piper Seneca 1, but it might as well have been a 747 to me. After a hundred or so hours in that aircraft, I found myself in the right place at the right time to be able to ferry some aircraft for an aircraft distributor and eventually on to a Piper Navajo job. As a ferry pilot, I would fly around the country with the power pulled back to economy cruise and everyone was happy. The aircraft burned less fuel on the trip and I logged more time. Even today, I still ferry aircraft when I can, just for the opportunity to fly someplace totally unscheduled.

Another option could be a flying club operation with a twin. You might have to fly with an instructor for a while before you can take the aircraft alone, but once you qualify, the rates will probably be relatively cheap. It will take some time and effort on your part to search out the bargains and opportunities.

Just when you find yourself with a few hundred hours of multiengine piston time under your belt and you start shopping for some of those flying jobs, you might run smack into another category of time someone will ask for: turbine time. *Turbine* is basically a generic word for jet engine time. The time can be logged in two forms, however. The first is *turbo prop time*, where the aircraft is powered by a jet engine connected through gearing to a propeller. This could be a King Air or Cheyenne, or an Avanti.

The other is *pure jet time*, in a Cessna Citation, Learjet, or Boeing 737. After I accumulated about 1,200 hours of turbine time in a turbo prop aircraft, I felt like pretty hot stuff. Then, when I went looking for a job, some of the ones I wanted asked for jet time. It seems a carrot will always be in front of you somewhere. As soon as you have jet time, someone will probably say they wished you had more time in that jet, or they'll tell you they wished you were type rated in some other kind of jet. It's just the way this industry works.

And Now for Something a Bit Different

Because the title of this chapter includes the word "ratings," this section may at first appear to have digressed somewhat. Every part of flying does not necessarily carry with it a new pilot certificate or rating as you'll read about here. Learning to control an airplane in unexpected situations is critical no matter what airplane you fly or the type of company that employs you.

Life Outside the Box

by Robert P. Mark
Reprinted by permission *Aviation International News*—
www.ainonline.com

Ask most professional pilots about either the USAir accident in Pittsburgh or the United Airlines crash in Colorado Springs when two Boeing 737s flipped upside down before impact, and the discussion often focuses on whether it was wake turbulence, a roll cloud, or a rudder hard over that caused the crashes. Make a case for survivability in either of these two accidents and people will raise their eyebrows and roll their eyes as if to say, "Are you kidding me?" Few experts, except perhaps the folks at Boeing and Don Wylie from Aviation Safety Training (AST), believe the pilots could possibly have survived a rudder hard over, if that was, indeed, the cause, as the NTSB claims. USAir flight 427's old analog flight data recorder did not track rudder position. The question that still remains, however, is couldn't these pilots have recovered their airplanes? The simple answer is they were never trained to recognize and recover from those kinds of upsets.

Aviators, especially those who climbed the professional ladder from the ranks of flight instructors, think they have a pretty firm grasp on aerodynamics. But, after spending a few days in the company of Don Wylie, most experts find themselves wondering how well they paid attention during their early flight-training days. Wylie, a former Air Force F-4 pilot in Viet Nam, saw a parting of the waves, so to speak, after the USAir crash. "You only have to listen to the transcript to grasp the significance of what happened, no matter what turned the airplane over. The final words of the captain to the First Officer are 'Help me pull.' The USAir Boeing was fully stalled when it crashed." Holding the wheel fully aft denied the crew any chance of a recovery. In the COS crash, the 737 was in a steep spiral all the way to the ground. According to AST, loss of control in flight has contributed to more transport category aircraft accidents than any other single cause in the United States and is a close second to Controlled Flight Into Terrain (CFIT) as a killer worldwide.

Wylie's two days of ground and flight training, called the Advanced Maneuvering Program (AMP), makes it clear that the outcome of an in-flight upset need not always be fatal (Figure 3-3). AST employs 21 instructors, including Wylie, at its hangar at Houston's Hooks Airport. This AIN reporter had the opportunity to attend a G-4 simulator session at Simuflite's DFW training facility that Wylie taught as part of recurrent training for two General Electric pilots, Nick Esposito and Mark Ripa. The simulator session was followed by a trip to Hooks for a two-hour flight beneath in a T-34 to see if I could

Figure 3-3 The military adds unusual-attitude training early in the training regime.

transfer my classroom instruction into stick-and-rudder techniques that might save my life. This pilot learned that while old flying habits can be tough to break, it could be done.

AST has trained over 1,900 pilots in the Advanced Maneuvering Program and counts a number of corporations, such as Abbott Labs, Merck, the U.S. Army, and Morgan Stanley, among its clients. Another customer, Bill McGoey, is GE's chief pilot of standards and training for a fleet of two BBJs, two G-4s, three Challenger 604s, and two helicopters at the company's Stewart NY base. "We've always thought upset training should be a regular part of our pilot recurrent." McGoey says this concept was reinforced when one of GE's early hangar partners purchased an aerobatic Decathlon and regularly trained crews in unusual attitude flying. "Because there was a great deal of press about possible control issues on the BBJ," McGoey says, "we wanted to make our passengers feel as comfortable as possible that we were properly trained for an upset. I found that selling this training was actually pretty easy. The pilots who have been through AST's upset program thought it was some of the best training they'd ever experienced."

Back to some aerodynamic discussions, the nucleus of Wylie's training. Had the USAir pilots realized their aircraft was no longer flying, reducing the Angle of Attack (AOA) would have been necessary to make the wing fly once more. The problem was, to those pilots, the view out the window was not one they'd been trained to see and their "muscle memory," as Wylie calls it, told them instinctively to pull to escape the ground, which turned out to be a fatal mistake.

Wylie believes, with the right training, they would have had a chance to recover. But, combine the untrainable startle factor of a real upset with passing, rote knowledge of unusual attitude aerodynamics that most pilots have and the outcome may have been sealed. According to Wylie, "We've been training a generation of pilots who don't use the rudder and have become systems operators, simply relying upon technology. Students also worry too much about complying with ATC instructions and maintaining altitude, even when the world is falling apart around them." AST emphasizes this last point by the flight-recorder playback of a Boeing 747 that rolled over at 35,000 feet after the pilots failed to notice an increasing AOA following an engine failure. With the aircraft upside down, the pilots asked ATC, which was completely unaware of the problem, for a lower altitude, rather than admit they were falling out of the sky.

Wylie says it takes some visionary thinking on management and on the part of pilots to begin flying outside the standard 1G box most have been trained in. "Some students have an edge, though," says

Wylie, "such as tail-dragger pilots—people who are not afraid to kick the rudders when necessary." Additionally, Wylie says helicopter, glider, ex-military, and even remote control (RC) pilots do a pretty good job of recovering from severe upset maneuvers.

After watching thousands of pilots try to fly themselves out of the dark corners of their aircraft's performance envelope, Wylie concludes, "Pilots fly like they've been trained." Fighter pilots learn to fly their aircraft to its performance extremes to gain a tactical advantage over their enemy. This just happens to translate into a succession of abrupt, high G maneuvers. On the civilian side, recurrent training reinforces a reluctance to work the aircraft's controls hard, even during an emergency, for fear of upsetting the folks in back. Wylie says it's time pilots learned how to "spill a little coffee" to save their lives and those of their passengers. "I've watched pilots who were properly trained, recover a 777 at 250 feet AGL when the upset occurred."

But recovering from an upset means more than simply understanding aerodynamics, for just as the upset is occurring, at a time when the two people up front need to be completely on top of what is happening and how to escape, the human brain slips into "hypervigilance," as anyone who has ever been part of an automobile accident will understand. Events appear in slow motion, as a person's vision begins to narrow and auditory exclusion commences. Adults scream for their moms and the brain reverts to any form of movement it believes will save it from extinction. Unfortunately, that "muscle memory" can force pilots to pull when they should push, or to kick the wrong rudder at the wrong time.

Another issue that regularly comes up in Wylie's classes is how poorly some crews are prepared for an upset in normal flight. He includes a slide of a typical transport aircraft cockpit with the foot rests prominently displayed. While the ability to rest their feet on the instrument panel represents success to some pilots, Wylie worries that many crew may not understand how vulnerable sliding the seat back and putting their feet up leaves them if the airplane suddenly rolls upside down. Comfort is one thing, but "the pilot should always have the ability to command full control of the yoke and rudders whenever they are needed," he says. "They should also be properly belted in." Wylie also worries that pilots are religiously trained to aim their aircraft around the sky following the flight director's commands. Wylie suggests all pilots review how to instantly unclutter their CRT display if a recovery becomes necessary, to avoid following a command that could lead to the ground. A company's CRM procedures should decide who turns off the flight director and yaw damper during an upset.

The Four Forces

Everyone with a pilot's license has heard the discussion about the four forces that affect an aircraft in flight: lift, drag, gravity, and thrust. But do we really have any practical knowledge in our heads other than those old drawings in the FAA Handbook with arrows spurting out of the aircraft at odd angles? Sure, 2Gs of force are applied to an aircraft in a constant altitude turn, which must be balanced by an equivalent amount of lift, but what does it feel like? And what does it feel like when you enter a 2G turn in one second rather than in the nice lazy fashion we've all been taught? And what about quickly rolling into a 70-degree bank that requires 3.3Gs worth of lift? The aerodynamic discussions in AST's classes, especially those that focus on an aircraft's resultant lift vector, make a pilot consider perspectives on flying they either forgot or must admit they never understood in the first place.

Wylie calls an upset, "an excursion that is both unexpected and unplanned for," no matter what the attitude. And, after training thousands of pilots, Wylie has seen some trends that show how little a pilot's experience level means when they flip over. For instance, when the attitude indicator shows brown side up during an upset, most pilots will instinctively kick the bottom rudder to try and get away from the ground, an act that almost guarantees a fatal termination of the maneuver.

If rudder induces yaw during a hard over to the left, Wylie teaches rolling the ailerons right. According to the way Boeing builds their airplanes, he says, the 737 has enough ailerons at any speed above stall to overcome a rudder hard over, if the crew reacts properly. But pilots must know enough to "push" as they reach a high angle of attack. Think how that integrates into all our previous flight training lessons. The airplane is rolling out of control because of a rudder hard over, or a wake vortex, with insufficient aileron to recover. Only one option is left to regain control of the aircraft . . . reduce the angle of attack. That means lowering the nose to unload the airplane. Once pilots enter hypervigilance, however, most are unable to push on the yoke. But, pushing on the yoke immediately increases the power to weight ratio as the wing begins flying once again, which means roll response to aileron input also increases dramatically.

Finally, the ground school sessions focus on AST's guiding light for pilots who find themselves upside down and unable to think their way out of the problem. Step on the sky! When the airplane rolls, punch the rudder on the blue side of the ADI to find the quickest route back to level flight. Little thinking is needed. If no blue is showing, simply (yeah, right!) punch the rudder in the direction of the ADI pointer to accomplish the same thing. In these noncognitive events, as Wylie calls them, when it is life or death, pilots always seem to push the rudder away from the ground coming up, even if the rudder is what brings the aircraft closer to the ground faster, which is why he calls it noncognitive. Sounds good, at least in theory.

Wylie emphasized again the importance of using the rudder smartly, although not aggressively. He prefers the term "managed application of rudder" during an upset, rather than anything that remotely sounds like stomping on anything. Some of the big transport builders and operators around the world have make it clear that pilots should avoid using the rudder during an upset to prevent setting up the oscillation some claim tore the vertical stabilizer off American 587 departing JFK. Wylie thinks that attitude is rubbish and proves it during the flight training portion of the course.

The G-4 Simulator

The advantage of practicing extreme attitudes in a simulator is that you'll never hurt yourself. The disadvantage of trying these maneuvers in a simulator is your brain instinctively knows you can't hurt yourself, which can render some portions of simulator training relatively ineffective. Unusual attitude recovery during recurrent training in the simulator is normally accomplished at 15,000 feet, where many pilots feel like heroes if they successfully roll an airplane through an upset. But, if that upset occurs a mile on final at 400 feet in the air, everyone on board will probably become a statistic.

Wylie's training combines aerodynamic theory with a trip through the simulator for some maneuvers to drive the classroom points home. McGoey says, however, "Most companies send their pilots out for two weeks of training each year, so adding another day or two for upset training can be a tough sell for some." Although unusual attitude training is required by the FAA during recurrent, pure upset training, such as the AMP program, is not.

The folks at GE were kind enough to let me look over Nick Esposito's and Mark Ripa's shoulders as they worked their way through Don Wylie's upset training add-on to their own G-4 recurrent. Nick and Mark are experienced pilots, although relatively new to GE. Neither one came with a military flying background.

Right after takeoff, ATC asked Esposito to turn the G-4 left 40 degrees for traffic. As he entered the turn, ATC asked him to expedite, which caused Esposito to naturally increase his angle of bank. Now, in a steep turn close to the ground, the left engine quit, which made for a pretty exciting ride as Esposito attempted to maintain control of the aircraft that was now trying to roll over on its back. The caveat was that Wylie had instructed Esposito not to use the rudder during recovery, but to fly with elevators and aileron only. At this slow airspeed and with the ground coming up fast, the ailerons were not terribly effective as Esposito tried to roll the aircraft level. But Wylie told Esposito to lower the nose, probably the last thing most pilots would try. As Esposito reduced the angle of attack and the ailerons became more effective, the Gulfstream began to roll wings level. Thanks to some expert piloting on Esposito's part, he managed to avoid the ground, despite the EGPWS yelling at him every second or two. It was an impressive demonstration.

Wylie next had Esposito set up for level flight at 20,000 feet and 280 knots. Esposito rolled into a few 90-degree banks, but did not pull or push on the elevators as he turned. To demonstrate how effective the rudder can be in controlling the lift vector and, hence, the aircraft, Esposito tried controlling the aircraft through the use of rudder only. The airplane was slow to recover, but it was definitely under control. Esposito next tried with no controls at all, but simply using differential thrust to point the airplane where we wanted it, a trick perfected by the United Airlines DC-10 crew a few years ago in Sioux City. The value of an alternative means to make the airplane fly was clear. We also tried some upsets in which Esposito recovered by rolling through the upset, rather than stepping on the sky, the recovery preferred by many pilots. The results were dramatic. At the end of the roll, the G-4's airspeed was through the redline and the G-meter said the wings had departed the airplane, despite how good it looked out the window. We also lost about 5,000 feet during the recovery. Wylie said the lack of a "wind in the wires" feel to the simulator is a liability in this kind of teaching.

Wylie had Esposito fly the aircraft near the top of its performance envelope, around 51,000 feet, to demonstrate more aircraft control issues. As the aircraft approached an intentional stall—something that is not too tough at this altitude where all performance parameters begin to merge—Wylie asked him to try and maintain altitude, the way we're all taught stalls and recoveries. When the stick shaker rattled the wheel, Esposito released some back pressure, which cost some altitude. As hard as Esposito tried to coax the aircraft through the shaker and return to his original altitude, it was clear that was not going to happen. Wylie's point was not even to try. Make the airplane fly by unloading it and simply realize you must trade altitude for control . . . and worry about ATC later.

The final trick Wylie showed everyone was from a five-mile final point with the G-4's gear and flaps down full. Wylie simulated a wake turbulence upset by rolling the aircraft over about 120 degrees and had Esposito recover. At a slow speed and in a dirty configuration, Esposito quickly learned that, when aileron and rudder are not effective enough, the only option is to push—or stop pulling so hard—on the yoke to get the airplane turned around. Wylie again mentioned his earlier 777 recovery example. After some more pretty fancy flying on Esposito's part, we landed safely after another upset even closer to the ground. Ripa then took the left seat and repeated the same exercises.

In all, each GE pilot probably spent an hour and a half in the left seat watching demonstrations of the aerodynamics Wylie had taught in ground school before trying to turn their knowledge into action. Esposito reacted positively to the training and added, "I think I need to use the rudder much, much more." Next, it would be my turn to integrate what I'd learned with my stick and rudder capabilities in the Beech.

The T-34 Flight

Back in Houston, the midday July heat was already almost unbearable at Hooks Airport as Don Wylie and I pre-flighted the Beechcraft T-34, N44KK. Wylie showed me the basics of how to shed the aircraft's clear bubble canopy if a problem arose, as well as how to pop out of the five-point harness and use the parachute if that became necessary, a checklist item that works wonders on the confidence of we straight-and-level guys. The T-34 uses a stick for control rather than a yoke, and it also has a G-meter to record the truth about past maneuvers.

Because AST likes to train T-34 pilots in pairs, we teamed up with another student, Charles Kerins, the owner/pilot of an Aztec, based in the British Virgin Islands and his instructor, Blake Thomas, in another of AST's T-34s. Because both instructors are ex-fighter pilots, we taxied out in formation to the end of Runway 17 Right. After a quick run-up, we took off together with a right turn northward toward Lake Conroe. Once we reached 3,500 feet and were clear of the Houston Class *B*, we tried some 60-degree steep turns to feel the 2Gs of back pressure necessary to maintain altitude. Wake turbulence is always a hot topic. With the smoke turned on from our formation's lead ship, we flew through the wake of the other T-34. Even though these are small, relatively light aircraft, the roll motion generated as we flew through the smoke was eye-opening. Then the two T-34s went their own ways for separate training sessions.

Don Wylie and I quickly moved into accelerated stalls. But the way Wylie teaches them is totally different from what I recall being taught or ever taught my own students. We were trained to fly them with power and ease the yoke back—albeit quickly—to feel the *G* forces and realize that an airplane stalls at a higher speed when the wing is loaded up. When Wylie demonstrated the first accelerated stall to me, we pulled 2Gs in about a quarter of a second and, for the first time in my flying career, I felt the G-forces that make an airplane stall at a higher speed. My neck quickly succumbed to the compression of the stall, but we were out of the maneuver as quickly as we'd entered it. I had Wylie try another, so I could watch more closely, as well as feel it. This time, it was clearly apparent not simply what 2Gs felt like, but how the aircraft instantly began flying again when it was unloaded with the release of back pressure, just as I'd noticed during one maneuver in the G-4 simulator. I tried a few of these myself and realized I had something new to teach my students.

Roll offs were next as we flew through some simulated runaway nose-up trim maneuvers. Wylie had me hold the stick back to feel the force necessary to maintain level flight while he cranked in five degrees of nose-up trim. When I released the forward pressure, the Beech shot upwards like a rocket. Using ailerons and rudder only, I rolled the airplane into a 70-degree left bank and watched the climb instantly stop. The forward pressure on the stick immediately disappeared as Wylie's explanation about the changing lift vector—now moving sideways—came back to me. The demonstration is to prove the airplane, although erratic, can be controlled with a runaway trim condition before the airplane rolls out of control. Forcing the stick forward to try to hold the airplane down could have exceeded the airplane's structural limits.

Another demonstration that would never have worked in the simulator is a wingover. It's not that we couldn't perform a wingover in the simulator, but we could never have recognized the nearly 4Gs my ex-F-4 instructor pulled during the turn. Until a pilot has experienced this kind of G-force, they'll never completely understand how disorienting this seat-of-the-pants feeling can be alone. And we never rolled much past 120-degrees before recovering.

Then came the G demo, which will put the fear of God into most any pilot who flies around with their feet planted firmly on the floor at cruise. If a 2G upset occurs, something Wylie says is pretty easy, the pilot can neither raise their legs off the floor nor their arms to the yoke to pull—or push—as necessary against the force of gravity to regain control.

Now it was the day of reckoning, time to try stepping on the sky and see if it made sense in flight. I thought I'd seen this during the G-4 ride, but my guess was, from the way Don Wylie kept yelling to, "Push, push," on the rudder, I wasn't quite fast enough. Wylie rolled the airplane into a right turn of about 160 degrees, not quite upside down, but close. I tried stepping on the sky as I pushed the stick to the left, but the airplane acted as if it were trying to decide what I'd asked it to do before it made a move. Don told me to "Push that left rudder next time." I thought I had. Another 160 to the left and I still wasn't fast enough to avoid watching the airspeed climb close to redline and almost force me to yank the stick back pulling a bunch of Gs.

Time to try another maneuver as I slowly felt myself beginning to liquefy in the Texas heat. We moved to spatial disorientation—it worked—to prove how easily even an experienced pilot can lose track of where they are when they begin to exit that warm fuzzy 1G world we all live in. At this point, I should have realized Wylie never, ever gives students things they've seen before, although he's extremely fair in trying to recognize a student's limits.

Scare them or make them sick and the learning is over. And, he says he's had those same results from pilots of all experience levels.

Recoveries from steep turns with aileron only prove the airplane can be brought back to level flight without slamming the rudder at all. But, what is truly eye-opening is when recoveries from those 160–180 degree excursions are made using both aileron and rudder. The airplane almost literally snaps back to attention. Wylie gave me another chance to step on the sky and, this time, as we were rolled just about inverted, I fought my desire to roll through and pressed—hard—on the right rudder as I pushed the stick right. The airplane did snap right back. "Now you're talking," Wylie chuckled. We tried a few more and the theory began to sink in. The only item on the checklist we had no time for was spins. We rejoined the other T-34 for landing. Once we returned to the hangar for a debrief, my wingman—or maybe I was his—Charles said, "The AMP worked for me. I froze the first time we tried a spin, though. If that had happened for real before the course, I guess I would have had no chance of recovery."

No doubt most of the training AST offers will make many pilots uncomfortable, to say the least—from the ground-school statistics and cockpit voice and flight data recorder playbacks, to the simulator upsets that sometimes end with that deafening rumble and flashing of red lights characteristic of a simulated crash, to the loss of vision and feeling that your head is headed for the bottom of the airplane as the blood races toward the pilot's feet in a real-life 3G maneuver. But, making people uncomfortable enough to be willing to spill a little coffee in the back may just be the only way they'll live long enough to talk about the value of this kind of training.

Admittedly, some of the maneuvers Wylie demonstrates are on the edge of statistical reality. But, if the big one happens, this kind of training might be the only safety net a pilot owns. After a pilot completes the course, they'll emerge with a healthy respect, as well as a newfound experience level, for how to survive an upset that will help offset that muscle memory everyone has used since their early student pilot days. Aviation Safety Training's AMP costs $1,000 per pilot as an add-on to a recurrent simulator session. The initial T-34 training costs $2,995, with recurrent training in the Beech set at $1,495.

Despite the fact that I nearly lost my lunch in the 98-degree Texas heat, I left Hooks Airport with a sense of awe that I'd just learned something I wish I'd known during the other 34 years I've been flying. I think AST should try stealing the old American Express card slogan for their training: "Don't leave home without it."

A Few Tips for Aspiring Female Pilots

Anyone can be a pilot, but not everyone will want to follow this career path. I've asked Sandy Anderson, a former vice president of Women in

Aviation International (WIAI) and a Boeing 747 Captain for Northwest Airlines, to speak to the ever-increasing number of women pilots, as well as the valuable networking opportunities available through WIAI. Captain Anderson's ideas and WIAI also have plenty to teach men about industry networking, and the WIAI convention each spring should not be missed for the wide range of aviation scholarships awarded alone.

Women Are Making a Difference

by Sandra L. Anderson—www.wai.org

Women are making a difference! There has been a dramatic increase in the number of women earning ATP and Commercial certificates in the past 10–12 years. Women are making a positive impact in flight instructing, charter flying, corporate and airline flying. If you are interested in any of these pilot occupations, it is important to gather as much information concerning pilot careers and opportunities with people in the industry. Aviation organizations, groups, conferences,

Figure 3-4 Women in Aviation International is a great association for both women and men.

air shows, and libraries are excellent sources of information concerning various pilot occupations. Some of the following tips may be utilized in all areas of aviation (Figure 3-4).

No matter what your occupational objectives are in the aviation industry it takes passion, assertiveness, and perseverance to achieve those goals. Each individual female brings her own unique perspective, skills, and capabilities to her particular occupation. To achieve your goals it will take passion in your heart, assertiveness of your soul, and determination.

Flight Instruction

To secure a job in flight instruction, find a reputable fixed-base operator (FBO), aviation school, or aviation university with good student enrollment figures. If you graduated from an aviation university, consider the possibility of remaining at the university as a Certified Flight Instructor (CFI). Consistent demand for aviation instruction can assist in securing your continuous employment as a flight instructor. Also, what is the opportunity for you to achieve additional flight instructor certificates and ratings, such as a CFI-Instrument and CFI-multiengine land or sea, while employed with a particular school or university? The more certificates and ratings you have as an instructor, the more valuable you are to your employer. Is your employer willing to help you achieve additional certification and ratings? If you enjoy teaching or working towards the next step in your pilot career, flight instruction is a rewarding profession and an invaluable experience.

Charter Flying

Jobs in the charter flying business vary greatly among companies and FBOs. The varieties include duty hours, benefits, maintenance quality, type of aircraft and avionics equipment, travel destinations, quality of training, and reputation of the company. Charter pilots enjoy an informal interaction between pilot and passenger. Passengers are greeted by the pilots and all aspects of the flight (itinerary, catering, and special requests) are reviewed with the customer. A charter business generally means day trips with some overnight stays.

Most charter flying companies do not require a four-year degree, but the better a pilot's education, the better their chances of getting hired. Some issues to consider when applying for a charter job are duty hours, "on-call" requirements, benefits, pilot attrition rate, day trips, overnight trips, and destinations. Make sure you have researched the charter company you are considering. Are the company's aircraft logbooks and records in proper order and has the company had any FAA violations or fines?

Airline Flying

Airline employment is highly competitive. You must ascertain the hiring requirements of the airline you are interested in prior to sending in an application and résumé. It is vitally important to talk with other pilots who have gone through the hiring interview process. Before an interview, learn as much as possible about the airline. Consider a membership in an industry organization that provides continuous information on what the commuters, regionals, and major airlines are doing.

Some areas to research are the airline's route structure and the pros and cons in working for this particular commuter, regional, or major airline. Does the carrier fly domestic routes only, or domestic and international routes? Will I have to relocate? Will I have to commute to my domicile? What is the pilot attrition rate?

Corporate Flying

Most corporate pilots look for employment in an area of the country where they'd like to live. The objective of a corporate pilot is to move people from one address to another.

Depending on the size and structure of the company and its flight department, corporate pilots may handle all aspects of the trip, from flight planning and weather to the destination transportation and all catering supplies. Pilots fly the same aircraft, usually with the same passengers and, occasionally, new customers. Corporate pilots also have the opportunity to fly in some of the most advanced technological aircraft available.

Corporate and airline flying have similarities and differences that need to be considered when deciding which avenue to pursue. For both, you need an excellent background, and preferably a college education, with a similar number of flight hours required for both—depending on the equipment the company uses. Some of the differences include scheduling—the airline schedule will be much more structured and the corporate aviation schedule much more varied. In addition, with corporate aviation, you become a part of the company's structure and culture, with opportunities for advancement within the flight department, as well as elsewhere in the company. Management and good communications skills can be important, even in entry-level jobs.

The WIAI conference is a unique opportunity for women to network with other female professionals who provide a wealth of aviation experience. In discussing issues and opportunities with female professionals, prospective aviation women can discuss employment opportunities, interviewing techniques, professional appearance, and make valuable career contacts. Many WIAI members are exceptional mentors, not only in their chosen profession, but also on human resource contacts. Additionally, the WIAI conference provides an opportunity for women peers to associate and network if they are looking to transition to another aviation opportunity, and to learn of employment opportunities through the various aviation companies that exhibit each year. Visit the WIA web site at www.wiai.org.

It's also time to begin looking at what a few representative airlines are thinking about when they begin the search for pilots. Here's a look at some 2007 airline hiring profiles.

Hiring Profile: Air Wisconsin

Minimums
- 1,000 hours total flight time
- 250 hours multiengine
- ATP written within one year of hire
- First class medical
- Valid passport

*Pilots must have a high school diploma or equivalent, provide proof of eligibility to work in the U.S., and must be able to read, write, understand, and converse proficiently in English, the common/universal language of aviation.

Domiciles
- Philadelphia, PA (PHL)
- Washington, DC (DCA)
- Norfolk, VA (ORF)

Compensation
- Wages are set by the collective bargaining agreement between AWAC and the Ai Line Pilots Association (ALPA).
- First Officer—$24.17 (first year), $35.38 (second year) per block hour
- Captain—$56.99 (first year) per block hour
- Per Diem—$1.40 per hour from time of check in at domicile until check out
- Minimum Pay Guarantee—75 hours per month for all pilots
- Guaranteed 12 days off per month
- Duty and Trip Rigs
- Pilots are paid the greater of one of the following:
  - One hour of flight pay for every 2.0 hours of duty
  - One hour of flight pay for every four trip hours
  - Three-hour minimum for each calendar day

Training
- No training contract or pay-for-training program
- Pay during training—2.5 block hours per day
- Two weeks Basic Indoctrination—Appleton, WI
- Two weeks Systems Ground School—Charlotte, NC
- Three weeks CPT & Sim—Charlotte, NC
- Single occupancy hotel rooms provided

NOTE 1: Resumes are only accepted for positions that are listed under our Current Openings. The résumé or cover letter must indicate the position(s) of interest. Résumés with no position listed or indicating "Any Open Positions" will not be accepted or retained and will be discarded.

NOTE 2: All employment is contingent upon passing a federally required ten-year background check and preemployment drug screen. If hired, you will be required to complete an extensive application that will be used to complete the background check and may be required to clear an FBI Fingerprint process. All statements must be accurate and, if anything is found to be false, it could result in disqualification.

Hiring Profile: NetJets

At NetJets, we take pride in knowing that our pilots are the best in the world. We work closely with FlightSafety® International to ensure that you get only the finest in training. Our operations and network of Flight Service Specialists (on-site meteorologists, experienced dispatchers, top quality vendors, excellent flight followers, and more) contribute to making all of your flights safer, more reliable, and more thorough. In addition, NetJets pilots fly schedules that don't leave them fatigued.

Domiciles
- Columbus, OH (CMH)
- Teterboro, NJ (TEB)
- Dallas, TX (DAL)
- West Palm Beach, FL (PBI)
- Los Angeles, CA (LAX)

To qualify as a NetJets Aviation (Citation, Hawker, Falcon, Gulfstream 200, Boeing aircraft) pilot candidate, you must have the following prerequisites:

- Airline Transport Pilot Certificate (Multiengine Land)
- Current FAA First Class Medical certificate
- 2,500 hours total pilot time
- 500 hours fixed-wing multiengine time
- 250 hours instrument time (actual or simulated in flight—excludes simulator time)

NetJets International (Gulfstream large cabin aircraft) pilot requirements:

- Airline Transport Pilot Certificate (Multiengine Land)
- Current FAA First Class Medical certificate
- 5,000 hours total pilot time
- 500 Gulfstream hours (preferred)
- Gulfstream type rating (preferred)
- Crewmembers must live in the Continental U.S. and within one hour of an airport with scheduled airline service

Hiring Profile: Virgin America

If you have passion, drive, and a guest-friendly spirit, you might be the kind of person we are looking for. Virgin America (VA) is seeking a special kind of person who will help define who we are. We'll supply the planes— brand new Airbus A320-family aircraft—and the headquarters in beautiful San Francisco, California. We want you to bring the heart and creativity to make this the most beloved airline in the sky. Together, we can create a company where inspired people will always love to work.

Your Role?

The captain will partner with the First Officer and In-flight team members in ensuring a safe journey and a positive VA experience for all. This position will safely operate the aircraft in accordance with all Federal Aviation Regulations (FARs), company policies, and procedures, while maintaining a lasting impression of friendliness and outstanding service, representative of the VA brand.

What you can bring:

Skills
- FAA Airline Transport Pilot (ATP) Certification
- Check Airmen or Instructor time highly desired
- 5,000 hours total time in airplanes (excluded: Helo, Sim, F/E time)
- Airbus type rating preferred
- Current FAA Class 1 Medical Certificate
- Must pass FAA-mandated drug test
- FCC Radio License
- Must possess a valid U.S. driver's license
- Current passport
- Legally authorized to work in the U.S.
- Must pass required federal background checks and VA pre-employment background checks

Experience and Expertise
- Excellent communication skills

- An established knowledge of Microsoft Office, Adobe Acrobat, and Internet e-mail functions
- Recent flight experience
- 2,000 hours total time as PIC
- 1,000 hours as PIC total time in turbine aircraft
- 500 hours on EFIS-equipped aircraft is highly desired

VA is a brand new airline . . . and a new airline brand. We're building a business by doing something we truly believe in: creating an airline people love. To get there, we're hoping to make the flying experience and guest service offering both outstanding and unique.

The VA work environment encourages initiative, rewards innovation, and values individuality. Our team is friendly and helpful. We believe in hospitality.

Perhaps, most importantly, we all crave the opportunity to be part of something bigger than being just another airline.

Timing and Process

We are looking for ten initial Cadre Check Airmen followed immediately by VA's first class of line Captains. During 2007, we plan to hire more than 100 pilots.

We will review résumés and invite qualified candidates to participate in preliminary phone interviews.

In-person interviews will begin September 14, 2006. Our first round of offers will be extended at the end of September.

Note: At press time, Virgin America was still awaiting certification from the Department of Transportation to operate in the United States.

4
The Job
Hunt Begins

So, you've spent a ton of hard-earned cash on a commercial certificate and instrument rating, as well as a multiengine rating and, perhaps, even a flight instructor certificate. You've managed to build some flight hours, and maybe some multiengine time, and now you think it's time to start looking for a way to bring in some money, instead of just writing checks. But where to look and, furthermore, once you do find some job listings, how do you plan to get that interview?

The difference between where people end up as pilots relates not only to where they train, but the environment in which they're immersed almost from day one. In Europe, there are few general aviation airplanes flying, which means many fewer pilots can afford to take a small airplane out for a ride on the weekend, which translates into far fewer people who experience flying until they've made a career decision. That's one reason ab initio training is more prevalent in Europe than in the U.S.

By contrast, the abundance of small airplanes in the U.S., despite many Americans' complaints about high-rental prices, means pilots can win a private license and afford to fly for 50–100 hours or more before making a career decision. For the purposes of this chapter, my definition of low time relates to a pilot with less than 500 hours total time.

A key factor in the search for any flying job is to tell as many people as possible that you're in the market for starters. Call classmates, old employers, even people you've interviewed with in the past. This can effectively multiply your own job search efforts many times. You never know when a friend might run into someone else who has an opening. I think there are two different kinds of people who look for flying jobs: those who look pretty much on their own and those who find someone to search for them. I've personally used both methods, and I think each has its own benefits and drawbacks.

Searching on your own means you're a pretty busy person, at least if you work the way I do, because someone looking for a full-time job already has one . . . looking for work. The search for a flying job (compared to talking about looking for a flying job) requires not only hard work, but also a succinct plan of how you'll reach this new goal. There's that word "goal" again. Funny how that word keeps creeping up, but I believe that, without clear-cut goals—a personal flight plan—you'll end up making emotional decisions that might not be in your best long-term interests.

Being involved with airplanes is always an emotional experience, so make sure you've some idea where you're headed before you go marching off toward a new job. As one pilot says later in the book, a goal doesn't necessarily mean flying only one particular airplane, but it can also include the need to decide the kind of flying that fits your lifestyle as well as the location you want to live. If you're relatively low-time, almost any flying job will probably be beneficial, as long as you can afford the cost. And, sometimes the cost can be dear, everything from moving a few thousand miles away to starting wages that appear to rival poverty scale. Only you can decide.

In this section, we talk more about a few job-search techniques. Some will be additional publications, some will involve firing up your computer to access the Internet, while others will be full-fledged pilot recruitment services anxiously awaiting your signup, so they can share their knowledge . . . for a fee. All of these can help you quickly accumulate a wealth of information about who owns aircraft and which companies are hiring.

Notice that I don't mention "not hiring." If a company owns aircraft, they all must hire . . . eventually. It's a matter of finding out who they are and planning your appearance on their doorstep at the proper moment. Some of the magazine classified sections contain job ads, while some are ads for training schools. But sometimes just talking to people at the training schools can generate ideas to help your plan take shape. Many of the training schools also use 800 phone numbers, so call and ask questions. If you train at a particular school, what can they do to help you find employment? In fact, that is something you should ask every school you train with.

What follows are some of the sources not necessarily organized by their importance. They are all important. The critical element of any job hunt is what to do with all the information you uncover. The Internet alone can be overwhelming without some way to organize the data.

Internet Job Hunting

Here's the deal.

A computer won't find you a job, but your chances of finding a job without an Internet connection and some savvy search skills are significantly reduced. But don't confuse online methodology as a substitute for good-old in-person networking. It's not. Online job searching is simply an evolutionary improvement in how pilots look for work.

Note: This is not a complete list of companies that offer help to aspiring pilots through traditional and online methods. These are simply the ones I have some personal experience with.

Information services for pilots began over 30 years ago when the Future Aviation Professionals Association (FAPA), as almost everyone came to know it, was first opened. Run by two airline pilots—Louis Smith and Gary Ekdahl—and a helpful staff, they saw a need not simply for airline interview preparation, but for all the collateral services, such as company background and résumé preparation for pilots in search of airline jobs within the United States.

Eventually, the company outgrew itself and FAPA shut down. FAPA's former vice president of marketing, Kit Darby, himself a United Airlines pilot, started Air Inc. (www.jet-jobs.com). Darby's company offers online-based resource materials, a print magazine—*Airline Pilot Careers*—and numerous reference books. Additionally, Air Inc. provides applicants a phone number for one-on-one telephone counseling about airline-specific interview questions. Air Inc. charges various fees, depending on the options the applicant chooses. Initial annual subscription costs vary: online-only access costs $180 per year, while online access and an initial set of Career Development system publications costs $230.

After FAPA closed in 1996, Louis Smith started FLTops.com (www.fltops.com) and provides material in online format only. At press time, FLTops.com's service charges a one-time fee of $29.95 and $4.95 each month thereafter. FLTops.com offers members and visitors free access to industry news compiled five days each week by FLTops.com writers. Archival news access requires FLTops.com membership. FLTops.com does offer members background information prior to an interview (but in an electronic format), as well as an easily searchable fleet database to learn which airline flies which aircraft. The company also offers a domicile database for each airline it covers.

Louis Smith said, "What makes us different is that FLTops.com is not based on the print model of an aviation information magazine. We only cover the Big 13 in an Internet format that pilots simply jump in to." But Smith was also vocal about what FLTops.com is not. "You won't get an individual evaluation and we won't give career advice. Pilots just want to know where the jobs are and we provide that information from the key people doing the hiring at the airlines. We've transferred a lot of the work to our customers. Pilots make a capital investment and, in return, they get a really good product at a really good price." FLTops.com also sends out a weekly bulletin. Just tonight, I learned applicants to American Eagle will be pleased to learn that the airline has modified its hiring standards to accept résumés from pilots with 500 hours total time if they have been working in a Part 135 company. What's that kind of information worth?

Smith's advice to applicants, "Networking is still #1 in finding that job. You have to get to know someone at the companies you want to work for. You can't beat someone telling a potential employer about you before you're hired." For more information, visit the web site at www.FLTops.com.

Processing applicants has always been expensive. One Internet-based service, Airline Apps.com at www.airlineapps.com, is attempting to help airlines keep those costs in line. This company can seriously reduce your workload, while it keeps your personal hiring materials in front of many carriers at the same time by acting as the direct application service for a number of airlines. The advantage to an applicant is that the electronic form only needs to be filled out once. After that, the applicant simply chooses the publication option of which airlines will receive the material. For an extra fee, the company will compare your résumé to those of pilots being hired at the various airlines. The initial service costs $54.45 annually, plus a one-time, $9.95 set-up charge.

I logged in to the service to learn more as I wrote this text and found the process pretty simple. The company lists the following airlines as participants: American Eagle, Comair, Continental, Frontier, Mesa, Spirit, and Trans States. It also lists Air Wisconsin, Air Tran, Atlantic Southeast, Atlas Air, Boston-Maine Airways, Cape Air Nantucket, Chautauqua Airlines, Express One International, Gemini, Grand Canyon Airlines, Island Air, Piedmont Airlines, Pinnacle Airlines, Polar Air Cargo, and World Airways as companies accepting résumés. Check the web site for updates, although information on the company itself at the web site is limited.

One element of the application process that cannot be left out is whether or not a pilot needs a college degree to apply. Many companies *recommend* one, but don't *require* a sheepskin. David Jones, former editor-in-chief at *Flying Careers* magazine (no longer published) and now a contributing editor at FLTops.com offered some relevant advice. "For younger pilots who are tempted to build time any way they can, rather than finish their four-year degree, they should know that the statistics show the number of pilots hired without a four-year degree is low. Occasionally, a few high-time pilots might get in without one, but they are few in number. The difference in flying and other jobs is that everything you do—on and off the job—can count for something, too. Don't do something stupid, like pick up a DUI. And, as always, your attitude is critical. A bad one will come back to bite you."

As with other jobs in the aviation industry, there is a cyclical expansion and contraction. Some of the online resources we listed a few years ago have disappeared. Many new job boards have sprung up as some of the larger services have contracted. AOL, for example, once was home to a huge aviation forum that has now ceased to exist. That's not necessarily bad because more sites can offer a wider range of views on any given hiring situation. But more sites also translate into more places to visit regularly to stay current on the topics being discussed.

No matter how busy you might be, put AVSIG on your list. AVSIG is an acronym that stands for Aviation Special Interest Group, and AVISG was one of the first. Originally a part of the CompuServe information service, the *AVSIG* message board migrated to the Web when CompuServe—like AOL—thought the group's core was too small to bother with. Although the web address might seem a bit odd (www.aero-farm.com), it's a place you must visit. You'll find people from around the industry ready to talk about almost anything aviation, such as flight training, history, books, airline jobs, or even air traffic control.

Mike Overly has been running AVSIG—www.avsig.com also works—since I can first remember going online, which must be around 1989 (Figure 4-1). In his own words, here's how Overly describes the site. "AVSIG works because everyone in aviation is invited to solve problems in the same place under the same house rules: no anonymous wisdom, no flaming, no shilling, no obscenity, but, otherwise, anything goes. We're probably the last forum on Earth to require real names for posting, but we continue to be amazed at how the level of accuracy and civility goes up when your name is stamped on every post. Accountability is everything in aviation, including online discussion. The real-life personal and professional friendships that develop in this kind of environment are just a bonus."

Figure 4-1 AVISG is one of the oldest virtual aviation water coolers.

Aviation Employee Placement Service (AEPS) (www.aeps.com) is only available online and claims approximately 7,000 member companies worldwide. In case you're thinking that seems like quite a few airlines to represent, you should know that AEPS does not work exclusively with pilots. Member companies range from airlines, to charter and corporate operators in search of mechanics dispatchers, and other aviation professionals. Company president Jim Dent, himself a former Pan Am pilot, claims some 56,000 members, about 9,000 of whom are pilots in search of cockpit positions. Dent said, "In our system, you can restrict who views your information. The companies who search our database also don't get unsolicited résumés and only see your résumé if you fit the qualifications they want." His advice to applicants is to "keep your information in our database as current as possible."

It does offer an easily customizable data record for a pilot's qualifications that new members fill out online when they sign up. Then, when a member company has a position to be filled, they simply put the necessary criteria in their computer and perform their own search. The benefit to both the pilot and the member company is they each get results that meet their requirements. If a regional carrier wants pilots with 2,000 hours total time, no records of pilots with less than that amount of time are sent, saving everyone a lot of time and energy.

AEPS also sends out a weekly aviation industry news update—via e-mail—published in conjunction with *Air Transport World* magazine. Whenever a member company performs a search and extracts the information on a member, AEPS also sends that member an e-mail update. Members can read in-depth profiles of the companies AEPS represents. AEPS charges a monthly membership fee—currently $12.95—but it does not charge for updates to the member database. Visit its web site at www.aeps.com.

Avcrew.com (www.Avcrew.com) is a compact, easy-to-navigate web site that offers corporate pilots and potential employers a place to find each other. And, that's what this is all about. Jobs are posted as they come in and pilots can post their résumés here. Elsewhere in this text, you'll see an example of the dialogue found on the interactive portion of AvCrews's site. It's a place to talk about the issues of the day as they relate to corporate flying. A recent question asked about the toll airline flying was taking on corporate flight departments, while another asked for opinions on fractional operations.

Richard Harris, publisher of AvCrew.com and an NBAA member, said, "AvCrew was the first site devoted only to corporate pilots. As employers learn more about the benefits of technology when searching for qualified pilots, this medium will become even more popular (to locate pilots)."

This small grouping by no means represents all the Internet sites of value to a pilot applicant. These are only a few of the formal job connection sites. At another location in the back of the book, you can find a simple list for you to indulge yourself. Many of the other valuable sites include some of the worldwide message boards where pilot types hang out: PPRuNe (www.pprune.com), Airline Pilot Central (www.airlinepilotforums.com), The Corporate Pilot Board (www.corporatepilot.com), and Aviation Job Search (www.aviationjobsearch.com), all offer great locations to begin information tracking. By now, you've also been exposed enough to the Internet to realize that sites come and go on an almost-daily basis. Be prepared to organize your material. A search a moment ago on Google under the phrase "Pilot Jobs" produced 13 million hits. Adding the word international helped. That search only produced 9 million hits.

Airline Pilot Central (www.airlinepilotforums.com) is one of those places I visit regularly simply because it is so chock-full of information from all over the world about flying, from dialogues between pilots about their latest adventures, to discussions about air traffic control and even banner towing, that I'm afraid I might miss something important.

Some of the Publications

Most all the news services have some presence on the Web because the cost of production is much reduced for publishers. But instant access to the Web translates into almost instantaneous updates to everything you'll see there.

One of the largest aviation information publishers is McGraw-Hill (and this is not a plug for our publisher, but simply fact). Some years ago, McGraw-Hill saw the opportunity to build a portal to act as a central clearinghouse for aviation information. That idea became www.aviationnow.com and is home to many of the publications that follow. Many are available in print and most are also available online through McGraw-Hill's Aviation Intelligence Network that links to aviationnow.com. McGraw-Hill offers a package price to the Intelligence Network that I highly recommend.

The print version of the World Aviation Directory (WAD) is about ten pounds of information on the aviation industry that you ever thought about or, possibly, even might have wanted to know. The *WAD* is a world-renowned directory to all things aviation, from airlines, to fixed-base operators, to flight schools, to parts suppliers, to government aviation agencies. And, best of all, the WAD covers the aviation industry worldwide. Here's a glimpse at some of the facts you can locate in this big volume.

From the Industry Trends and Statistics section:

- Worldwide Carriers Systemwide Traffic
- U.S. Majors and National Carriers

- Domestic Financial and Traffic Summary
- Financial Indicators
- Expense Indicators
- Fleet Analysis
- U.S. Major Carriers
- U.S. National Carriers
- Worldwide Carriers
- U.S. Major Carriers Aircraft Operating Expenses
- Domestic
- Systemwide

One section lists the name of every airline flying in the world, as well as its home base. These can be easily cross-referenced to the air carrier list, which lists the address, telephone, and FAX number for an airline, as well as the major executives of that carrier and the current fleet size, something that might come in handy during an interview.

You'll be suitably impressed with the information that's also available on aircraft manufacturers and charter companies. Finally, if you know the name of a particular aviation person you're trying to track down, the alphabetical list could be just what you need. A current WAD (it's released twice a year) is not cheap, about $200 per year, but probably no better source of information is available in book form. Most library reference sections carry a *World Aviation Directory*, so you can take a look before you shell out your hard-earned dollars.

Many of the following publications offer not only classified sections to steer you toward more information for your job search, but also stories about what's happening within the industry, such as a new airline start-up or a corporate flight department opening or closing that might influence your decisions.

Most of these magazines are also advertising-driven, which should translate into a bit of skepticism on the part of the reader (especially flight schools that promise a job after graduation). Be certain you've verified the claims any school admissions person gives you before you sign up. As with the rest of this book, you may find an interesting perspective during your climb up the aviation ladder, whether you're a beginner or near retirement. Tell us about it for future editions at rob@propilotbook.com.

Professional Pilot Magazine

Based in Washington D.C., *Professional Pilot*— know within the industry as *Pro Pilot*—calls itself the monthly journal of aviation professionals with spotlight articles on airline and corporate operators, as well as technical articles on state-of-the-art electronics and flying techniques. The slant is definitely toward professional pilots—more corporate pilots than airline— although newly rated pilots can certainly benefit from the educational aspects of the magazine.

A recent issue of *Pro Pilot* contained these stories: the state of aviation in Ohio, a pilot report on the Canadair RJ jet, an avionics article on Flight Management Systems, a profile on a hurricane-hunting NOAA pilot, a story on alcohol abuse, and a piece on Martin State Airport. Regular features include an IFR chart-refresher quiz and basic aviation news.

The magazine's classified section recently listed various flight training schools that might not advertise in other publications, as well as a few listings for help under the jobs-available section.

Contact *Pro Pilot* magazine at 3014 Colvin St., Alexandria, VA 22314, or 703-370-0606. This publication is not available on the newsstand. E-mail *PRO PILOT* magazine at editorial@propilotmag.com or visit its web site at www.propilotmag.com.

Flying Magazine

Flying magazine claims to be the world's most widely read aviation magazine. Its emphasis is on smaller general aviation, although many articles highlight the newest aircraft and equipment in corporate, airline, or—occasionally—military aviation. This magazine uses a number of regular, highly respected columnists, each with an interesting perspective on what's happening in the industry.

The magazine's classified section has grown, and it now covers seven full pages. In a recent issue, no less than 46 flight schools were advertised. The classified section also included 13 ads for employment information, everything from one that asked for low-time pilots to one specifically aimed at corporate flying jobs.

Recent articles in *Flying* magazine included a story on flying jets as a single pilot operator, a look at a fractional ownership program for the Cirrus Design SR-22, and a piece on how well a new Light Sport Aircraft (LSA) might work for training private pilots. *Flying* is available on the newsstand, at most libraries, and at www.flyingmag.com.

Business & Commercial Aviation Magazine

Business & Commercial Aviation magazine is written for the crews of corporate aircraft, although many of the articles cover subjects of interest to all pilots. Recent stories covered accident-prone pilots, icing certification, the London City Airport, and the aviation user-fee controversy.

BCA has a solid monthly aviation news section too. The marketplace section is small, but many ads of various categories are scattered throughout the magazine. *BCA* is part of the McGraw-Hill collection of aviation magazines. Subscription information is available at www.aviationnow.com. *BCA* is *not* available on the newsstand.

Aviation Week & Space Technology Magazine

Av Week, as this magazine is known, is the weekly bible of what's happening in all facets of the aviation industry worldwide. With editors scattered around the globe, you can expect the most comprehensive description of aviation events that could affect your career. Coverage of the general aviation end of flying is a bit skimpy at times, but the magazine more than makes up for it with their coverage of airlines, military, and corporate aviation. If you follow aviation, this publication is something you must have. Their classified ads often relate to major flying jobs, along with employment services in the industry. *Av Week* is also part of the McGraw-Hill collection of aviation magazines. Subscription information is available at www.aviationnow.com.

Air Line Pilot Magazine

Air Line Pilot magazine is the voice of the Air Line Pilots Association (ALPA), the major union for airline pilots at 39 different airlines in the U.S. and Canada. Some of the stories tend to look at the political interests of a union with more than 60,000 airline pilot members. The classified section is pretty short on anything of interest to aspiring pilots. What makes this magazine particularly interesting is the insider's view—from a union perspective—of what's happening within the airline industry, such as how new FAA regulations could affect a pilot's job.

Recent stories in *Air Line Pilot* magazine included a look at the life of a regional pilot, a study of the ALPA group at ATA airlines, a feature on fatigue, and another on the age 60 limitation that ALPA opposes, as well as regular comments by ALPA's president. *Air Line Pilot* magazine is not sold on the newsstand. Much of the content is available at www.alpa.org/.

Air Transport World Magazine

Air Transport World (ATW) magazine is the place to learn about issues relevant to the airline industry, but from a management perspective. The magazine's editors focus on U.S. and international carriers each month, with a wide variety of topics ranging from a highlight piece on at least one individual airline per issue, discussions of how companies finance aircraft purchases, new engine performance, and the aircraft manufacturing industry itself. I'd add a subscription to *ATW* simply to offer you a different point-of-view from the managers at some of these carriers you might be interviewing with; for example, their stance on labor issues.

ATW also offers readers a news flash system available through the web site. Contact Air Transport World subscription at 216-931-9188. *ATW* is located at 1350 Connecticut Avenue N.W., Suite 902, Washington, DC 20036. Find *Air Transport World* magazine at www.atwonline.com.

Aircraft Owners and Pilots Association (AOPA) Pilot Magazine

Aircraft Owners and Pilots Association (AOPA) Pilot magazine is the publishing voice of the 410,000-member strong Aircraft Owners and Pilots Association. This magazine is focused on general aviation, but in the past year, it has added an interesting and useful feature called "Turbine Pilot" to explain the intricacies of flying jets and turboprops to people who are used to piston engines. This feature alone could make the magazine worth the price to a relatively new, yet aspiring professional pilot. The *AOPA Pilot* magazine is offered as part of the annual membership dues (currently $39) from AOPA. If anything more is to be said about AOPA, besides the fact that nearly 60 percent of all active U.S. pilots are members, it's that the $39 brings so many extras, you'd be crazy to pass it up.

Recent articles included flying with over-the-counter medicines, lost communications' procedures, a report on Cessna's Caravan, and a look at what's doing at Lycoming and Continental Engines. The classifieds tend to be rather lean, but they would still be a good source for training school ideas. Contact AOPA at 1-800-USA-AOPA or at www.aopa.org, where the content also is well worth the 39 bucks it costs to be a member.

Airline Pilot Careers Magazine

Airline Pilot Careers magazine is published each month by Air, Inc., mentioned earlier. The Air, Inc. editors sift through thousands of pieces of information and produce plenty of forecasts for the industry.

Each month, *Airline Pilot Careers* magazine highlights at least one airline (and usually two) and tells all: the history of the company, hiring requirements, fleet size, pay ranges, pilot comments, and tips on making it to the first interview. Recent articles include a cover story on United Parcel Service (UPS), how to find international flying jobs, and how pilots can be conned by investment scams.

To keep applicants aware of the qualifications of pilots being hired each month, *Airline Pilot Careers* offers a new-hire feedback section, where recent pilot hires relate their experience levels at the time they were given the nod.

Here's a listing from a recent issue, in which I switched my name with the real new hire. Rob Mark, 35:2, 20/100 vision, United. B-727 FE, 7,150 hours, 1,000 jet, 3,000 turboprop, ATP, FEw, CFII, MEI. B.A. degree. Cargo. LR-36, DA-20, BE-99, PA-31, Burbank, CA.

Broken down, this means the pilot was 35 years, two months old, with 20/100 vision. He was hired by United as a B-727 Flight Engineer with 7,150 total hours logged, of which 1,000 hours were in jets, 3,000 in turboprops.

At the date of hire, this pilot held an ATP certificate and had passed the Flight Engineer written exam. He also held the instrument and multiengine instructor ratings, as well as a B.A. degree. He flew mostly cargo in Learjets, Falcons, Beech 99s, and Navajos. He lived in Burbank, California. *Airline Pilot Careers* magazine is not available on the newsstand. Contact Air, Inc. at www.jet-jobs.com.

Aviation International News Magazine

Aviation International News magazine is the big news magazine. This magazine is not only big in the sense that's it's chock-full of news about airline and business aviation, but it's also a solid news-hub and an information storehouse about the business of aviation such as fixed-base operator and airport. And, oh, yes . . . it's also big, physically. The *Aviation International News* format is glossy-magazine style, but twice the size of a regular magazine at 11" by 17" (Figure 4-2).

While the purpose of *Aviation International News* magazine is to be a news vehicle, *AIN* allows for much longer articles than other magazines.

Recent stories included a Pilot Report on the Citation Mustang (reprinted in this book), as well as a story about NetJets Europe pilots aim to unionize and a look at the production snags that have recently plagued the Eclipse VLJ. The news stories make this magazine important, especially because there's no classified section. *AIN* offers readers a twice-weekly e-mail News Alert publication with breaking industry clips. Subscribe at http://www.ainalerts.com/ainalerts/.

Figure 4-2 AIN is one of the premier aviation news magazines.

AIN also writes daily publications for most of the major aviation trade shows around the world, including NBAA, Paris, Dubai, and the HAI helicopter event, and it has recently begun a TV show at those events. This magazine is not available on the newsstand. Contact *Aviation International News* at www.ainonline.com. AIN TV can be found at www.AINTV.com.

The Flyer

This tabloid is normally devoted to smaller general aviation aircraft. On reading a recent issue, I looked through its "Pink Sheet," which is *The Flyer*'s classified pages, and found ads that could be of interest to a new pilot looking to build time. Under the "Help Wanted" section, I saw, "Pilots fly 1,001 hours per year with great pay. Alaska offers seasonal and full-time employment. . . ." No promises, but an opportunity in a place that a future professional pilot might not look. Also, *The Flyer* can be an excellent source of ads for the parts necessary to keep that airplane of yours in top shape. This tabloid is the sleeper of the publication list, so be sure to check into it. Contact *The Flyer* at 1-800-426-8538 or at http://www.generalaviationnews .com/.Air Transport Association.

Air Transport Association's Smartbrief

Inclusion of an aviation association might at first seem odd, but the Air Transport Association (ATA) offers readers a news service that must be a part of any pilot applicant's daily read routine. Called *Smartbrief* (www.smartbrief.com/ata/), each day ATA delivers at least a half dozen stories pulled right from the news wires or the front pages of major news organizations around the world. Some, such as the *Wall Street Journal*, may require an online subscription to read the entire story.

Aviation for Women Magazine

Don't be fooled by the name. *Aviation for Women* is a magazine for professional pilots and professional pilot wannabes, as they are sometimes called, as well as many other facets of the industry. The print magazine of *Women in Aviation International* brings readers current features that keep them up-to-date about women in the aviation industry. Articles feature women who have made aviation history, professional development ideas to help you in your career, and current-topic articles. A subscription to *Aviation for Women* magazine is included with a WIAI membership.

Additional key features include scholarship information with details on requirements and application forms, as well as organization and chapter news about key local and national issues. Because the scholarship information is for men and women, the price of a WIAI membership is well worth the opportunity to compete. Find out more about it at www.wiai.org/.

AOPA Flight Training Magazine

AOPA Flight Training magazine provides constant coverage of the flight training industry because the motto here is "back to basics." The issues are designed not only for the student of various ratings, but also for the instructor. *AOPA Flight Training* offers a host of resources for student and private pilots, such as flight and knowledge test-preparation quizzes and detailed maneuver explanations. Instructors will find a ready list of all necessary logbook endorsements and sample lesson plans.

AOPA Flight Training magazine produces an excellent national list of flight schools, which Editor Mike Collins was kind enough to let us reproduce as Appendix A. The magazine is owned by AOPA and you can find out more about it at http://flighttraining .aopa.org/.

The *Wall Street Journal*

No publication subscription list would possibly be complete in your search for a new position without considering the *Wall Street Journal (WSJ)* (Figure 4-3). The *WSJ* is published six days a week and can be delivered right to your door each day or purchased at most newsstands and drug stores.

Nowhere else can you read the kind of up-to-the-minute reporting on issues that hit home to you if you're in the job hunting and, hence, the interviewing mode. This in-depth coverage of emerging topics is the *WSJ*'s competitive edge over many of the other publications here. While the *WSJ* is certainly a business newspaper, the quality of the reporting is excellent and normally portrays quite balanced views, even on labor issues.

Because so much of the business world is tied to the realities of air transportation, the industry seems to appear as a regular topic. Recent stories have included in-depth looks at the struggle at FedEx between management and its pilots' union, a major series on labor in America, great coverage of the strike at Northwest Airlines, the involvement of the unions in United Airlines' recent dismissal of its company president, and how much job hunting for pilots has moved overseas.

Subscription information can be found by calling the *Wall Street Journal* at 1-800-JOURNAL. And don't forget that a subscription to the print version also gives you a discount on a subscription to the customizable online version of the Journal, as well. I can't imagine not also subscribing to the online *WSJ*. www.wsj.com.

Figure 4-3 The *Wall Street Journal* online is a critical resource in the hunt for a flying job.

Trade a Plane **Magazine**

This may, at first, look like another of those magazines that will make pilots scratch their heads, especially when they notice the canary yellow paper on which the magazine is printed. Visit *Trade a Plane* magazine at http://www.trade-a-plane.com/index.shtml, and you'll be drawn into a publication that sells things, mostly airplanes and accessories. But, you'll also run into a wonderful employment section for those hard-to-find jobs . . . the ones that help you build time in out-of-the-way locations. "PILOTS FOR FAR 135, ME (C310) cargo, requires 1100TT, and multiengine rating, OH, WV, KY, CT, NJ, NY, great schedule, great pay," or " PROFESSIONAL PILOTS FOR Part 135, Birmingham, Dallas, Charleston WVA, and Mobile bases. Must meet 135 IFR requirements, Cessna 208 Caravans, and Piper Saratogas."

Airline Business **Magazine**

Quite honestly, *Airline Business* magazine was a new publication to me, which first appeared in my mailbox as the text for this book was being written. *Airline Business* magazine is a competitor to *Air Transport World* magazine, except *Airline Business* is produced in the United Kingdom.

Recent issues included stories on Polar Air Cargo and the recent alliance among American Airlines, Canadian, Qantas, Cathay Pacific, and British Airways, as well as an excellent commentary piece on the outlook for the world economy after a recent meeting of the International Monetary Fund (IMF). Visit its web site at www.airlinebusiness.com.

Flight International **Magazine**

If there is a second version of the aviation bible that is *Av Week*, but from a European perspective, that publication would be *Flight International* magazine. Available in both a print and an online edition, *Flight International* has been read around the globe for 90 years. What makes a subscription to *Flight* extremely valuable is the often quite different perspective to the same world issue and its effects on the industry (Figure 4-4). *Flight International* has also begun offering feature podcasts.

Flight International magazine is also famous for the cutaway drawings that look at the inside bits of aircraft back to the 1930s. A complete archive is available online at http://www.flightglobal.com. The *Flight* web site also regularly looks at technical details for military aircraft, space vehicles, helicopters, and boasts considerable job-hunt information. A sample of the technical depth of the magazine can be seen in their study of the A-380.

Figure 4-4 *Flight International*'s web site offers access to many valuable resources.

AvWeb

AvWeb is the first of the new aviation news sites on the Internet that seems to have become successful. This online aviation news site is owned by a larger publication group, which also controls more than a dozen other magazines in various industries.

You can find an interesting twist on the day's aviation news, as well as the availability of regular feature-story podcasting here, but don't look for job ads. You can also sign up for another service to send you a twice-weekly newsletter. Another item you won't find on too many other web sites yet is a little orange rectangle that looks like this. This is an RSS feed symbol that makes it easy to subscribe to a blog (covered in Chapter 8).

FLTops.com

This is the news and information site for FLTops.com. Online news is updated five days a week and is free to Internet users who are content to browse the recent headlines. The searchable archive of past news articles, as well as the additional resources FLTops.com provides, is available only to subscribers.

FAA Public Affairs List

One of the best—and least expensive—methods of staying in touch with what is happening in the aviation industry is by subscribing to the FAA's list for news releases. Send an e-mail message to this address: listserv@listserv.faa.gov. In the body of the message, type subscribe-faa-newsrelease (your name). Visit the main FAA web site at www.faa.gov and click the news button.

A copy of every press release the FAA sends out will be delivered automatically to your e-mail account. You'll receive an e-mail from the FAA confirming you've been added to the list.

And don't leave this section believing that aviation publications are the only places jobs are advertised. I've certainly found them in the "Help Wanted" section of the *Chicago Tribune*, the *Denver Post*, the *New York Times*, and, of course, the *Wall Street Journal*.

NTSB Mailing List

Similar to the press release list from FAA is the version run by the National Transportation Safety Board (NTSB), at http://www.ntsb.gov/Pressrel/pressrel.htm. Not only will you find a listing of detailed aviation accident reports at the main NTSB site, but subscribing here will give you copies of the latest aviation safety recommendations and opinions by the Board, which is a recommending body to the transportation industry as a whole and the FAA specifically for aviation.

If I were interviewing a pilot applicant, I can't think of a better place to find a few questions that might give me some insight into a pilot's background than on this site. Would you be prepared if an interviewer were to ask about the latest aircraft icing recommendations the Board had released?

The Résumé

As you begin looking for any flying job, you need something to announce yourself as a professional pilot in search of work. You need a résumé, in addition to a snappy cover letter specifically tailored to each target company. The thought of putting together a résumé makes some pilots cringe, but you might as well become used to it if you intend to remain in aviation or nearly any other profession (Figure 4-5).

A number of excellent books on résumés are currently available in the bookstore, or at Amazon.com, but be careful about choosing any of them. They aren't normally designed for pilots, who often possess some unique

Paul Allan McCartney - 1201 Abbey Road - Herndon, VA 20169
Phone: 703-730-3015 - Fax: 703- 730-6904 - cell: 703-479-2281
Website: www.bizpilotvirginia.com - e-mail: bizpilot@comcast.com
ssan: 291-69-0034

Objective: Career pilot with an international corporate flight department

Certificates and Ratings:
Airline Transport Pilot, Airplane Single and Multi Engine, Land
Certified Flight Instructor, Airplane, Instrument, Multi Engine
Aircraft Ratings: HS-125, G450, CRJ
FAA Class 1 medical
Restricted Radio Telephone permit

Flight Experience: **Total – 7276**

PIC	4185	Turbine	4705
SIC	2604	Instrument	702
Night	1857	International Ops	1307

Aircraft Specific (hrs): CRJ 1487 HS-125 1257 G450 1050

Aircraft Specific Schools:

Flight Safety Recurrent	G450	11/06	(Part 91/135)
Flight Safety Recurrent	G450	12/05	(Part 91/135)
Flight Safety Recurrent	HS-125	09/04	(Part 91)
Flight Safety Initial	G450	12/04	(Part 91)
American Eagle Initial	CRJ	10/97	

Work Experience:

Sallie Mae, Part 91/135 corporate flight Co-captain Gulfstream 450 Dulles, VA	2004 to present
Fred's Box Company, Part 91/135 corporate flight Co-captain Hawker 800 - Raleigh, NC	2000 to 2004
American Eagle, Part 121 scheduled air carrier Captain, CRJ, Dallas, TX	1997 to 2000
American Eagle, Part 121 scheduled air carrier First Officer, CRJ, Dallas, TX	1995 to 1997
American Eagle, Part 121 scheduled air carrier First Officer, ATR 42, Chicago, IL	1992 to 1995
Quick Air Charter, Part 135 charter Captain, First Officer, C-421	1990 to 1992
Sam's Flying Service, Charlotte, NC Flight Instructor, Part 141 school	1987 to 1990

Education:

University of North Dakota Bachelor of Science—Aviation Management	1984 to 1987

Figure 4-5 Sample pilot résumé.

talents that don't always fit into a standard résumé format. Most of the major aviation information services, such as FLTops.com and Air Inc., are affiliated with people who will write a proper pilot résumé. Many will design your résumé from scratch or critique yours for a fee.

In case you decide to go it alone and write your own résumé, here are a few tips you need to consider. First, neatness does count. And this isn't simply because Human Resources (HR) people like to grade résumés, it's because you're trying to sell yourself through your résumé , a document that could be your initial contact with a company. Make it easy on the receiver to locate the information they care about, which means keep it short—one page, if possible—and don't make the type size too small.

Every résumé needs a stated purpose or goal. This not only offers you an opportunity to state your goal—a career pilot position—clearly, but it also gives you the chance to tailor your resume specifically to the company you are applying to, "A Career Pilot Position with Continental Airlines." Word processors make this kind of targeted marketing of your qualifications easy, but with one caveat. Be sure that you have looked at all the qualifications on the résumé before you send it out. If you send a résumé off for a corporate job, I'd remove the reference to having passed the Flight Engineer Written with a score of 98 percent. And, don't forget to put it back in when you send it off to an airline that requires the FE written to apply. Try creating different versions of your résumé for different targets.

The exact layout is up to you, but it must contain at least all of your personal contact information, the certificates you hold, your flight times, education, and other activities that might have an influence on the hiring decision, such as awards or membership in outside organizations, especially those in which you held leadership positions. The truly personal information—age, height, weight, marital status, and so forth— is extremely controversial. By law, an interviewer cannot even ask your age or marital status. My view is to leave the rest off as well. As a final check on the finished document, why not offer it to another pilot for their comments?

If you're through with the writing, make certain you verify that all the dates make sense, too. Plan to print your résumé on a good-quality laser printer, although many of the new inkjets also do a great job. Use only a bright white paper to make the type stand out easily. No matter what, make a backup copy of your résumé and save it somewhere safe.

Before you send a digital version of your résumé to anyone, try sending it to another e-mail account as a test. This offers you an opportunity to see what your résumé will look like when someone else opens the document. One of the quirks of computers is the formatting can sometimes change during transmission, and this is good to know in advance of e-mailing your résumé.

Worth mentioning, too, is that the entire résumé concept is in a state of flux now, simply because many airlines and some corporations are using services to process online forms. Not only does using a service cut down on paperwork, but it makes it easy for a potential employer to qualify them as they arrive. They might see 35 electronic résumés and decide they only care about people with more than 2,000 hours total time. Hand-sorting papers takes hours, where a simple database filter is all that's needed with digital data.

Another new concept is the pilot web site, which essentially offers a potential employer an opportunity to download a copy of your résumé, look at your photo, and click on more detailed explanations of your background, if they choose. This is a customer-focused search world. Your job is to offer an employer the information they need in the format they asked for. But, it doesn't mean you can't try something that makes you stand out a bit. A three-page web site is pretty inexpensive today. If you haven't done it yet, visit www.networksolutions.com and register a domain with your name—www.johnsmith.com or some easy-to-recall combination of your name. And, for heaven's sake, don't build a web site with a silly domain like "worldsbestpilot.com." You'll make yourself look foolish.

Certified Flight Instructor (CFI)

Now it's time to focus on one of my favorite subjects: flight instructing. If you considered some of the issues we've discussed so far, you might already have your Certified Flight Instructor (CFI) rating. If you've not yet earned yours, you might want to give the idea some thought. With a CFI certificate, you'll be eligible to approach a flight school about a position with a small amount of logged flight time. I would begin by calling the local flight schools in your area and simply asking for the Chief Flight Instructor. Ask for a few minutes of their time or set up an appointment to stop in and visit.

It won't be a secret that you're a low-time pilot because the vast majority of new flight instructors are. Many flight schools see this as a benefit. New flight instructors not only have a great deal more enthusiasm for the job, but they are also much more familiar with the subject matter because they've just completed the courses. I would always try some local schools first. If the school is not interested in a new full-time instructor, ask if they'll consider you for a part-time teaching job. If you really want the job, I would try just about anything to reiterate your desire to work. The worst they can say is no. And no only means no right now. Then, try them again in a month.

If the local flight schools don't prove fruitful, you need to decide whether an out-of-town position is a consideration. In aviation, I've found the best jobs always seem to be somewhere other than where I live. For a while, I was lucky enough to work for an airline based where I lived. Unfortunately, the airline eventually went bankrupt and I was faced with a major decision . . . to move or not to move. I decided not to move and, a year later, I was forced to change my mind. Certainly, you can sit around and wait for a position to open at a local flight school or you could just bite the bullet and start looking elsewhere.

One method you could use is to pick up a copy of *AOPA Flight Training* magazine's annual directory of flight schools or the one produced by the University Aviation Association at http://www.uaa.aero/.

While both lists are primarily designed as a guide for students seeking flight schools for their own training, it only makes sense that these schools need instructors. Flight schools always seem to have a turnover, as pilots move on to bigger and better things, so the job is to locate the school that needs another CFI (Figure 4-6).

The directories are broken down by states, so if your sights are set on Florida, you still have over 100 schools to try. If Colorado is more to your liking, the number is less, but it's still a hefty 25. The directory also includes

Figure 4-6 Flight instructors must be proficient in simulator operation.

full addresses and phone numbers to ease your job search, so set your printer to run off a few dozen more résumés, put some money in the kitty to pay for the higher telephone bill this month, and get started.

A final word about flight instructing, a place where nearly 25 percent of my total flight time evolved. Personally, I enjoyed it, even though, at times, it was grueling work during heat waves and freezing Midwestern winters. The money has never been wonderful, although a few ads we've shown here will convince you that all schools are not created equal.

The owner of a flight school I know spoke to me at lunch about CFI wages when I asked why they were still so low at times. "Because CFIs are willing to work for dirt money" was his reply. "They don't actually care about the money. They only want to log time. Many would work for free." That stunned me a bit, but I later realized he was probably right. My point is you should realize your need to log time should not screw up the job for someone who does not want to log time as their primary purpose in life. And, if we don't begin trying to turn instructing into a real profession by acting like professionals and demanding a decent wage, there isn't going to be any aviation industry some day. The argument may sound, to some, comparable to the voting rationale about what difference a single vote could possibly make to anyone, but as we have seen in the early part of the twenty-first century, a few votes can make a huge difference. Save some of our industry for the next pilot. Demand a decent wage in trade for sharing your knowledge and enthusiasm for aviation.

Some readers might wonder why a story that might seem so dated appears in an updated edition of a book such as this. The answer is simple. The story that Scott M. Spangler (now editor of *EAA's Sport Aviation* magazine) tells is as true today as when he wrote it.

Critical Shortage

by Scott M. Spangler, Editor, reprinted by permission
AOPA Flight Training magazine

June 1998—Aviation is facing a shortage of specially trained pilots who are critical to its survival, and few people realize this shortage looms.

Aviation is facing a shortage of specially trained and certificated pilots, who are critical to its survival. This frightens me, but what's even scarier is that few people realize this shortage looms. Who are these pilots that aviation cannot live without? Quite simply, they are the poorest paid and the most maligned of all professional pilots. They are certificated flight instructors.

It shouldn't have to be said that if there were no flight instructors, there would be no pilots. Yet, the aviation industry and pilot community take CFIs for granted. Young instructors are branded as "time builders," who teach only to acquire the flight time needed to get a "real" job as an airline or corporate pilot. And, with more than 67,000 pilots holding current flight instructor certificates, we always have plenty waiting in the wings to replace those who do manage to hitch a ride on the airline gravy train, right?

But how many of those 67,000-plus CFIs actively teach? Reliable estimates put the number at less than 15,000, or fewer than 25 percent of all CFIs. Why the disparity between current and active? No one can say for sure, but here's a safe bet: Most of the pilots who earn a CFI see it as a waypoint—not a destination. Some CFIs have full-time jobs unrelated to aviation and teach a couple of new students a year. Far more formerly active instructors have moved on to "real" flying jobs, but they keep their CFI current as a matter of pride.

Of all the certificates and ratings a pilot earns, the flight instructor certificate is the most difficult. The fact that the FAA recognizes pilots who possess the knowledge and ability to teach the art and science of flying says a lot about their commitment to their aviation education, and letting it lapse would mean throwing away the time and effort it took to earn it.

The problem the aviation community faces is a declining number of active teaching professionals. Here are some additional indications of the growing CFI shortage. In 1997, the airlines alone hired 11,396 pilots. During the same period, the FAA issued 3,958 new flight instructor certificates. That disparity can't continue for long before we feel its effects.

At this year's Women in Aviation International convention in Denver, not one of the 150 participants at a career-planning seminar raised a hand when the speaker asked how many wanted to be a flight instructor. Not one.

Can you blame anyone for not wanting to be a flight instructor? My son's piano teacher gets paid more ($20 for a half-hour lesson) than most CFIs, and he doesn't have to earn—or renew—any type of certification to teach.

Whether or not people in aviation admit it, "the industry" caused the shortage by its treatment of CFIs—low pay and low status—and only the industry can solve the problem. How? Here are some ideas the participants (all 25 of them) touched on during a session at the recent National Air Transportation Association/Professional Aviation Maintenance Association "Super Show" held in Kansas City.

Pay flight instructors more money for the essential service they provide. Too often, flight instructors bear the brunt of general aviation's price competition. Ironically, this works against getting more people to realize their flying dreams. People aren't stupid. They pay $40 an hour for their kids' piano lessons, so they naturally assume they'll pay more for a teacher of flying. When a flight school or instructor apologetically says they must charge $25 an hour for flight instruction, these smart newcomer-students are going to wonder why it's so cheap—and whether they're getting what they're paying for.

Naturally, instructors must give good value for the money they're paid. Flight schools, the employers of most flight instructors, should encourage giving good value and support it by paying for or offsetting the cost of additional training, classes, seminars, and participation in programs such as the National Association of Flight Instructors' Master CFI program.

Paying CFIs more will encourage instructors to make a career of flight instructing, which should benefit students and the flight training industry. Let's face it, a high rate of CFI turnover costs everyone money to fund such things as training new instructors, insurance, and maintaining and building a strong customer base.

In addition to paying flight instructors a living wage, we—all of us who fly—need to treat CFIs as what they are—professional educators. Give them the respect they deserve. And we'd better start now because the shortage is real, and it's getting worse every day. Think about this. When we have no more flight instructors, our flying careers will last only until our next flight review. How much is flight instructing changing? A recent job ad posted by Western Michigan University is trying to entire CFI's with salaries near $50,000.

And here's a look at how a new flight instructor fit himself into the aviation industry.

Profile: Neal Schwartz, CFI, 340 Hours Total Time

Senior at Duke University, Majoring in Economics

When *Flying* magazine editor Richard Collins called Neal Schwartz on the phone, Schwartz was shocked. But shock was quickly transformed into excitement as Collins told the college student and licensed private pilot that he'd won the Sporty's Catalog Scholarship that would bring him $15,000 to help pay for his additional ratings.

Right from the start, Schwartz attacked the flight school search problem like any good student of economics would. "I didn't want to spend the money inefficiently, so I did tons of research on flight schools," Schwartz recalled.

"I found that Part 141 schools were not necessarily better than Part 61, just more expensive," Schwartz said. "You're really only as good as your own personal study habits. I happen to think that pilot education is more a function of the pilots themselves than the school. The airlines really don't care where you got your ratings from. I went to a Part 61 school and felt it was a better education because I was much more involved in the decision process from the beginning."

Schwartz knew he wanted to fly almost right from the start. "I knew how fast an F-14 flew when I was ten. I wrote my college acceptance essay on learning to fly since I picked up my private when I was 17. I paid for that by being a lifeguard during the summers." Schwartz now dreams of flying an F-16 for the U.S. Air Force.

For a man who is only 21, Schwartz has many valuable insights to share with other pilots. "Most people completely underestimate the value of networking. Pilots often don't want to be involved in the business end of looking for a job. They feel they've spent their money and expect a job to just come to them. I've met pilots by just walking up and talking to them. Sure it's a little intimidating, but pilots love to talk about themselves."

Schwartz also felt that the Internet has played a key role for him in staying in touch with other pilots. "AOL and its message boards have been incredibly helpful," he said. "I got a job flying an airplane across the country from a message I found online. I'm very comfortable and believe you can weed out the goofs online pretty easily by the questions you ask and their responses. I've found the Student Pilot Network, AVWeb, and the Landings sites [all mentioned in the chapter on useful web sites] to be very helpful as well."

But Schwartz differs from some pilots when it comes to the issue of a formal education. "I think an aeronautical science degree is a copout," he said. "It's important to have something to fall back on. What can you do with that degree if you need to? You need to have the whole college experience. I think you can fly on the side, and read the books, and be just as good as someone from Embry-Riddle."

Finally, Schwartz tells aspiring pilots to "get a job at the local airport. The money is terrible, but you're always surrounded by pilots, people who are already doing what you want to do."

A CFI'S Job Is an Important One

On the first day of work as a CFI, you might easily be saddled with three new primary students and a commercial applicant. Obviously, the job is to prepare these students for their check rides and ratings, but there's certainly more to your job than that. No one should begin instructing without being aware of the awesome responsibility they hold. Because of you, or in spite of you, a student will eventually take to the skies alone, based on what you've taught them. Teach them well. Make certain your student's brain contains the knowledge they need to keep them out of trouble. Think about them before you send them up solo. Can this student handle an engine failure? How about 360s on downwind from the tower? A strong crosswind that suddenly appeared? You are your students' role model. You can make or break their career with your attitude and your style.

I've never told this story to a soul because I was too embarrassed, but many years ago, I wanted to be an airline pilot and attended a large state university with a flight school program to begin the work to make my dreams come true. I was 17 years old. Perhaps you'll keep this story in mind as you start off on your adventures as a flight instructor.

My flight instructor was a young man of only about 21, but he obviously knew a great deal more about flying than I, so I settled myself down to learn. His name was Dick and he was known as a screamer. (No, I didn't change his name. Perhaps he'll read this someday and realize what an idiot he was.) He didn't teach by presenting and reinforcing; he taught by yelling until students either really understood or were intimidated enough to say they understood.

In the airplane I first flew, a 7FC Tri-Champ with one seat in front and one in back, Dick taught the practical portion of flying the same way. He yelled! Even worse, however, in the cockpit, Dick would hit me from behind if I didn't perform correctly. I still remember my first landing with him; it's like it was yesterday instead of 35 years ago. On final, I guided the aircraft fairly well, but he screamed at me the last few feet, "Flare . . . Flare . . . Flare!" Wham! We hit the runway. He just kept yelling flare on final, but he'd never explained the concept of the flare. The man was a Class A jerk. Unfortunately, at 17 years old, I couldn't handle it. I finally left school, convinced I was the idiot, and didn't fly again for five years.

You heard about the bad instructor. Here's how another instructor saved a career that had not yet begun. As a sergeant in the U.S. Air Force, I was assigned to a Texas Air Force base that happened to have an aero club. At this time, I had not been inside a small aircraft for nearly five years. Looking back on this, it was pretty obvious to most everyone, except me, that I really did want to learn to fly . . . badly. I began hanging around the aero club, but not going in. I just stood around outside, like some lonesome pup, looking at the three Piper Cherokees the club owned.

One day, an instructor happened to come out as I was walking around one of the planes. "You a pilot?" he said. "No," I said. "No, I'm not." "Sorry, you just sort of looked like a pilot." He smiled and walked away. About a week later, I was back. Just snooping. The same instructor walked out of the hanger. "Boy. For a guy who isn't a pilot you sure hang around here a lot. Why don't you just join the club?" "No. I don't think I could do this," I said. He just stared at me for a minute. "Who told you that? Some really nice instructor?" "Well, not exactly," I replied. "Hi," he said. "My name is Ray." I introduced myself and he told me he was about to fly the Cherokee out to check the VORs.

"Want to go along?" I only waited half a second before I said yes. Ten minutes later, we were climbing westward out of Austin, Texas. "Want to fly it awhile?" I took the controls and felt the airplane move to my inputs. I turned, I dove, I climbed and turned, and climbed again. When we landed, Ray asked me if I wanted to join the club. I did, and I never stopped flying again. Thanks, Ray, wherever you are. You saved a flying soul that was almost lost. Good teachers not only teach; they influence lives. Use your power wisely.

Time Builders

Once you have your certificates, don't be surprised if a potential employer checks out your logbook and says, "Thanks. But call us when you have a little more time." For the most part, I can guarantee you this is going to happen. Plan to grin and bear it, but take a look around for some of the flying jobs that perhaps aren't as glamorous as you might think at first. Let me share a bit of personal experience with you on a couple of the ways that I built quite a bit of my time over the years.

Ferrying Aircraft

One method is by ferrying aircraft. I still ferry airplanes around the United States. These very words, in fact, are being written from a splendid Florida hotel where I've arrived after bringing one twin-engine aircraft down to trade it for another to take me back to Chicago.

Ferrying airplanes began for me about 15 years ago, just after I'd received my flight instructor rating. I didn't need the flight instructor certificate for this kind of flying, but the opportunity just happened around that time. I was instructing at an airport near Chicago, and after many months of hanging around everywhere at the airport, looking for a break, I got one. At the airport restaurant one afternoon, another pilot pal of mine introduced me to a young lady who just happened to run an aircraft rental firm (that, unfortunately, has long since gone out of business). During lunch, I mentioned that I had recently picked up my CFI and hoped for a professional pilot's job someday if I ever got lucky and could make the flight time requirements. She looked over at me very casually and said, "I need someone to ferry some airplanes around for our company. Would you be interested?" I almost choked on my lunch!

The first airplane I flew was a Piper Arrow. I think I must have had all of about 10 or 15 hours of retractable gear time, but they didn't seem to mind. I flew to St. Petersburg, Florida, on the airlines, got a checkout, and made my plans for the trip back to Chicago. The experience was valuable. Not being too weather smart about Florida, I didn't know that leaving Florida in the middle of a hot summer afternoon was not the greatest of decisions, but I checked weather and left anyway . . . visual flight rules (VFR), because the airplane had only one VHF radio and an ADF.

I spent the next three or four hours dodging showers, thunderstorms, and ever poorer visibility. By the time I reached the Atlanta area, the weather was terrible and, finally, I managed to land at Charlie Brown Airport, thanks to a radar steer from a kind Atlanta Approach controller. I landed just in time to learn that the field had gone IFR in rain and approaching thunderstorms. The next morning, the field was clear, with two miles visibility and fog, so I asked for and received a special VFR clearance from the tower and departed northwest bound. Once clear of the Atlanta TCA, I called Atlanta Center for VFR advisories. The rest of the trip offered me the chance to talk to more towers and centers. All together, I logged about 8.5 flying hours. I gained experience, as well as the time, which helped to make up for the fact that the pay was pretty poor. It did improve as I became more valuable to my employer.

Before that month was out, a conversation with another man gained me another ferry trip. This time, I was off to Miami to bring a Cessna 150 back to Chicago. From that trip, I gained 11.5 more hours in my logbook.

What made this whole venture great was the way it sparked my enthusiasm to find more ferry work. Now, with two long trips out of the way, I was starting to feel more confident about my abilities. I began circulating my freshly printed business card to other operators at other airports. I added an answering machine to my phone just to make sure I didn't miss any possible trips. Another month later, I made a trip to Colorado Springs in a single-engine airplane and came back in another. I ended up flying this trip twice. One trip, I logged nearly 20 hours round trip, the next about 16. It all counted toward increasing those total hours.

Just in case you think this is all glory and fun, though, let me give you a real-life example of just how a typical ferry trip ran recently. It began with the phone call. "Hi, Rob. It's Jan. Are you free for a two-day trip tomorrow?" There's seldom a great deal of notice in this game. "Sure," I said. "What's the trip?" "First of all," Jan said, "You're going to catch a ride with Tim in the Mooney over to DuPage to pick up a Cessna 421. You have 421 time, don't you?" She was relieved when I said yes, because I possibly could have been one of a dozen pilots she called who were either busy or not interested in the trip. "We'll want you to leave as early in the morning as possible. Actually, you'll be taking the 421 to Naples, Florida, to drop off Chris, who's going to look at a Mooney. Then you'll be flying the 421 back up to St. Pete." "Sure, no problem," I said. We hung up and I called my wife to tell her what was happening.

Before the phone call was complete, my call waiting beep told me something had changed. "Hi Rob, it's Jan. Change of plans. Why don't you get here around 10 A.M.?" "Fine," I said. The next morning, all packed and ready to go for a two-day trip, I checked my bag. All my IFR charts were current. I don't carry Jepps anymore because the NOS are easier. When they expire, I just buy new ones like the VFR charts. If I don't fly IFR for a month, I don't end up spending money on charts I don't use. The next item is a book. Never take ferry flights without a book. You never know when you'll find time to read. Finally, clothes. It's a two-day trip, but I pack for three . . . just in case.

Before I get ready to leave the house for the airport, I call in. "How's the trip coming? Everything okay?" "Hang on just a minute, Rob," Jan says. "Be here around 11 A.M."

"Elevvvven . . . Ah. We're going to have a tough time getting all the way to Naples and back up to St. Pete if we don't get going 'til almost noon. I have to be back here Friday morning." "Okay," she said. I show up and we head out for the ride to DuPage to pick up the 421. After my ferry pilot pays the bill, I do a thorough preflight; this is important! Never fly a ferry flight without a thorough preflight. This is where ferrying airplanes becomes serious work. Most ferrying work is to transfer aircraft from one place to another so they can be sold. Sometimes the aircraft will be sold before you take off; sometimes you'll be the first contact a potential buyer has with the sales company and the aircraft. Always remember that it's much like buying a used car. Some people take meticulous care of their machinery, while others are lucky they change the oil . . . ever.

Take a look at the logbooks to trace the history of the machine, as well as whether the aircraft is legally capable of IFR flight. Your duties as a pilot involve your being sure of the aircraft you fly. Don't take someone else's word for it. If you are ever ramp checked by an FAA inspector and the aircraft is not legal to be flown for one reason or another, your license is on the line. Check the aircraft for oil or hydraulic leaks. What shape are the tires in? Many sales occur as the airplane approaches the annual. The current owner might not want the expense, but there could be more problems.

When I preflighted the 421, it looked fine, as did the logbooks. I noticed, though, that the right engine was fairly high time. It had just about reached its TBO. That was definitely something I wanted to keep an eye on, but the aircraft had passed the last annual, so it was legal to fly. After I sat down in the cockpit, I took a few minutes to refamiliarize myself with the 421 cockpit before I started running the before-start checklist.

When I reached engine starts, I hit the button for primer and starter, and the left prop growled for about a half turn and stopped. The battery was dead. I called for a Ground Power Unit (GPU). The first GPU didn't work, so a second was brought as a replacement. Once the aircraft was started, I taxied slowly, checking ground steering and making sure all the electrical systems were up. Remember, I still didn't know what had flattened the battery.

After a VFR takeoff (I wouldn't have left instrument flight rules (IFR) with an almost-flat battery), I watched the engine gauges closely. Things looked normal during the short 15-minute flight to the next airport at DeKalb, where I arrived at about 1:30 P.M. So much for a crack-of-dawn departure!

At DeKalb, I waited for the other pilot, who was supposed to be there when I arrived. He wasn't. I checked weather toward Florida while I waited, and I learned that thunderstorms were building near Chattanooga and Atlanta. A few funnel clouds had already been seen at St. Petersburg, thanks to a stationary front about 40 miles from there. I just shook my head and sat down to wait. If my passenger didn't show up soon, the chances of making that kind of distance would be slim, not even counting the state of the weather, which was becoming more exciting by the hour.

The problem with just waiting while the destination weather gets worse is that your anxiety level tends to rise. My passenger didn't arrive until 4 P.M., 5 P.M. Florida time. We launched by 4:20 P.M. and made it as far as Birmingham, Alabama, where we stopped for fuel and a potty break. We would have been a bit closer to south Florida if the thunderstorms had not made us deviate west. Along about Chattanooga, where the cumulo bumpus really began, I realized that the aircraft's radar was out to lunch. As I did the walk around at BHM while the fuel was being added, I found an oil leak in the right engine. Not just streaks, but a fair amount dripping off the inboard side of the cowling.

It was now 8:30 P.M. Florida time, and the flight was definitely ending right here, right now, until a mechanic could check out the leak, which would not be until the next morning. Don't fool around with this kind of thing. If you're going to ferry airplanes, you must know when to cry "Uncle." To tell you the truth, the fact that the radar had rolled over and died had already made the decision for me—I definitely wasn't flying after dark. There would be no way to see what was ahead unless I tried to avoid the areas of lightning, and I'm not that brave!

The next morning, the mechanics found a loose fitting; they tightened it up and replaced some of the oil we'd lost. By 10:30 A.M., I was on my way. Notice I said I. My passenger had left. He needed to be in south Florida earlier, and he caught an airliner out, so all I needed to do was fly to St. Petersburg. That took about 2 plus 20, right up to the hangar of the waiting buyer.

By this time, I was almost a day later than I had planned. Good thing I had two days' worth of clothes. The mechanics would be inspecting the aircraft before the buyer decided to buy and said they might be finished by noon the next day. Other commitments were now conflicting with this late time frame, and I called the sales office to let them know it could be a problem. They weren't happy about it, but it happens. These days, there are a lot of freelance ferry pilots, with very few on salary to anyone. This is a benefit to the sales company because they only pay for you when they need you, but it does, sometimes, cause headaches.

By the next morning, I was starting to get concerned because of my commitments. If I brought the buyer's aircraft back (a Piper Seneca), it would take seven hours from St. Pete back to Chicago, but a ton of thunderstorms were between Florida and Illinois. The two-day trip was now in its third day, looking at a possible fourth. Luckily, the salespeople realized the dilemma before I even mentioned it again. The 421 stayed in St. Petersburg, as did the Seneca for the next pilot. I took the airlines back to Chicago and made my meeting with an hour to spare. Of course, I did manage to add almost six hours of C-421 time to my logbook.

Is ferrying airplanes for you? In this little scenario, I tried to give you a look at some of the bad along with the good. Ferrying aircraft can be a great adventure and a heck of a lot of fun. I've gotten to fly many different kinds of aircraft into places I'd often only heard of. Becoming involved in this kind of work tends to have a cumulative effect, too. Once people know you're around and available for a trip, they tend to use you.

More than once, I've been called by more than one firm to fly trips on the same day. Then there might be times you might not fly any for a month, so while the work can certainly be interesting at times, it definitely is not steady. A friend of mine and I both ferry for the same company, but I seem to be called more often than he is. This is where being assertive helps, I think. I call the scheduler about once a week, just to say hi. I'm convinced that's why I'm called out more often.

The pay for ferrying aircraft is hardly union scale. The pay is based pretty much on a couple of things: how badly the company needs you or an airplane somewhere else, and how good a negotiator you are. I remember ferrying airplanes years ago for $50 a day, plus expenses. Today, the rates are higher, but you have to stand up for what you think you're worth. Just be aware that some other people might not value your services as highly as do you.

The only regrets I have are having missed some of the really great trips. Last winter, I was set for a trip from Chicago to Tacoma, Washington, in a C-172. It probably would only have been about 12 or 15 flying hours, but the scenery would have been great. I was weathered out. The other trip I wanted was ferrying a couple of C-421s from London, England, across the Atlantic back to the Midwest. Stops in Iceland, Greenland, and Labrador would have made for quite an adventure.

I don't think I would ever call ferrying airplanes boring, as long as you realize your limits—just how far you're willing to go in what kind of an airplane and into what kind of weather. If you ever run into a ferry job where the contractor seems to care more about his machines than about your life, there's only one solution. Run—fast!

As we look at the state of the aviation industry today in the early twenty-first century, there may not be as many opportunities to ferry aircraft around. More and more ferrying jobs also require some turbine time as those types of aircraft become more common. Also, contract piloting companies have emerged on the Internet to offer solutions to a company that needs to move an airplane. You need to search those out.

Banner Towing

Banner towing turned out to be more interesting than I first expected. Let me tell you how I found the job. This might have been just luck—I don't know—but, as I was wandering around a local airport, I happened on an airplane of a slightly different model than one I had once owned, a 1968 7ECA Citabria that I logged some 600 or so hours in during the few years I owned it. As I was wandering around this airport, just being generally nosy, which means talking to people, I happened to see another Citabria painted in the same scheme as my old one.

Because tail-dragger aircraft are such a rarity these days, I walked over to look at it and found the aircraft to have an unusual array of what looked like chicken wire strung beneath the aircraft from wing-to-wing. The tail also had some sort of unusual hardware attached near the tail wheel. As I found myself wondering what it was all for, the owner walked up and we started talking. I mentioned I had owned a similar aircraft.

He told me the hardware on the back was for the tow hooks to grab aerial banners, while the chicken wire arrangement was a night sign, which from the air, would look much like the moving marquee at a bank that tells you the time and temperature. The contraption that ran the night sign was a big steel box that sat between the pilot's feet and in front of the control stick. I guess the owner figured out I was OK when he heard how much tail-dragger time I had logged. All totaled, I probably had about 1,000 hours at that point in time. He told me he'd be looking for a new tow pilot in a month or so, and to give him a call if I was interested. I must say I made a regular weekly pest of myself. I called just often enough to let him know I was interested. He finally called back to say the training would be in a week, and he asked if I could make it. I was there in a flash.

Because I already had time in the airplane, the training process was greatly reduced. The ground school lasted an hour, I think, while Barry explained the main parts of both the underwing night sign and the tow system. Considering what the job was—aerial advertising—the equipment was quite simple. For a banner tow, the pilot connected three long lines, each with a hook on one end, to the hardware at the back of the tail wheel.

The hooks, as well as the rest of the rope, were pulled back inside the cockpit. As the pilot needed another hook and line for a new banner, they simply opened the door in flight (better be buckled in), and tossed the rope and hook clear of the aircraft. It then swung free about 15 feet below and behind the plane. Each time you were through with the banner, you'd pull a lever in the cockpit, and both the banner and the rope would drop free back to the ground. Sounds easy.

If there was a tough part, it was picking up the banner. But I'm going to let my friend Sark Boyajian—another banner tow pilot—tell you how that all works.

Profile: Sark Boyajian, Banner Tow Pilot

Banner towing is basically outdoor aerial advertising that sends personal or business messages. I am authorized to fly anywhere in Chicagoland and surrounding states.

Banners behind the aircraft can be as long as 125 feet. Each letter attached to that sign is 5 feet high and 3 feet wide. The maximum sign I will tow, then, is about 40 characters, although some banner-tow companies, like those in Florida, might tow a sign with as many as 60 characters. I've never had to refuse a message because there always seems to be some way of abbreviating the words.

The aircraft I fly, the Cessna 150, is the most-efficient and least-expensive one to fly in a banner-tow operation. To pull signs longer than 40 characters, I'd need an aircraft with a more powerful engine because of the drag the big banner produces on the aircraft, and such an aircraft could be much more expensive to maintain.

The aircraft towing the banner is normally flying about 50 to 60 m.p.h. If I fly much faster than that, the nylon banners behind the airplane can rip. The banner itself is weighted on the bottom to make certain it flies correctly behind the airplane. There's also a small tail chute on the end of the banner. Without that, the banner would continually twist and spiral, making it impossible to read from the ground.

The method I use to pick up the banner is unique because we never take off with the banner already attached to the airplane. It would drag on the ground. Imagine the sign lying on the ground flat. A short pole at the beginning of the banner, where the message starts, has a 250-foot rope attached to it. That rope is then stretched across the ground and, eventually, strung between two 10-foot-high poles, much like a line strung between two small goal posts. Attached to the tail of the airplane is a 30-foot steel cable with a grappling hook on it.

Once the aircraft has taken off, I release the grappling hook, and it trails along behind the airplane. I try to fly the airplane close to the ground, normally about 20 feet high, with the hook trailing in back, and aim between the goal posts. Just before I reach the posts, I pull the nose of the aircraft up quickly. The hook in back swings much like a pendulum on a clock and, hopefully, grabs the rope from between the posts. If so, I fly away with the banner attached.

I know when the banner is attached because I'll feel a bump and feel the airplane slow down as I'm pushed forward from the drag that's been added by the banner. I climb the aircraft out steeply to prevent the banner from dragging along on the ground. Eighty percent of the time I pick up the banner the first time, but on really windy days it may take three, four, or even five times to pick the banner up because of the way the hook swings around.

I also never land the airplane with the banner still attached. Again, it could get torn up on the ground. I tell the tower I'm coming in to drop the banner, and I begin a gradual descent over the airport to around 100 feet. When I am near the ground crew, I pull a lever in the cockpit that releases the tow cable, and the banner floats down to the earth.

Weather is also important when I tow a banner because, sometimes, the forecasts are not correct. The federal regulations say we can't be closer than 500 feet to any clouds or any closer to the ground than 1,000 feet in a congested area. When I look out the cockpit window, the visibility must also be a minimum of five miles.

We will, obviously, not fly in thunderstorms or freezing rain, but we will fly when it's raining, in light snow, or even when the winds are strong. The wind doesn't affect you as far as getting up into the air, but when you are up there on a windy or gusty day, I am going to be tossed around because the airplane has this long rudder. The wind just pushes me up, down, and sideways, particularly in the Loop area. There have been a couple of days when I was flying that I had to go back to the airport because the airplane was almost uncontrollable from the way the winds blow around the downtown skyscrapers. As far as the readability of the banner to the customer, it might be twirling a little more in the wind, but it's still quite stable.

During the evening hours, when you would not be able to see a banner, I can advertise with my electric aerial billboard from about 1,000 feet above the ground. It's a wire grid strung beneath the aircraft from wingtip to wingtip. When it's extended, it measures about 40 feet long and about 8 feet wide. The grid is wired with between 200 and 300 light bulbs, about a foot apart. The machine with the message is in the cockpit and, when I run that, the message appears on the grid, beneath the aircraft, like the moving marquee sign at a bank. I can also vary the speed at which the message moves. The advantage of the billboard over a banner is I can change the message while I'm in the air and stay over, say, Soldier Field for three hours at a time with different messages.

The maximum I fly is eight hours a day, but that long a day is rare. For instance, when the Bears are playing, the flying time is about 2-1/2 hours or possibly 3-1/2. This gives me time to pull about two banners in an afternoon.

This job has a lot of pressure. From the first moment I speak to a customer until the job is completed, I have to be aware of the weather. I am always looking at my watch, wondering if I will take off on time. Will there be a problem? Will I be able to get back home safely? But it's still a great job.

I fly almost everywhere. I'll fly around a house, a boat in the lake, a ballpark, or even the Hancock building. The banner plane circles the target—the Hancock building, for example—in a counterclockwise direction, because that's the way the banner is written—left to right. The normal distance I fly is about a half a block away from the building. If I fly farther away, the letters are too small to read.

I flew past the 95th Restaurant at the top of the John Hancock Center with a banner that said, "Jane, will you marry me? I love you, John." My policy is if the woman says no, the guy doesn't have to pay.

Pipeline Patrol

Pipeline patrol jobs are tough to locate, but they certainly can help you build time—quickly! The days are long, and the airplanes or helicopters are sometimes small. One company I know of employed a pilot in a Mooney to begin in Minnesota on Monday and follow a natural gas pipeline (visible from above by certain markings unique to pipelines) for almost 800 miles down through Louisiana. To avoid the risk of missing something on the ground, the flying could neither be performed at high speed nor at high altitude. I've seen pipeline patrols that finished five states away from where they began, only to watch the pilot quickly take off on in some other direction to focus on pipelines three states east.

The flying speed in the Mooney for this work was around 80 knots, so the flight took a long time. Altitudes were often at 500 feet or even less in uncongested areas. Realize, too, that pipeline or powerline patrol is not always flown in good VFR weather. One powerline pilot told me that special VFR can be a way of life for this kind of flying. It can be dangerous, but you certainly will see a great deal of the country. When the pilot in this flight finished his patrol, he'd turn around and fly the reverse track along the route to his starting point again. One of my friends found his pipeline job by word of mouth, just hanging around the airport and telling everyone he met that he was looking. You might also try calling local utilities. Their public relations departments should be able to tell you if that company uses aircraft to patrol their lines. If not, they might be able to steer you in the right direction.

Freight Flying

In case you're wondering why I'm bringing up freight flying as a somewhat alternative job when everyone knows the pilots at Federal Express or UPS are flying big aircraft—DC-10s, and Boeing 747s—let me say that those jet carriers represent only a small portion of the freight carried in this country. Literally hundreds of small charter companies are flying mail and small packages, or even boxes of nuts and bolts to where they need to be—usually to small towns or medium-sized cities. Often these freight routes are flown in older aircraft specifically selected for their capability to haul large amounts of cargo, but not necessarily in conditions some pilots might find ideal.

A company in Chicago flies night freight in Beech 18s and DC-3s, both old radial-engine aircraft from the era of World War II. The reason they use these aircraft is they're like flying Mack trucks. They'll fly with just about anything that can fit into the fuselage cargo doors. An *old freight dog* (that's what they call them), Mark Goldfischer (now a DC-9 first officer for a national airline), told me what he recalled about his time flying cargo.

"I was employed at Zantop International for almost four years. Believe it or not, I first learned of Zantop from an old magazine story, while flying canceled checks in Florida and California. Through the article, I realized they would consider me with my 2,000 hours total at that time; I was 25. I was also parking next to Zantop at the Atlanta freight ramp most nights. Using this opportunity, I spoke to a couple of pilots, and they said to get the Flight Engineer Turboprop written out of the way, and then call Zantop. I did, and I landed an interview and the Flight Engineer job on the Electra (L-188).

"I spent my first year and a half flying out of Willow Run Airport in Ypsilanti, Michigan. The operation involved transporting civilian freight to most of the major cities in the United States. Zantop had a fleet of 23 Electras at the time, the largest Electra operator in the world. The Electra is a wonderful airplane from a pilot's perspective. To this day, she's my favorite. Lockheed made her as strong as a workhorse, with plenty of power. I've talked to many pilots throughout this industry who have flown the L-188, and I always see a twinkle in their eye and a warm smile on their face when they think of the days when they used to pilot her around the skies. Whether flying people or boxes, she's a real treat. Zantop also operated DC8s and Convair 640s.

"Flying freight takes a toll on your body and health, though, because so much of the work takes place at night. Whether you make $25,000 or $125,000, it doesn't matter; your body doesn't know the difference. I can honestly say that 50 percent of the time in the saddle, I was fatigued, and 25 percent of the time I was too tired to be flying. That was the worst part of the job. Inevitably, when your body's clock struck bedtime, it was time to go to work. Of course, I've said nothing of all the time spent eating out of machines, trying to stay awake by gulping coffee in absurd quantities, and all the other wonderful luxuries that come with the freight industry (Figure 4-7).

"On the other hand, direct clearance to your destination, less weather to worry about, and uncluttered radio frequencies and airways are part of the good side to hauling boxes when the rest of the world is sleeping. Hotel time? Yes, plenty of it. For the most part, package pilots spend all too much time in the 'pilot prisons.' One year, I spent 250 out of 365 days in hotels! Ouch! Lucky for me, I didn't have a family back home.

"Yet, through all the sham and drudgery, somehow I still managed to have a lot of fun. What do I miss the most? The guys. My fellow freight dogs.

"For some reason (and I'm sure I'm not being totally impartial) freight pilots seem to have more varied and vivid personalities than their people-pushing cronies do. Freight pilots are more humble and likable as people. For some reason, their egos don't get overinflated with huge salaries and flirting flight attendants. They do their job silently and safely, and they go home without much ado about anything. They don't get enough credit for a job that's done as well as any passenger pilot who works for some major airline. I miss their kind.

Figure 4-7 Freight pilots often see a sunset and a sunrise in the same flight.

"The schedules vary, depending on your company, base, and equipment. At Zantop in Ypsilanti, I worked three to four days a week. Weekends were mostly free because most freight companies didn't move freight the whole weekend. Zantop played the game a little differently, though, because, on Saturday, they'd spend the money to commercial flight you back and out to whatever city on Monday in time for the Monday night hub. Most companies don't foot this bill. A typical show time would be 4:00 P.M. If you had to come in and go out on the same hub, that would really take the wind out of your sails. The wait time was about three hours. You could either go to the Lazy-boy lounge for sleep or to Denny's for eats. Of course, you get to know Denny's menu real well. And sleep was always that kind of unsatisfying, dirty sleep."

Military Contract Flying

Goldfischer had this to say about military contract flying: "Logair was different. I spent my last year as FE in Warner Robbins, GA, flying freight for the Air Force. Schedules were a week on and a week off. Actually, Logair was where you wanted to be. You were home almost every night. I left for two reasons. One was, of course, to further my career at Pan Am, and, two, was because I had simply had enough of freight flying. It was starting to affect my health, which I couldn't accept. I would definitely recommend a freight job to build time; it's a neat side of the industry to experience and learn about. As a career? Well, for some, maybe. It's up to the person. I can only speak for myself in that it wasn't meant to be. For some, it could be rewarding. Incidentally, my qualifications at time of hire were CFII, MEL, A&P, turboprop FE written, and a four-year B.S., Aviation Technology, from Embry-Riddle."

Flying for AirNet: A Time Builder or a Career?

An Interview with Craig Washka, Director of Training

In April, 1974, Jerry Mercer became AirNet's first pilot, in addition to being the sole team member. Little did he know then that AirNet—originally called PDQ Air Service—would grow from a single-pilot, single-airplane company into one, 25 years later, with 118 aircraft, 150 pilots, and 1,200 employees nationwide. That original aircraft—N1814W, a BE-58 Baron—still flies for AirNet.

The concept was simple. Banks needed canceled checks moved between cities as quickly as possible to reduce the interest they'd pay on those checks—the float, in banker lingo. Although electronic banking has changed the need to carry canceled checks, AirNet is using the fleet for small package freight and traditional aircraft charter. Mercer provided that opportunity with an ever-increasing fleet of piston aircraft—Navajos, C-310s, and Barons—and jets—Lear 35s. AirNet is now the largest transporter of canceled checks in the world. "Our niche market is delivering time," said Craig Washka, AirNet's director of training.

One of the best parts about flying for a company like AirNet—Starcheck on the radio—is that pilots will be out building PIC time almost from the first. They can move up rather quickly, too. "In a year and a half, a pilot can move from piston captain to First Officer on the Lear 35," said Washka. "In fact, we have the largest civilian fleet of Learjets in the world. Upgrades are based on a seniority bidding system. Because of attrition (about six to eight pilots per month), AirNet is always looking for new pilots. "We have never had a furlough either," Washka added.

Pilots will log 1,000 hours or more a year at AirNet, flying about six hours per night, four nights per week. They could be based at any of

AirNet's 70 bases scattered around the continental U.S., from which they fly some 585 flights each night during the week and 150 flights on weekends. AirNet pilots are all salaried based on their position—piston Captain, Lear First Officer or Lear Captain. All salaries are based on the number of years a pilot has worked for the company. A piston Captain starts at $18,000 per year (see other AirNet pay scales in Appendix D), with a $3,000 raise after one year. Washka adds that "there is a $1,500 sign-on bonus, as well." All training is paid for by the company, but pilots do sign a training contract—a promissory note—to help AirNet recoup some of their training costs if the pilot leaves the company early. For a closer look at AirNet's pay scale, see Appendix D.

"We know that many of our pilots use us as a stepping stone to the majors, although many of our pilots are career-changers," Washka said. "We just expect them to be involved in a challenging career on the back side of the clock. Short of going to war, this is the most challenging flying around."

A typical day—or night—for an AirNet pilot begins around 9 P.M., like run number 501, for example. The pilot leaves AGC (Pittsburgh, Allegheny County) at 9:10 P.M. and flies one leg to CMH (Columbus, Ohio), where they'll help unload and reload the aircraft and be off again by about 1 A.M. for CLE (Cleveland, Hopkins). There, they lay-over until about 4:15 A.M., when they depart again for CMH and turn the aircraft around to depart by 5:30 A.M. for AGC. Upon landing at AGC, the pilot signs out for the night about 7 A.M.—10 hours of duty and about 3.6 hours of flying.

Interested? Here's what you need to consider—a minimum of 500 hours total time, with a multiengine and instrument rating. An ATP is not necessary. "Initially, our interns screen the résumés," says Washka. "That resume must be put together well. Those interns are like hounds and look for spelling errors. More than two pages and you can forget it, too. We like to learn about a pilot on a single piece of paper and respond to every résumé we receive, currently about 250 per month.

"If the pilot meets our minimums and the résumé is well-laid out, they receive an application. If an applicant lives close to an American Flyers location, we send them there for a preliminary assessment, which we call PASS (Preliminary Assessment Selection Service) Phase 1. There, they take a written test to learn their basic knowledge level. If the applicant scores 70 percent or better, they go on to a simulator check in a Frasca 142 to see if they can fly to commercial pilot standards. Eighty percent of the hiring decision is weighted toward the simulator ride. They also take a PPS Personality test, which is a tool to help us make a decision, but it has been deadly accurate. If they pass all of this, we invite them to Columbus for an interview.

"The interview lasts about 20 to 30 minutes, and we do it in the middle of the night. We are trying to learn who they are. Do they have knowledge of the aircraft they're currently flying? We also ask them some situational questions and whether they'd like flying at night. We'll tell them on the spot, or certainly within a few days, if they're hired. When we give them a class date, they give us a $300 deposit to hold their spot. We give them the money back after training is completed through bonuses, moving expenses, per diem, and so forth."

Training at AirNet—like flying there—takes place in the middle of the night. Ground school runs from 8 P.M. to 6 A.M. for a month. "This is where some pilots might decide that flying all night is not for them," Washka said. "Pilots are not hired until they pass their Part 135 check ride at the end of training." All initial AirNet training is accomplished in a Baron, and the Frasca 142 and AST 300 simulators. AirNet pays for a pilot's lodging in a large apartment complex near the airport, where they learn to work together in training teams.

"While they're in training, we observe how well they work together," Washka adds. "We expect the strong ones to pull the weaker ones along. The pilots who go home every weekend usually don't make it." After training, pilots fly together with a Line Captain before being set free to fly alone. That could be two days later or a week. Most PICs are on their own within a week. "It's like a whirlwind, night PIC and IFR right off the bat," said Washka.

Want a few tips for getting hired at AirNet? "Be prepared and research the company," Washka says. "Make sure you review the regs, weather, multiengine procedures, and the AIM. You need to fly a simulator before the interview, too. If a pilot is not current on instruments, they're kidding themselves to think they'll get through. Here, attitude is everything," he adds. "Some applicants we've met are great pilots, but they don't have the right attitude. Remember, too, that part of the interview is me asking whether you have any good questions about the company or the job. A pilot with the right attitude and good stick skills can succeed here."

Washka concludes, "The major airlines realize our pilots are talented. We hire so many." Do your homework about AirNet at http://www.airnet .com./Careers/careersFrame.htm.

The Jobs: A Look Outside the Country

We began this chapter with wondering where the jobs can be found? The answer is: jobs are everywhere. They're hiding on neighboring airports, perhaps your own or one 75 miles down the road. What you see in this book is that some flying jobs are found on web sites run from 10,000 miles away and offer work even further away.

Two services that come to mind are Rishworth with offices in Washington state, Sweden, and New Zealand, and IASCO in California. Rishworth is at www.rishworthaviation.com, while IASCO is found at www.iasco.com. And, here's a quick look at some of the employment opportunities outside the U.S. that Rishworth is featuring. Remember, all these pilots started somewhere and, at one point, were wondering if they, too, would even pass their private pilot check ride.

Recent Employment List

Fokker 100 Captains—Europe—Duration of contract: 3–12 months

B767–300 Captains—Seoul—Duration of contract: 2 years, renewable

B737 EFIS & NG Captains—OKAY Airlines—Tianjin, China—Duration of contract: 2 years, renewable

Embraer 170/190 Crews—Chennai (Madras), India—Expected start date: ASAP—Duration of contract: 2-year renewable contract

CRJ900 Captains—Skopje—Duration of contract: 3–5 months

Atlas Blue A321 Captains and First Officers—Marrakech or possibly Agadir, Morocco—Duration of contract: 12 months, renewable

Korean Airlines—B744 & B777 Captains—Anywhere on the KAL network—Duration of contract: 5-year term, renewable

A320 First Officers—India—Duration of contract: 2 years, renewable

ATR42/72 TREs & Line Captains—Vietnam—Duration of contract: To January 2008, renewable thereafter

ATR72 Line Captains—Can fly up to 65 yrs!—India

EMB-135 Instructor—S.E. Asia—Duration of contract: six months

J41 Captains and First Officers—Dubai—Duration of contract: Permanent

SAAB2000 Captains—Europe—Duration of contract: six months with possible extension

5

The Regional Airlines

It is time to set the record straight. "Commuter" pilot is out, and "regional airline" pilot or simply "airline" pilot is definitely in. If you're flying a B-737 or an Airbus, you might wonder why anyone cares, but to a regional pilot, "commuter" pilot just doesn't tell the right story anymore (Figure 5-1).

In years past, a commuter pilot was someone who drove little propeller-driven airplanes—Aztecs, Navajos, King Airs, and the like. Those eventually evolved into Metros, Dorniers, Shorts, and Beech 99s, hardly little airplanes, but still the name "commuter" pilot stuck. With that commuter name came an impression of the pilot—someone young, inexperienced, and flying little scooter airplanes—certainly not something to encourage much respect in major airline circles. But, as Bob Dylan once said, "The times they are a changin.'"

For some time the norm for regionals was a loud, cramped, 19-seat turboprop. Today, only a single company produces a 19-seat airplane. Regional airplanes arrive with glass cockpits and sophisticated flight management systems to squeeze out every ounce of performance. Because the vast majority of regional airlines now code-share with a major carrier, the overall trend of the '90s is to provide a class of service that closely resembles that of the regional airline's major partner.

A current market forecast says, "Regional airlines are already a key element in the strategy of majors and flag carriers, and they will continue to demonstrate strong growth in the long term." The forecast, recently released by Bombardier/de Havilland, also states that, "The number of seats offered by regional carriers is expected to grow at an annual average rate of 3.7 percent . . . with delivery of 7,420 aircraft in the 15- to 90-seat range . . . over the next 20 years."

Along with the increased need for aircraft to meet demand comes a requirement for crews to fly these aircraft, not only in the U.S. but in Europe, China, and India. One industry expert said the concept of a pilot

Figure 5-1 An ERJ and an ATR await takeoff at Manchester Airport in the U.K. (Courtesy Ian Schofield: www.jetphotos.net.)

shortage over the next few years does not go far enough. He says calling it a staffing meltdown is more accurate.

The most recent fact sheet from the Regional Airline Association (RAA)—www.raa.org—defines a regional as a "short- and medium-haul scheduled airline service connecting smaller communities with larger cities and hub airports, operating 9 to 78 seat turboprops and 30 to 108 seat regional jets." RAA says, currently, some 70 regional carriers are in the U.S. The European counterpart to RAA, the ERAA—www.eraa.org—also represents 70 airlines in Western Europe. ERAA saw a nearly 7 percent rise in total traffic in 2005 over the previous year. U.S. passenger enplanements were 150.9 million in 2005 (134.7 million passengers transported in 2004) and are up more than 100 percent since 1995.

In addition to overall passenger growth, of interest to pilots is the size of the fleet. On January 1, 2005, RAA reported the fleet size at 2,757 aircraft. Fifty-nine percent of the aircraft were turbofan (jets), with 10–19 seat turboprops covering 9 percent of the fleet, 20–30 seat turboprop: 4 percent of the fleet, and 31–70 seat turboprop: 12 percent of the fleet. Fewer than ten-seat aircraft made up 16 percent of the fleet. In 2005, 97 percent of passengers were carried on jet-powered aircraft, either turboprop or turbofan. Over 14,900 regional airline daily departures took place in 2005.

New sources of regional airline expansion are expected to come from new city pairings, supplemental service on existing major/national airline routes, increased frequency on existing regional airline routes, and route transfers from major/national airline partners. Industry- and FAA-growth predictions through 2017 call for passenger enplanements to reach 241 million, revenue passenger miles to grow to 139 billion, and the fleet to increase in size to 3,851.

RAA believes the major constraints to regional airline industry growth include the financial difficulties of the major airlines, limitations of the air-traffic control system, regulatory cost burdens, lack of runway capability at some airports, and airline scope clause restrictions.

Along with increased seating capacity comes increased range as regionals opt for small, pure jet aircraft. Certainly a less well-publicized aspect of pure jet aircraft is curb appeal. Some passengers think a jet is safer, more comfortable, and, overall, a better machine to fly on. A recent Bombardier study says as many as 100 new city pairs might materialize from agreements between the United States and Canada, routes that would be perfectly suited for the regional jets, such as the Canadian RJ, EMB-145, and the Dornier 328 jet.

Comair, a Delta Connection company, is an all-jet regional. But the new turboprop regionals soon to be showing at major airports everywhere will give the jets, and even the majors, a run for their money when it comes to treating passengers in style and safety.

Some regional aircraft are brand-new designs, while some evolved from earlier models. And some fall somewhere in between. The Saab 2000 looks like a super-stretched version of the Saab 340. The Fokker 50 looks like an F-27. The Canadair RJ looks just like a stretched Challenger business jet. Yet, while these aircraft resemble earlier models, their performance places them in quite another category. The RJ seats between 56 and 70. The Saab 2000's 360-knot cruise is almost 80 knots faster than the smaller Saab 340 and the fastest of the regional turboprops. The Fokker 50 slices through the sky nearly 40 knots faster than the F-27. The Dornier 328, the ATR, and the Embraer Brasilia are all new designs, while the Beech 1900D is a new version of the 1900 first introduced in 1984.

If you've viewed the inside of a modern airliner such as a Boeing 737-700, a B-777, or even an Airbus 320, you'd feel right at home in a new-generation regional airplane. In Europe, in fact, the Boeing and Airbus aircraft often fly the routes as regional aircraft, alongside the newer Embraer and Canadair regional machines. Almost every regional airliner leaving the factory these days incorporates a glass-cockpit design that brings to these aircraft the efficiencies of advanced flight management, engine, and avionics control systems, which, until recently, were the domain of only large air-carrier aircraft.

Regional carriers operating the de Havilland of Canada Dash 8, for instance, also expect to begin certification of crews to Category II landing standards in the near future. The new 35-seat Dornier 328 will use a sophisticated Honeywell SPZ-8800-integrated avionics system to give the crew a simple, yet precise, answer to whether climbing to FL 250, even with a headwind or staying low, is more efficient for the best fuel burn. And despite the new numbers of regional jets flying around the world, the fluctuation of fuel prices has kept aircraft like the Dash 8 and the ATR-42 and 72 on the front burner as regional airline profit centers.

This all means that much of the back- and brain-breaking work involved in regional operations, such as eight or more instrument approaches a day, will be reduced. Not so long ago, regional carriers were flying large aircraft—like the Shorts 360, which weighs in at 26,000 pounds, or the 46,000-pound F-27—by hand on 12- to 14-hour duty days because the aircraft had no autopilots. Many of the 19-seat aircraft, such as the BAe Jetstream 31 and the Dornier 228, have the more common two VORs, an ADF, and a transponder for continuous operation. When was the last time you heard of anyone flying a 12-hour day by hand in a B-737?

As cruise speeds for turboprops continue to rise, and as more and more jet aircraft join regional airline fleets, the typical trips that regionals fly are changing dramatically. No longer will all regional airline pilots be found banging around in the bumps and weather at 6,000 feet and 200 knots. Many regionals are moving into the big leagues with the aircraft they fly. But these new airplanes are going to force changes in the way people—such as airline managers, FAA officials, and fellow pilots—view regional operations.

Code Sharing

In 1992, the major airline industry was still attempting to pull itself out of one of the worst economic times in its history—until the chaos after 2001, that is. They'd lost about $10 billion, more than all the combined profits of all the airlines since commercial flying first evolved from the old biplane mail-carrying days.

There has been one bright spot in the airline industry, however, and that's at the regional level. If you were to look at the financial profitability of three of the most successful airlines, you'd find one of them is Southwest Airlines, which flies Boeing 737s. The other two, Atlanta-based Atlantic Southeast Airlines (ASA) and Cincinnati-based Comair, are regional carriers. At the time, these two regional carriers were not simply meeting their payroll; they were making money hand over fist. Another carrier on the East Coast, Atlantic Coast Airlines, was also making a tidy little profit as a United Express Carrier. (Atlantic Coast eventually tried to operate independently from United as Independence Air and failed.)

One major reason for the success of the regionals has been their code-sharing agreements with their major partners. In the early days of regional airlines, back when these airlines were flying fairly small, poorly equipped aircraft, the commuters—as they were called then—became involved in agreements with major airlines in a "I'll scratch your back if you scratch mine" kind of deal. The major airlines, such as American, for instance, contracted with a small carrier to provide a feed from the smaller cities to American's jets at a hub location like Chicago O'Hare. It was much cheaper to run a 19-seat turboprop from Peoria to Chicago than it would be to fly that route with one of American's MD-80s.

To make sure the regional aircraft were flying as full as possible, American allowed these regionals to use American Airlines flight codes in the computer reservation system. A travel agent could book someone through from a large city to a smaller one and fly on American all the way (or at least what passengers *thought* was American Airlines all the way). Quite a few passengers were shocked to exit a large American Airlines jet to learn that the remainder of their trip would be aboard a turboprop or small jet that only looked similar to an American Airlines aircraft.

For the airlines, at least, this seemed a match made in heaven, for a while. With the profit margins in the airline industry as tight as they are, some of the commuter carriers were unable to survive, and carriers such as American would often awaken one morning to find out that one of their code-sharing companies had closed up shop the night before, leaving hundreds of American Airlines passengers stranded. The airline knew this couldn't continue, so American began buying the code-sharing regionals themselves. At least by owning these carriers, the airline could be certain of controlling its partner airlines.

What made some of the regionals so successful is that their cost structure is considerably less than the major airlines. As you saw earlier, while the price of regional airliners is certainly in the millions, that's a drop in the bucket compared to a $250 million price tag on a Boeing 777. The other major factor, besides the relatively low cost of purchasing these regional aircraft, is the salaries paid to regional pilots is much lower than salaries at a major airline.

Code-Sharing Facts

Fifty-three code-sharing agreements between regional and major/national airlines were in place with seven code-sharing regionals being wholly owned by major/national airlines. Three code-sharing regionals were partially owned by major airlines, with 43 code-sharing regionals in marketing agreements. Overall, 99 percent of regional airline passengers traveled on code-sharing regional airlines in the U.S. during 2005.

Regional Pilot Pay

- Average Starting Pay—$22,080
- Average Sixth Year First Officer Pay—$36,628
- Average Junior Captain Pay (Sixth Year)—$64,689
- Average Senior Captain Pay (18+ Years and largest aircraft)—$94,080

Data courtesy www.FLTops.com

Regional Jobs

It's a fact of aviation life that most low-time pilots will have better luck hiring on with a regional than on a major carrier (Figure 5-2). This is not a suggestion of the only way to win an airline job, if that's your goal, but it's certainly the road most traveled toward a job at the majors.

FLTops.com President Louis Smith said, "The regionals are expected to hire nearly 18,000 new pilots by 2018." How tough is the job search at the regionals? The path to the cockpit door is similar to other flying jobs. Persistence often beats everything else. One hiring expert said another

Figure 5-2 The Embraer ERJ 190 series has become a significant part of the fleet mix for airlines, such as JFK-based Jet Blue Airways. (Courtesy Embraer Aircraft.)

reason many are not hired at the regionals is they often realize how quickly the hiring standards can change due to marketplace fluctuations. The regional hiring strategy—the strategy to be hired anywhere—translates into beginning the communications with the airlines you're interested in early. You don't need to wait until you precisely meet every qualification. And, once you do file the paperwork—whether a handwritten application or an online version—update it. Merely sending in your application and waiting for the phone to ring won't get you an interview.

Although, at press time, United is currently not hiring pilots (they are hiring flight attendants, however, which is a good sign), they seem content with an annual update. Some regionals want to hear from you every six months. I'm not saying you need to make a pest of yourself, but you do need to let the airline know if you're interested. If you have a friend already working for the carrier where you'd like to work, ask if they'll write a letter of recommendation. I've never heard of a company yet that didn't like to see a note from one of their own employees—a known entity—talking about a potential employee. It just makes good sense. If you don't have a friend at an airline, make one. But, most important, let the airline know you're interested. The key is to find the right balance between too anxious and too laid back.

The following company no longer exists. During a series of tough times for parent company ATA, the ATA Connection simply disappeared. Perhaps because I enjoyed meeting the people at this airline and found their energy inspiring, perhaps, too, because I lost an airline at Midway Airport, I present their story here. Lessons and experiences in this airline are common to the regional industry and I believe they make this piece worth your time.

Chicago Express—The ATA Connection

If you stand on any ramp at Chicago's Midway Airport on a clear day and face northeast, the view of downtown Chicago's monoliths of commerce is breathtaking, with the Sears Tower, the Amoco building, and the John Hancock building being some of the more prominent. As aircraft after aircraft taxi by, however, you vaguely begin to realize you've seen that cityscape somewhere before. But where? Then, a Chicago Express Airlines' Jetstream 31 crosses a taxiway headed for the B concourse and you make the connection. There, painted in teal and blue on its tail, is that same image of downtown Chicago that has served the company as both a logo and a marketing tool since Chicago Express' first flight in August 1993.

The airline's radio call sign "Windy City" seems appropriate, too.

Chicago Express is the brainchild of Michael J. Brady, the airline's chairman. The company is also privately owned by the Brady family, so no financial data is available. Brady is the former boss of Memphis-based Express One, the Northwest Airlink regional he sold in 1997. Chicago Express was a subsidiary of Phoenix Airline Services, which also ceased to exist at the time of the sale. Brady has now taken a much more active role in the airline's operation because it is the only aviation company he owns.

The idea for Chicago Express evolved in 1993 from the void in service left when the original Midway Airlines went bankrupt in November 1991, transforming Chicago's second airport into a virtual ghost town. It also left many smaller cities with no air service to Midway Airport, a place that had emerged as the airport to fly from when price was a major travel consideration. While Southwest Airlines captured most of the longer haul markets after 1991, the 200- to 300-mile segments were virtually abandoned. Chicago Express was developed to fill that gap. The company is now also directly linked—through a code-sharing agreement with American Trans Air. ATA recently announced a major ramp-up of service at MDW from 6 gates to 12, with flights increasing to 36 daily in 1998 and to 90 by 2004. Doug Abbey, president of AvStats & Associates in Washington, D.C., believes that "American Trans Air is committed to growing at Midway, and that can only help Chicago Express."

Scott Hall, Chicago Express's vice president of operations, said, "We have a long-term contract to provide connection service with ATA in Chicago and, perhaps, Florida at some later date. We provide the service and get paid a flat fee per flight operation. This takes a bite out of the cyclical nature of this industry." Chicago Express flights are listed alongside ATA's own, essentially making the smaller carrier an invisible partner to the larger. Code sharing is not new to Chicago Express Airlines, however. The company had a limited code-sharing agreement with the new Midway Airlines. In fact, the plan called for moving the entire Chicago Express operation to Raleigh Durham in 1996. "We were spooling up, thinking about taking on Midway at RDU," Hall said. "We even had ten pilots in training on the Saab 340. Then it all came to a grinding halt." That halt precipitated the first furlough in Chicago Express's short history as ten pilots on the bottom of the seniority list lost their jobs. All were quickly offered work at Express One Airlines.

Hall believes his company's bond with a major carrier was inevitable. "It is very difficult for an independent carrier to exist without a mainline partner. The public believes that our link with ATA is an endorsement [of our company]. The switch to all Part 121 carrier operations has also helped our company's image." Doug Abbey believes that, "with American Eagle going to Jets at O'Hare and United Express somewhat up for grabs there, too, Chicago Express may have some unique opportunities at MDW. Midway is not slot-constrained like O'Hare."

Chicago Express operates 12 BAe-3100s. "Chicago Express has gone against the commonly held dogma that the 19-seat business is dead," Abbey added. They believe there are still good markets out there. It is incumbent upon all regionals to raise the bar on passenger service.

Hall added, "People are demanding cabin-class service and our load factors are up."

These factors have the company looking at larger aircraft that they hope to be operating by the end of 1998. That decision—possibly on SF-340s, EMB-120s, or Do-328s—will be made jointly with ATA management. Chicago Express had a short stint with two SF-340s a few years ago as the plans with Midway Airlines took shape, but it returned those aircraft to the lessor in 1997 when the deal fell apart. The 340s could again be the logical choice for upgrading Chicago Express since, as Abbey said, "Brady was one of the largest Saab operators, and Saab is very much in the used aircraft market."

Chicago Express currently has 56 pilots on the payroll, but that number is expected to rise during the rest of 1998 as fleet size increases. Hall added that "we will hire 25 to 50 this year," a number he believes is quite conservative. "That number could go higher if we add more cities and larger aircraft," Hall said. "We're after pilots who are outgoing, who will pitch in, and who are very professional on procedures," reflecting the necessary one-on-one customer service aspect of 19-seat aircraft crews. Despite the upbeat nature of the Chicago Express–ATA link, Hall said, "we have not yet decided to change our paint scheme."

Opportunities for Chicago Express through its connection with ATA also spell opportunity for pilot applicants, especially because, Hall said, "Our attrition is high. Of the 56 currently employed, those below seniority number 16 have only been with the company since April 1997. We are losing two to three pilots per month right now, mostly to United, Delta, and ATA."

The company has a gentleman's agreement with ATA to interview Chicago Express pilots who have at least three years' experience with the carrier. That agreement may be formalized at some point in the future, but has already succeeded in a move to the majors for nine pilots in the last year. The tenure of pilots at Chicago Express is about three years before most leave for a major airline position. Currently, the pay scale does not extend past the four-year point anyway. Matt Guyso, a Chicago Express Captain, said he particularly liked the upgrade training when he was a First Officer. "As you progress in experience, the Captains would definitely expect more and more from you."

Chicago Express's only hub is Midway Airport from which they fly to Des Moines, Iowa; Madison and Milwaukee, Wisconsin; Grand Rapids and Lansing, Michigan; Indianapolis, Indiana; and Dayton, Ohio. Every other leg includes a landing at Midway Airport. Crew domiciles are located at Chicago-Midway, Dayton, Des Moines, Grand Rapids, Indianapolis, and Milwaukee.

The company performs all maintenance—except for small line items—at their base in Grand Rapids, as well. Essentially, no flying occurs on Saturday, except from the Indianapolis and Milwaukee bases. Each outstation is staffed by three crews with no reserves. "Everyone knows they need to be at work," Hall said. "If they think they are going to miss a trip, we expect them to call ahead" (so the company has time to make other arrangements).

The company puts a great deal of faith in their pilots and makes attempts to keep them happy, especially, as Hall said, "since, you'll work hard here, for not a lot of money." Crews get together at the outstations each month and build their own schedules. Management's only concerns are that the lines are legal and all trips are satisfactorily covered. A failure among outstation crews to agree on a schedule would cause the company's default schedules to take over, something Hall said, "Hasn't happened yet." Management is said to be quite flexible on trip swapping, as well.

Another major benefit for Chicago Express pilots is the aviation education they receive operating at an airport as busy as Chicago Midway. Toss in the Chicago weather environment that changes a couple of times a day and it's little wonder Chicago Express pilots are being scooped up by the majors. Tom Collins, a First Officer, felt his education was also enhanced by the Captains he has flown with. "Circumnavigation of thunderstorms is the most interesting. The Captains know what needs to be done, but want to see if I know what to do. They give me a lot of leeway in my decisions." Collins also believes his experience dealing with passengers has been important to his career. "Smiling and helping them relax is important. I like to deal with the people."

Chicago Express is nonunion, although Hall said pilots may have been approached by the Teamsters. "We believe we are doing a good job for the pilots right now. A lot of our success comes from the fact that we listen to our employees." The company offers pilots a 401K, although the airline does not match those contributions, and also provides passes and jump seat privileges on other carriers. Major healthcare is a part of the working environment at Chicago Express. A new crew lounge at Midway was also opened recently.

Pilots typically fly about 70 to 75 hours per month. Reserve line holders will see about ten days off per month, while regular line holders can expect anywhere from 12 to 16 days off. Depending on their domicile, pilots might sit on reserve for three to four months before getting a regular line of their own. Pilots are all salaried at Chicago Express, with First Officers starting out at $14,941 per year and a Captain at $20,950. Any flying time over 80 hours makes crews eligible for bonus pay. Pilots may choose to pick up additional flying time for anywhere from $60 to $160 per day extra, depending on their cockpit seat. They also receive $15 a day in per diem when they're on an overnight.

What does it take to work for Chicago Express? Hall said the minimums for pilots who do not come through a FlightSafety ab initio–type training program are 1,100 total time and 100 multiengine time. That is down from the 1,200 and 200 that was the norm just a year ago. A commercial pilot's certificate with multiengine and instrument ratings is also required. Hall added that "We've taken people from the Vero Beach [FlightSafety] program with as little as 300 hours total time. The average has been about 320 hours. We've taken about 40 pilots from there so far." First Officer Pete Beckmeyer arrived at Chicago Express "as a former company intern with only 280 hours total time, 70 of that in multiengine aircraft." Rich Vergaru, now a Captain, had "1,200 hours of total time and just under 100 hours of multi," when he started with Chicago Express, while Matt Guyso, also a new Captain, had "1,400 [hours] total and 250 multiengine."

While an ATP is not a requirement, having the written out of the way would be a plus, because it is necessary for the upgrade to Captain. Upgrade time is currently a mere three to four months at Chicago Express. Hall said that "a recommendation from another pilot carries a lot of weight" in the hiring process. "It's always easier to hire someone who you know something about. We've even had recommendations come in from pilots who have moved on to other airlines."

Experienced pilots have also been applying to Chicago Express through FlightSafety, said Beth Thornton, FSI's new-hire manager. "We've had Captain-qualified candidates go through." Additional programs to funnel new pilots to Chicago Express have also been initiated with American Flyers. The company is also investigating another program with the Prescott, Arizona, campus of Embry-Riddle Aeronautical University.

When FlightSafety receives a résumé, they scan it to be certain it meets the minimum requirements. If it does, the applicant will be contacted to arrange the next step in the process, an evaluation at a nearby FlightSafety training center. Thornton said, "There are three major parts to the evaluation. The background history, the simulator evaluation, and a written job assessment." The background history collects all of the pilot's certificates, past job record, and letters of recommendation. The written job assessment measures a pilot's general abilities and personality traits, and it takes about an hour and a half. It does not require a psychologist to administer or evaluate the assessment.

The simulator evaluation "looks for a proficiency level on instruments equivalent to ATP standards," Thornton said. "Pilots are briefed before the ride and on the results immediately after. They fly in both the right and left seat for the evaluation. We pair them up two pilots per session because we want to see how they interact with other crewmembers." Thornton said the entire simulator evaluation process takes about three hours. She believes one of the keys to a successful simulator ride is "being instrument-profi-cient" before the applicant takes the exam. Pilot applicants are responsible for their expenses to and from the FlightSafety location they choose, and they should plan one full day for the evaluation. Arriving the night before would be a good bet because the work begins promptly at 8 A.M.

Applicants will find there is a wide range of full-motion Level *C* or *D* sim-ulators they could be asked to fly. Some of these include the King Air 200, the Metro 2, or the Beech 1900D. Captain-qualified candidates may be asked to fly a Citation, Gulfstream 4, Boeing 727, Lockheed Jetstar, or Learjet. Thornton added that applicants are not expected to have a thorough under-standing of these aircraft, but they are expected to be able to demonstrate the basics of good instrument flying. Successful completion of the evalua-tion does not guarantee a job, however; Vergara was made a conditional offer of employment when he satisfactorily completed his simulator ride.

When a pilot successfully completes the FlightSafety evaluation, their files are forwarded to Chicago Express for the final in-person interview. Hall often conducts these interviews with someone from HR, as well as the company's Chief Pilot, Don Terrell. "We sit down and review the pilot's résumé and their work history to begin with," Hall said. "We ask a number of questions about Jepp plates for the technical interview and discuss a number of crew scenarios to check their judgment. We hire about half of the pilots who make it to the interview." Collins recalled that the interview also included questions about FARs and the AIM, as well as HR questions, such as "Why do you want to work for this company?" For him, the process of getting hired "took two and one-half months from start to finish." The air-line expects the more experienced pilots to explain the regulations and operational questions with considerably more depth than less experienced First Officer candidates.

Once hired, a pilot heads for FlightSafety's St. Louis facility for ground school and simulator training. Chicago Express is a pay-for-training airline so pilots should expect to receive a bill for approximately $9,450 for the Jetstream course, not including living expenses. Financing can be arranged through banks that already have an agreement with FlightSafety, but there is no guaranteed financing. Matt Guyso financed his training through "Wells Fargo Bank, at a rate that was much better than a credit card." While Guyso made a down payment to lower his monthly repayment costs, he added that "they would have financed the entire thing."

FlightSafety training includes about six days of indoctrination and crew resource management, two weeks of systems, and two weeks of cockpit procedures and simulator training. Training also includes eight simulator sessions of four hours each, where each student receives time in the flying and the nonflying pilot position. Normally, the pilot has an hour or two of training in the aircraft before they reach the line to begin their new job.

Guyso completed the flight training portion of his education by traveling to a Chicago Express outstation one night and waiting for that station's aircraft to return from the day's work. He and a Chicago Express instructor then went out and "flew one session of three hours." Initial operating experience is comprised of about 10 to 20 hours before the pilot begins their first regular Chicago Express trip. Guyso added that "FlightSafety had really excellent systems instructors. They would work with you if you were having trouble. They had a very positive attitude. We briefed before we started each day and at the end of the day, as well. We also worked in the cockpit procedures trainer every day." Vergara said, "The training at FlightSafety is intensive. Everyone parallels it to sticking a fire hose in your mouth. I suggest being highly instrument competent."

A typical schedule for a pilot based at an outstation, such as Milwaukee, might run like this. Showtime for the morning shift is a sleepy 4:30 A.M., with the first trip leaving Milwaukee at 5:15 A.M. for Midway. A quick turn finds the crew returning for three more round-trips to Milwaukee before they go off duty at 2 P.M. Other versions might include a two-hour layover after that first leg into Midway, followed by a round-trip to Indianapolis, and then a return to Milwaukee for the day's end. A typical afternoon shift would show at 1:40 P.M. for the first leg to Midway. That's followed by a round-trip to Dayton, and then back to Milwaukee to sign out about 8 P.M. Out of Chicago, some crews fly standup overnights that show in the early evening. They then fly one leg out, as the last flight to an outstation, only to be the first trip out in the morning back to Chicago. These can make for a night that is pretty short on rest.

If you're looking for a career as an airline pilot, nothing can be quite as handy as a type rating and a thousand or so hours of turbine PIC time. The management team at Chicago Express understands they are not yet at a point where the carrier will be a career for most pilots. Collins said, "I can't say enough good about this company." Guyso added, "I would recommend Chicago Express Airlines as a place to work. I think you're going to see good things happen here. Management understands that most of the pilots will eventually want to move on."

Current Chicago Express pilots had a few tips for others who might want to follow in their footsteps. Collins said, "There just is no way to prepare for the written exams at FlightSafety. Also, things happen in ground school pretty fast. You're thrown into this transition from pistons to turbines that I thought was hard to get through. I recommend buying *The Turbine Pilot's Flight Manual*, by Gregory Brown and Mark J. Holt. It made it all much easier to understand." Pete Beckmeyer said, for him, the best part of flying at Chicago Express is that "The Captains still remember when they were First Officers. The Captains are always encouraging me to do better each day."

Pete Beckmeyer's Career; Being in the Right Places at the Right Time

Pete knew he wanted to fly when he was eight years old. That's when his dad used to take him to the Dayton air show. The family would pull the back seat out of their old car and sit on the ground and watch. Beckmeyer recalls "things flying fast and making lots of noise."

As a teenager, Beckmeyer talked to people about getting into one of the service academies, believing the military to be the only real way to become a pilot. "I went to Ohio State and joined the Air Force Reserve Officer Training (ROTC) program. When I was a freshman, I learned that six seniors who were promised flight training were not going to get their wish. When I later transferred to the University of Cincinnati, I decided to forget ROTC."

Beckmeyer stayed involved in aviation by working one summer as a gate operations employee with Delta Airlines at Cincinnati. "That's where I heard about Embry-Riddle. Friends told me if you went there, you could pretty much write your own ticket in aviation. A lot of the gate agents at CVG jokingly told me to get out of aviation while I still could. I didn't and graduated from Embry-Riddle in 1996 with a degree in Aeronautical Science." But, it was what happened to Beckmeyer in the summer before graduation, in 1995, that became significant to his aviation career.

"I interviewed for an internship at Chicago Express Airlines at Midway Airport and got the job. I did a lot of word processing and worked on manuals. I worked with Scott Hall (now VP of Operations) or anyone who had anything for me to do. I got to do some jump seating and visit the maintenance base during that summer. Scott told me if I got my CFII and took the FlightSafety's Turbine Transition Course, they would give me an interview. He was anxious to have people who work well within the company."

Beckmeyer stayed in regular touch with Hall after that summer internship, the first critical step in the networking game. "I would send him an e-mail every so often. I also took the FlightSafety's Master Pilot Program. That was operationally oriented," to prepare for that first job. Unfortunately, when Beckmeyer called Hall in late 1996, things at the airline did not look good. "They'd just furloughed some pilots." Beckmeyer called Hall again in early 1997. When Hall said he was leaving the carrier, it looked to Beckmeyer as if his first intense networking sessions might be for naught (Hall was gone only a month before returning to Chicago Express). But, all this time, Beckmeyer had been looking on his own for flight instructor jobs. Beckmeyer had, in fact, interviewed with Sporty's Flight Academy, when the Chief Pilot from Chicago Express called and offered Beckmeyer a job based on Hall's recommendation. He quickly accepted. "Perhaps I was in the right place at the right time," Beckmeyer thought.

Beckmeyer believes—now that he's an experienced First Officer with 720 hours total time under his belt, half of that turbine—that "the people at Chicago Express give others the chance to learn a ton of great stuff from the Captains." He recalls, though, that "training was the most intense thing I've ever been through. In the beginning, you are just so overwhelmed, you hardly have time to think. At the end of my first day on the line I was exhausted, but it was a good exhausted kind of feeling."

Beckmeyer knows that, without that push from his internship with Chicago Express, despite having to pay for his training, "this kind of job would have been years down the line for me." He'd interviewed with American Airlines for an internship during that summer of 1995, but realizes now that not getting that job was probably the best thing that could have happened to him because "American Airlines would never have hired me at this point in my career."

Beckmeyer is now based in Dayton—an hour's drive from home— which gives him more time to spend with his wife and their 11-month-old daughter. Beckmeyer also gets to say hi now and again to those gate agents at CVG who remember him from the days when he also worked on the gates. "They told me they knew they'd see me again someday, but in a pilot's uniform."

Regional Airline Training Grows Up— That's the Good News

The thunder's crash was not overpowering, but it was loud enough to be heard over the whine of the turbine engines that hung on each wing. The flashes of lightning around the departure corridor prompted the Captain: "Tell the tower we want to line up on the active runway to check the radar before takeoff." The First Officer complied. The Captain knew well the tell-tale signs of a passing cold front, but decided the risks were acceptable for the takeoff. The gusty wind outside threw the rain against the cockpit glass.

The Captain told the First Officer he was ready and, after receiving clearance, advanced the power levers. As the heavy aircraft picked up speed, the wind rocked the aircraft about. At the First Officer's VR callout, the Captain pulled back on the wheel and called for gear up just as the fire warning light and bell on the left engine sparked to life. Outside, the wind seemed to blow even harder as the crew struggled to keep the machine in the air, using the drills they'd been taught over and over.

But the lack of full climb power and near max takeoff weight, combined with the high-density altitude this night, quickly showed their effects. As the airspeed began to decay, the Captain instinctively lowered the nose of the aircraft, but in vain. The airplane's left wingtip was first to make contact with the ground, heaving the aircraft violently to the left so quickly that the two pilots saw only a blur of lights outside before the red explosions were added to the violent rocking. In the final seconds, the crew knew that nature had won. Then everything stopped.

What Happened

"Let's talk about what just happened here," the instructor said as he turned up the lights in the simulator cockpit. The crew was still recovering from the savage, although nonfatal, crash they had just experienced. Many of you might have realized this story began in a Level C simulator. But, most of you didn't know that, this night, the crew training here was from a regional airline, testing the crew's skills against one of Flight Safety's EMB-120 Brasilias.

The level of sophistication in regional airline training has changed drastically. Checking out in a regional airliner used to include performing a stall series in the aircraft in the middle of the night, often in actual IFR conditions. Another favorite was pulling a power lever back with the gear still in transit to simulate a V1 power failure.

For a number of reasons today, however, training at the regionals is catching up to the majors with state-of-the art, full-motion simulators that allow an instructor to demonstrate a complete range of emergencies and unusual situations with no risk to the crew or aircraft. Just how dangerous training in the aircraft can be was demonstrated, yet again, in December 1991, when a Business Express Beech 1900 crashed during a middle-of-the-night training flight, killing the three pilots on board. A BAe 3101 Jetstream belonging to CCAir also crashed during a training accident.

What made the Business Express crash doubly ironic, though, was that the carrier had just signed an agreement with Flight Safety International. The training organization, based at LaGuardia Airport in New York, will perform all of Business Express's turboprop crew training in simulators. Business Express recently began acquiring British Aerospace BAe-146s, training for which will be conducted at the British Aerospace Washington Dulles training center.

Another reason for a more rapid conversion to training programs like those of major airlines is a certain amount of what FlightSafety Manager of

Product Marketing Bruce Landsberg calls "societal pressure." Airline passengers who fly a regional-size aircraft painted in a United, USAir, or American paint scheme usually believe that the training standards demanded of these crews are the same as those of the pilots flying the MD-80 or B-737 they just connected from. Until recently, nothing was further from the truth, because regional training standards were approached with a more hurried pace—"a rush-them-in and rush-them-out kind of training," as one pilot put it.

Previously, most regional carriers—often still called commuters—were operated under Federal Aviation Regulations (FAR) Part 135 rules that grew into "scheduled air carrier" regulations from the ranks of the on-demand charter services. As the size and capacity of aircraft grew, though, they often passed the 30-seat or 7,500-pound useful load cutoff that transformed those carriers into Part 121 operators.

Two Training Curriculums

This caused a great deal of turmoil because it required some airlines to provide training departments and records that could cope with the differences between the two training curriculums. Some that still flew both sizes of aircraft had to operate both programs simultaneously as a money-saving move. The FAA doesn't currently allow a Part 135 carrier to train to Part 121 credit standards because the agency believes this policy might be in conflict with the FARs themselves.

In December 1991, the Regional Airline Association (RAA) petitioned the FAA, requesting an exemption to Part 135 to allow regional air carriers to substitute pertinent sections of Part 121 to improve quality control and cut down on duplication. In their exemption request, RAA cites the FAA's own words: "FAA recognizes that the airman and crew member training, checking, and qualification requirements of Part 121 will always meet or exceed the requirements of Part 135. This is consistent with the recognition that Part 121 affords the highest standards of safety in civil flight operations." The FAA eventually changed the regs to remove most regional carriers from Part 135 governance and place them under the stricter guidelines of Part 121.

Elsewhere in the petition, RAA makes another significant point when it "recognizes the growth and maturing of the regional segment of the scheduled air carrier industry . . . because carriers operating under Part 135 are acquiring airplanes of increasing sophistication and are upgrading their training programs to take advantage of improvements in flight simulator capabilities and training techniques." FAA is currently reviewing RAA's petition, which the Air Line Pilots Association (ALPA) views as a major step forward for this segment of the industry.

Why have regional carriers waited so long to embrace simulators as the means of training their crews, even though majors have used simulators for decades? Most regional airline managers give the reason quite matter-of-factly: cost! But, training costs aren't always black-and-white issues. When Navajos, Metros, and Bandeirantes made up most fleets, taking the airplane offline at night, and sending an instructor and a couple of trainee pilots out to fly when the airplane would, otherwise, have been sitting on the ground, made good dollars and sense.

Training in Aircraft

One of the problems inherent in late-night training is the aircraft to be used for training would often end up at an outstation late at night, requiring the training crews to position themselves away from base, awaiting the airplane's arrival, which was often a large time-waster.

The training sessions were seldom good learning situations because they were conducted in the middle of the night, when most people's brains are in the sleep mode. If the training crews broke something on the airplane during the night, that aircraft would often be unavailable for the first flight in the morning from that station, causing untold conflicts back at the hub. Then, too, maintenance would sometimes take a back seat to the need to upgrade a First Officer or two.

Consider the overall quality of training. Simulating a good engine failure is pretty tough when the student sees the instructor reaching up to pull a power lever back, even if the movement is covered up with a piece of cardboard. When the instructor, trying to simulate a fire about to eat a wing, reaches over to hit the fire warning test circuit, the student knows deep inside that it's only pretend.

In fact, most aviation-training experts believe that only about 25 percent of the emergencies and unusual situations can be simulated in the aircraft itself. Before the simulator, many other kinds of aircraft problems were only talked about.

Finally, as if all these other items were not enough, the strength of the cost-effectiveness argument truly loses its impact in the safety aspect. How does an airline determine the cost of the lives of an instructor and two or three pilots, not to mention the loss of just one aircraft while training?

Some airlines saw simulator training as relatively inefficient, though. One carrier was giving its new First Officers 10 hours in the simulator, and another 5 to 10 hours in the aircraft before they took their Part 135 Second in Command (SIC) ride. The company subsequently learned it could train new First Officers to take the ride in the same 5 to 10 hours in the aircraft and save the cost of the simulator entirely.

To bean counters, the elimination of the simulator would seem an easy place to cut costs, but doing so certainly raises the question of quality. A pilot simply does not emerge as proficient after 10 hours of training as they do after 20 hours. The question then becomes, "Are companies trying to install competent First Officers in the right seat or are they merely training pilots to pass a check ride?"

Today, more and more airlines are training their crews to proficiency—until their knowledge of the concept has set in—whatever reasonable amount of time that takes. Another problem with the old-style cram method is this: even if you can force-feed the pilots' brains to pass the test, their grasp of the material two weeks after the ride is minimal. This is fine when you're a cook and can open a book if you need to, but it's deadly when the right prop overspeeds in an airplane.

Simulators' Cost Effectiveness

British Aerospace Manager of Marketing and Business Development William Grayson outlines one of the best economic as well as safety reasons for carriers to train in simulators. "When training in an aircraft, only one pilot at a time receives credit. In a simulator, both pilots train as an effective team and receive credit for the same flight. So, really, the aircraft would have to be twice as cheap to operate as a simulator" to be truly cost-effective.

Although some carriers are still sending First Officers to one class and Captains to another, the ability to train in a true crew concept is a benefit that shouldn't be minimized. With a simulator, regional crews can now learn flying and nonflying pilot duties the way they would happen in the aircraft on the line.

While the cost of a regional airline simulator is about $10 million, a few of the regional's major airline code-sharing partners view the cost as worthwhile. USAir purchased a Dash 8 simulator now installed at its Charlotte, N.C., training facility for use in USAir Express training. AMR Corporation operates an ATR-42 simulator at American's Dallas/Fort Worth training center.

In 1978, a United Airlines DC-8 with a landing gear problem ran out of fuel and crashed near Portland, Oregon. United management decided this would never happen again "at any cost." The cockpit resource management (CRM) program was the result. While this program has been a part of major airline training for years, it's only just beginning to make its way into the regional airline system. Both Comair, a Delta Connection carrier, and Piedmont, a USAirways Express carrier, use CRM as a part of their regular training program.

The regionals seem to be looking at CRM and installing the program on their own timetables, instead of merely reacting to an FAA mandate. Because carriers with an active CRM program have lower overall accident rates, instituting such a program should also reflect positively on a carrier's insurance rates.

For the pilots, CRM is a win/win situation because they get specific human-factors training to help them cope more efficiently with problems on the flight deck. ALPA's chief accident investigator at Comair, Captain Mitch Serber, said, "The airline's CRM program will soon be linked with a line-oriented flight training (LOFT) program to add additional feedback to the cockpit crew training loop."

One regional airline pilot, however, calls the CRM program in the regional system merely a buzzword to keep the FAA off the carrier's back. He said the reason his carrier instituted the program was because management believed "the system is safe and the equipment is reliable, so most accidents must be caused by pilots." Other pilots said their CRM programs were sometimes no more than a short video, or an even shorter speech, from the Chief Pilot.

Outside Training

Tough economic times often bring innovative new programs to an industry. One controversial program that has gained momentum recently is the management of an entire regional airline's training program by a professional training organization, and not by the airline itself. A case in point is the December 1991, Business Express agreement with FlightSafety International. Any pilot now interested in employment with Business Express is automatically referred to FlightSafety, which conducts all the initial screening of new hires. FlightSafety then determines which applicants will be referred to Business Express.

FlightSafety also conducts all new-hire, aircraft-specific training for Business Express. But the real sting in this program comes from the $8,500 bill that the applicant receives for that training. Opinions differ as to how or why a young, low-time pilot applicant would want to pay for their own training. A leading reason seems to be that, while the current supply of pilots far exceeds the demand, an airline can pretty much call the shots: Want to fly? Pay the bill!

The strongest motivator to the airlines, though, is the cost savings, because an airline no longer needs to fuss with the paperwork or cost of running a training department. This kind of "shoe-on-the-other-foot" program also allows the airlines to recoup some of the money they lost over the past five to seven years, when pilot after pilot left the regionals for the majors, taking their valuable training with them after a year on the line.

Another thorn in the side of experienced pilots is this: even those type rated in a particular airline's equipment are tossed into the same pool with the inexperienced pilots. Experienced pilots must pay to get hired. All the airline does is sit back and look at the fully trained applicants that FlightSafety sends them. (This training procedure is sufficiently unorthodox that ALPA's President, Captain Randolph Babitt, has instructed the Association's Collective Bargaining Committee to review the pros and cons of such training and to make suitable recommendations for dealing with the issue during work agreement bargaining sessions.)

FlightSafety conducts full-service training for four other airlines in the United States and occasional initial screening service for Atlantic Southeast Airlines (ASA). While this program might make company accountants smile each time they look at the money they save on a class of new hires, many of the pilots we interviewed believe the airlines that use this pay-for-training system are courting disaster. While no instances have been recorded where the quality of pilots flying the line has significantly diminished because of this kind of program, some ALPA pilots believe that day might be just around the corner. Again, this speculation is just that—a prediction of what might come.

"I think (the FlightSafety Program) looks good on paper," one pilot said. "These crews coming from the FlightSafety Initial Training program might really know the checklist and what to do if the antiskid fails, but these new training programs don't historically address the weakness of an aircraft. It takes an airline ground school, run by airline people, to address these kinds of things, the day-to-day problems you run into on the line—that's situational awareness. An instructor who might have come from teaching the Mooney program last week is not going to be able to teach a new-hire how to cope with an aircraft's drawbacks and how to be an airline pilot."

While many international airlines do put low-time First Officers into the right seat as the FlightSafety program is doing, "Those pilots sit in a classroom with, say, Lufthansa for two years before they jump into the airplane," one pilot said. "Then they sit in the right seat for years before they can upgrade." He emphasizes what he thinks is a potential for tragedy: "Some of these new First Officers could be moving over to the left seat with some very low total times as well as experience levels." ASA and WestAir are two carriers that eventually decided against turning their entire training departments over to FlightSafety after initial discussions. We shouldn't let the opinions of some cast aspersions on all, but we should also not let these predictions pass unconsidered.

One regional pilot sees the following scenario possibly unfolding because of what he believes is a potential conflict of interest: "Imagine an airline telling a pilot that even though that pilot paid for and passed the training, performance was marginal, but that the airline is going to give him or her a chance anyway. If the airline tells the pilots it expects them to really stay in line, this could really set the tone for how those pilots will react to a great many things in his or her future flying.

"Some examples might be the inability to say, 'No, I won't go out and fly around those areas of thunderstorms,' or 'No, I won't fly an aircraft that has not had proper maintenance.' Some of these kinds of pilots just won't have the good sense and experience to make good, sound decisions, even though they technically meet the requirements. The company owns their soul, but I think having pilots that won't give them any resistance is just what these airlines want."

Because the airlines can save vast amounts of money with this kind of program, pilots are likely to see more, rather than fewer, of them. Whether the preceding predictions do come to pass, however, only time will tell.

The Schedules

Flying schedules at the regionals tend to be similar to the majors, yet different. At a regional airline (Figure 5-3), you could find yourself with only ten days off per month if you're low on the seniority list poll. That means you're flying basically five days a week. But, I know a new Boeing 737 Captain at a major airline who is only receiving 12 days off per month. What you'll find, however, is that it's not how much time off you receive, necessarily, but how that time is organized that can make or break your lifestyle.

If you have split days off or your trip ends late one day and begins early on another, your time off can seem even less than it is. Asking to see a copy of a monthly bid line would not be out of line during a final interview. Depending on the carrier, and the scheduler, I have also seen regional airline schedules with lots of flexibility, sometimes offering many lines of 14, 15, and 16 days off each month.

Seniority

A constant subject at any flight department, be it corporate or airline, is seniority. Basically, seniority is based on who showed up first at a corporation because the number of pilots they often hire is small at any one time. At an airline, your seniority is not only an important thing. As one pilot said recently, "Seniority is the only thing." Seniority at an air carrier determines which aircraft you fly, how work schedules are organized, when a pilot is eligible for upgrade to Captain, and even when you can take your vacation.

Figure 5-3 Regional airline aircraft are as sophisticated as the equipment used by their mainline partners. (Courtesy Embraer Aircraft.)

Seniority numbers at an airline are usually established during initial training. The oldest member of the class is normally assigned the lowest seniority number. Today, some airlines are also choosing seniority numbers by a lottery system within the class. The seniority system is quite easy to understand.

Here's how it works. When two pilots bid the same schedule, the pilot with the lowest seniority number is the winner. If a new hire class is assigned different aircraft, the more sophisticated will usually be assigned to new hires with the lowest class seniority. If a pilot wants to live in a particular domicile city, and they have a lower seniority number than you, you'll be stuck in a city that's not your first choice until your seniority number is low enough to be able to hold that city. Just in case you were wondering, your seniority number does change with your company longevity. Each time a senior pilot leaves the company, a new pilot beneath him on the seniority list moves ahead to take over the old number. If you're given the option in a hiring situation of a later class date, but the aircraft you want, I'd take the earlier class date. Remember, seniority is everything.

I spent three years of my flying life at a regional carrier and, for the most part, I enjoyed it. There's no doubt that the days are sometimes long and the pay could certainly be better, but the people were great. At an airline like United, where seniority numbers are currently near 10,000, knowing the people—other than the immediate ones you work with—is often impossible.

Regional Pilots: Lifestyles of the Not-so-Rich and Famous

Reprinted courtesy *Air Line Pilot* magazine

A few years ago, when I was still flying an EMB-120 Brasilia for Midway Commuter, I was at a party that was fun and uncomplicated. Because some of us did not know each other well, we started talking about our jobs. When my turn came, I said I was a pilot. The faces of the people around me brightened noticeably—to most people, no matter how you cut it, hauling people around in airplanes is still a pretty glamorous line of work. Quickly, the air was filled with other people's lively stories of being trapped for hours on a runway somewhere waiting their turn in line for takeoff or a trip filled with violent "air pockets."

As the conversation progressed, one of the men asked me, "What do you fly, a B-737 or one of those DC-9 jets?" My reply was, "A Brasilia." "What's that?" another asked. "It's a 30-seat turboprop aircraft. Very fast. Very nice machine," I answered him. "Oh, it's one of those little puddle jumpers, huh? You wouldn't get me on that thing. I didn't know real airlines flew those little airplanes anymore."

I found myself becoming defensive as I related how efficient the airplane was, and how much money our commuter subsidiary had made for Midway Airlines, not to mention how similar flying a large turboprop was to flying a jet. "Are you trying to get on with one of the regular airlines?" the man asked. "I don't know. Maybe I should," I said. This was not the last time I would find myself questioning my job as a regional airline pilot.

Two months later, I rode jump seat on an ALPA-crewed B-767 to Los Angeles. As I stood in the cockpit doorway and introduced myself, ALPA card in hand, the two pilots welcomed me aboard. The conversation prior to the "before starts" was animated and lively. These two pilots obviously knew each other well and enjoyed flying together. Passing through 25,000 feet on the climb, the conversations began about the state of the industry and, at one point, the Captain asked what aircraft I flew. "A Brasilia," I told him. "Oh," was his only response, but quickly I noticed the atmosphere in the cockpit change. The rest of the trip, I was almost completely excluded from the cockpit chatter. At the end of the flight, I said good-bye and thanks, but the two pilots never even looked back at me. They only waved and said, "See ya."

My sense of alienation sometimes arose from contact with regular people and sometimes from other pilots. Once, a Midway DC-9 Captain tried to keep me from using the crew lounge at Midway Airport in Chicago by saying it was only for the regular Midway pilots. I showed him my ALPA card, and he moved aside.

Am I What I Fly?

I began trying to make sense of what I was seeing. I had worked hard to reach 4,000 logged hours with an ATP. Yet some people seemed determined to judge me by the kind of airplane I flew, not the job I did—kind of like saying you must be someone special if you drive a BMW.

Worst of all, I began to second-guess myself, especially because my paycheck showed what value the airline placed on my service. When I started flying for Midway Commuter as a First Officer, my annual salary was $11,700—that qualified me for food stamps! Because my wife had a good job, I was one of the lucky ones economically. But now, to add insult to injury, the people I had always thought of as peers seemed to look down on me.

I took the whole thing rather personally until I began talking to other Brasilia crews and, later, crews at other regional airlines. Pretty much everywhere, I learned, the crews who fly for regionals are regarded by many as the poor little brothers and sisters of the airline industry: by management, the flying public, and, sometimes, by other pilots. I started asking more questions.

Dash 8 First Officer Steve Varinsky summed it up: "Flying for the regionals is sort of like being in the minor leagues. Everyone looks at us that way—from the flying public to other pilots and even to other airline employees. Even when you travel nonrevenue, they look at you sometimes and say, 'Oh, you're only a commuter pilot.' I think this whole thing started a long time ago with people's perceptions of air taxis and commuters as companies that flew little six- and eight-seat airplanes."

Today, though, the differences between the service a regional carrier provides and that of a major are disappearing. Many regionals are now operated under Federal Aviation Regulations Part 121, just like the majors. In fact, services are purposely being blended together to give the public, in former AMR Chairman Robert Crandall's words, "a seamless flight." In this way, passengers will believe they are flying on one airline's airplanes from start to finish, even if they begin on a United B-767 and complete their trip on a WestAir Brasilia painted in United Express colors.

Today, too, shooting an ILS approach down to minimums in a Dash 8-300 with 55 passengers on board takes no more or no less effort than it takes for the crew of the DC-9-10 with 63 people on board that follows right behind. The newest regional aircraft are large, sophisticated, and quick. Some regionals today are even operating pure jet aircraft. But how regional crews are treated at the company level is where the major differences appear, on every subject—from pay and benefits to scheduling.

Many regional airlines are reaping huge profits at a time when their major carrier partners are losing their shirts. One of the most successful, Atlantic Southeast Airlines, gleaned a $32 million profit last year. Comair, a Delta Connection partner, soon to introduce jet service with the Canadair RJ, put a $2.9 million profit in the bank for just one quarter. Airline managers sometimes possess a bargaining chip because they have a captive audience in their cockpits. Where are their pilots going to go during tough economic times if a glut of unemployed pilots are still looking for work?

Low Wages

According to recent ALPA studies, the average annual wage for a First Officer at the lowest paid of all the ALPA-represented regional carriers, AMR Eagle Simmons Airlines, is $16,104. Even if you use the kind of simple logic most pilots love and say that this job is only half as tough as flying a 75-seat airplane, you'd find the Simmons pilot's wages are about 80 percent less than the right-seat driver in that 75-seat jet. Imagine raising a family on $335 per week.

Jetstream Captain Bob Phelan says, "We're looked upon more as slaves by the company: Just get out there and do your job. They think we should be paying them instead of them paying us." Shorts 360 First Officer Pete Trimarche says, "On a scale of one to ten, I'd rate my personal satisfaction with my job as a nine. But I'd give my quality of life a three. I graduated from college four years ago, and I still really can't support myself on my income." Dash 8 First Officer Evelyn Tinkl says, "Professional pilots should not be put in situations where they qualify for government-assisted food programs."

What Is Fair?

A number of regional pilots were asked what they considered fair money for their work. Brasilia Captain Ken Cooksey says, "No one expects a

Brasilia pilot to be paid MD-80 pilot wages, but pay should have a realistic floor, based on the number of seats in the aircraft and proportional to what other pilots are paid."

To some major airline pilots, the pay issue may seem to be only an irritant. But, beneath the surface may lie a dragon soon to be faced. As airlines look for new, innovative ways to cut costs, operations that violate current scope language may emerge.

Northwest, USAir, and Delta plan to continue removing jet aircraft from unprofitable routes and replacing them with high-performance, high-capacity turboprops operated by regional airline crews working for C-scale wages. In fact, using new Canadair RJ jets, Cincinnati-based Comair plans to begin searching out new, low-cost routes on its own, which don't even involve a feed with its partner, Delta Air Lines.

Of course, regional pay can be corrected, but whether it will be is another matter. One Comair pilot says, "Management's always yelling about how ALPA contract proposals will put the company out of business, but Comair has been profitable for 16 straight quarters." Comair's CEO David R. Mueller earns more than $500,000 annually in wages, with stock options bringing his yearly compensation to more than $1 million.

A 25 percent pay raise for each Atlantic Southeast pilot, right now, would cost the company an extra $3.5 million per year. Obviously, that is a cost Atlantic Southeast, with its $32 million profit last year, could afford. In Europe, Sabena Airlines regional DAT hired ex-Midway Airline's Brasilia Captains in 1992. The airline started these U.S. pilots as Captains at a beginning wage of $55,000 (U.S.) per year, plus cost-of-living adjustments.

Obviously, even though regional pilots love their work, their quality of life is often questionable at best. The hours are long, and the lack of duty rigs makes a pilot's time cheap, leaving companies little motivation to build more productive schedules. When regional airlines experience crew shortages, they just lengthen the duty days for the pilots currently employed. ALPA's President, Captain Randolph Babbitt, and Regional Airline Committee Chairman, Captain Stephen Ormsby, testified vigorously in 1992 on Capitol Hill on behalf of flight and duty time rules for regional pilots.

Although many of the pilots interviewed said they used ALPA benefits, most say money is a major limitation. A Business Express pilot says, "I'd like some of the insurance programs; but as a First Officer, I just can't afford them." But, many pilots believe the ALPA programs already in place are extremely valuable. Brasilia Captain Dan Ford says, "I consider my ALPA dues as a 'free lawyer' card. Just one use of an ALPA attorney, and I'll have more than paid back the dues I paid over the years."

Most regional airlines have minimal retirement programs, if any at all, leaving regional airline crews and their families to their own fate at age 60.

A Simmons Shorts 360 pilot has seen the handwriting on the wall with retirement programs at other airlines. He has "talked to too many guys who have lost out with their airline retirement plans, those *A* and *B* plans we all seem to dream about. I want an account with my name on it and my money in it."

Changed Domiciles

Express Airlines I pilots recently arrived at work to learn their airline's two main domiciles had suddenly increased to no fewer than 40 minidomiciles. With just a few days' notice, hundreds of pilots and their families were told to uproot themselves in a gypsy-like move or find a new job. Because Express Airlines I deemed the move necessary to improve its financial health, it did not consult the union.

Often, regional managers, at carriers where total pilot numbers and master executive council (MEC) strength is reduced, use a common technique to deal with problems. "If you don't like the procedure, there's the door." Other airlines use the "go ahead and grieve it" stance, knowing full well that while the grievance wheels are in motion, managers may still call the shots the way they choose.

Overall, regional pilots want to be thought of as just airline pilots. But a complaint sometimes voiced is the lack of respect they feel from their jet-pilot brothers. Saab Captain Mike Sigman says, "The rudeness of some major airline pilots is what really burns me. I used to fly a DC-9, but I don't anymore. Just because we are flying something smaller and slower shouldn't make other pilots look down their noses at us. We are all professional pilot members of the same union. ALPA National treats us as equals."

But the fault for regional carriers' troubles cannot all be placed on the backs of management. Years ago, the regionals were staffed with pilots who were often unfamiliar with the daily operations of working with a union. Today, though, more and more regional pilots are recruited from former airlines, such as Eastern, Midway, and Pan Am. While many of these more-experienced ALPA pilots understand the union, they report that many of the younger pilots still do not take an interest or, at times, do not seem to have any idea of what ALPA is even on the property for, other than a pay raise.

Help from the Majors

Indeed, the education process for regional MECs is a tough one. But, here is where the resources of the major airline MECs could be of great value to the regionals—through the experience these senior ALPA pilots could pass on to members flying at the regionals. The Delta MEC Central Air Safety Committee has greatly assisted the MECs of its Delta Connection carriers, such as Comair, for instance.

Can regional pilots see light at the end of the tunnel? Yes, but change will be slow in coming, something that may be tough for some pilots to swallow. Atlantic Southeast MEC's former chairman, Captain Cooksey, says that while the pilots asked for a great many items they didn't receive in their recent contract, "ALPA National felt we did pretty well for only our second contract. The fact that, in August 1992, the president and vice-president of our airline asked to meet face to-face with our MEC and Negotiating Committee was a major step forward in labor relations at our company."

At WestAir—now out of business—then MEC Chairman, Captain Walt Blore, said, "If I accept what the company says at face value, WestAir has a very promising outlook, although the company is already asking for some concessions in our upcoming contract." And, contrary to what some regional pilots believe, many major airline pilots do know and understand the problems their regional airline comrades face.

B-727 Captain Ken Adams says, "I have a great deal of respect for a regional pilot's job. We all share the same sky and we have to stick together." He believes that smaller aircraft need as much protection as the larger ones do, whether it's in equipment or duty rigs. Adams would like to see all regionals operated under Part 121, which they now do.

Flow-Through Agreements

One Delta pilot says that regional pilots are "doing one hell of a job; and in the future, all the majors will need some sort of agreement with the regionals if they are to continually replace their retirees with experienced crews. The military is just not a good source anymore." Prior to their bankruptcies, both Midway and Pan Am had flow-through agreements with their regional partner airlines, which not only allowed pilots to move up, but also gave those pilots a reason to remain at the regional level until their number came up, reducing retraining costs for those airlines.

The MECs of USAir and its wholly owned regionals (Henson, Pennsylvania, and Jetstream) have formed a joint committee to try to formulate a flow-through agreement. Regional pilots do not seem to be asking for special treatment, just equal treatment and a fair day's wages for a fair day's work. But the regional pilots can't bring about all the necessary changes alone. They need the help of their major airline ALPA brothers and sisters. If regional pilots don't receive that help, all of us in the airline industry may be doomed to repeat many of the hard lessons that ALPA learned over the last quarter century.

Regional Flying Today

Is the life of a regional pilot better or worse today? In late 2006, I'd have to say it has become worse. The contracts at many regional airlines have been ravaged by the management philosophy of the times: if you can't negotiate with employees, take them to court and let the judge toss the contract out. This new leverage has been used at both the majors and the regional carriers. Older airline folks will remember this kind of tactic first began in the airline industry—thanks to Frank Lorenzo—some 30 years ago. Shortly after Lorenzo took control of Continental Airlines, he had the pilot's contract tossed out in court.

A news clip on FLTops.com said this about Comair, once a giant in the regional airline industry.

"The Comair pilots, represented by the ALPA, voted to authorize their union representatives to strike if their contract is rejected in bankruptcy court. The 1,500-plus Comair pilots overwhelmingly supported the union's strike authorization ballot with more than 93 percent of the responding pilots voting in support of the measure.

"'We continue to negotiate with Comair management in an effort to reach a consensual agreement,' said Comair MEC Chairman, Captain J.C. Lawson. 'However, management appears to have decided that the fate of our contract should rest in the hands of the courts, rather than at the negotiating table with the pilots who have contributed so much in the success of this airline.'

"'Make no mistake, the pilots will not tolerate company-imposed pay and work conditions,' Lawson added. 'Today's vote should send a clear message that our pilots are united and we are ready to take all appropriate steps in defense of our working agreement.'

"Over the past year, Comair pilots have made significant sacrifices to keep the company financially solvent."

Hiring Profile: American Eagle

Here's a look at what American Eagle has to say, in early 2007, about its hiring requirements. One source said in March 2007 that American Eagle was so short-staffed that it had begun asking employees to refer friends for pilot positions and that the carrier expected to hire 700 pilots during the remainder of the year.

To be considered for an interview with American Eagle, Flight Officer applicants must meet the following minimum requirements:

- Commercial Pilot Certification with multiengine and instrument ratings
- Total fixed wing time to exceed 500 hours
- Multiengine fixed wing time in excess of 100 hours
- A current flying job (12 months of experience)
- Current FAA First-Class Medical
- IFR Currency
- FCC license
- Possess the ability to work in the U.S.
- Possess the ability to travel in and out of the U.S. to all cities/countries served by American Eagle (a valid passport is required)
- Ability to relocate
- Ability to work weekends, nights, shifts, holidays, and overnight trips
- Minimum age 21
- Flight time requirements are commensurate with experience (Figure 5-4)

Figure 5-4 An EMB-145 regional jet (Photo courtesy Embraer Aircraft.)

Preferred Requirements—Because of the competitive nature of the pilot position, the following qualifications are common among Eagle's most successful applicants:

- Actual instrument time 10 percent of total time
- Six months experience with a Part 121 or Part 135 operation
- ATP Certificate or successful completion of valid ATP written test
- Hours flown in most recent 12 months to exceed 500 hours
- Fixed wing multiengine turboprop experience
- College degree

PAY: Pilots are paid based on the number of flight hours flown each month. Our pay scales are negotiated by the ALPA and are competitive for the regional airline industry.

TRAINING: American Eagle's tuition-free, safety-based training program for pilots combines rigorous classroom study, systems training on computer workstations, and flight training on state-of-the-art flight simulators. Free room and board are also provided.

RELOCATION: American Eagle pilots are assigned to one of our seven bases:

- Boston, MA
- Miami, FL
- Chicago, IL
- New York City (LGA)
- Dallas/Ft. Worth, TX
- San Juan, PR
- Los Angeles, CA

    *Relocation is at the employee's expense.

RESERVE: New pilots are assigned reserve availability periods, in which they are required to report to their domicile within two hours. Reserves can plan on 10 to 12 days off a month, and they will be on reserve for the first 6 to 12 months of employment.

BENEFITS: As a regional affiliate of American Airlines, American Eagle employees enjoy the comfort of stable employment, competitive medical and dental benefits, and a company matched 401K plan.

TRAVEL: American Eagle offers employees and immediate family members one of the best travel programs in the airline industry. Employees are eligible to enjoy unlimited travel to American Eagle and American Airlines destinations, for themselves, their spouse, and their dependent children, for a small service charge. Employees are also eligible to receive a limited number of travel passes that may be used by friends and nondependent family members, for a service charge.

TO APPLY: If you are already a member of Airlineapps.com, simply click here and add American Eagle to your Job Targeting list, and then complete American Eagle addendum section information, if applicable.

New users may select single application to apply at no charge or choose to join the AirlineApps system as a member, if desired. New applicants will be taken through an easy-to-follow, step-by-step process for completing your application.

As we go to press, one company PSA Airlines, a US Airways Express carrier, has reduced its hiring minimums by eliminating the total flight time requirement in favor of the need only to possess a valid commercial pilot certificate.

6
The Majors

If you tell people you fly for a living, they seem to assume you fly for a major airline (Figure 6-1). Let's look at this segment of the flying world, which seems to be such a career magnet. First no matter what other pilots might say, the size of the aircraft they fly is important, at least to them. All pilots have big egos—it seems to come with the territory. But, what you'll find is that flying the biggest aircraft might not always be the best career direction.

Following the News

Earlier, I spoke of the volatility of the airline industry in this country and around the globe. It sounds as if this is a rare occurrence, but it is all part of that great circle of life to the aviation industry. When the first edition of this book was written in 1996, we were looking at recent airline losses, such as Midway, Pan Am, Braniff, Eastern, and others. When I wrote about the industry in 1993, Delta and American planned to lay off pilots. Northwest Airlines was on the verge of bankruptcy, while America West and TWA were already in bankruptcy. And, Continental Airlines had just recently emerged from Chapter 11.

There is simply no substitute for staying on top of the news about this industry. Doing this can help you make the decisions you must if your pilot career is to avoid some of the minefields that could be ahead. You must subscribe to a news service somewhere along the line, whether you read the actual newspaper each day or look at a monitor connected to your computer.

As I'm updating this book—in early 2007—the airline industry seems poised to emerge from some of the pandemonium carriers have experienced since the beginning of the twenty-first century. Today, Delta and Northwest Airlines remain in bankruptcy. US Airways emerged from bankruptcy a year and a half ago only to merge with America West. The US Airways name survived and America West is gone.

Figure 6-1 Qantas Boeing 747 waits its turn for takeoff at Sydney Airport. (Courtesy Peter Egglestone.)

In November 2006, the new US Airways made an offer to purchase Delta, although the deal was eventually abandonded. This seemed to signal the beginnings of the great wave of consolidation so many analysts said would be necessary for the carriers to get back on solid financial footing. The board of directors at Australian Qantas Airlines unanimously approved a bid by a private equity firm to take that airline toward a more profitable future. Sources say Qantas plans to purchase as many as 70 new airplanes once the deal is completed.

And, back here in the United States, the reactions to the Delta buyout have been mixed, with the Air Line Pilots Association (ALPA)—represented pilot groups essentially beating the deal to death. Delta management believes it can think its own way out of bankruptcy without any outside help from US Airways. So what did US Airways do? They sweetened the deal, but even additional inducements proved unacceptable to the Atlanta-based airline.

But, of course, while US Airways has been keeping Delta busy, Northwest Airlines has hired some investment groups to help them decide precisely what direction their future could or should be. The rumor mill has been ripe with suggestions that American Airlines might be looking at Northwest.

And to make certain that the smaller carriers are not left in a lurch, the CEO of JetBlue—in a speech to the NY Wings Club in late 2006—said his airline is also on the lookout for the right mix. It could be a marketing alliance or it could be a merger. Who would fit? JetBlue east means they'd need someone in the west. How about Frontier, a company with a strong base in Denver? Perhaps.

Having recently emerged from bankruptcy as well, United Airlines has been on the prowl for a complementary partner. This week, the second largest airline in the U.S. reported it has entered initial talks with Continental Airlines. United also expressed interest in merger options with Delta.

Finally AirTran, the once considered upstart discount carrier in Atlanta, made an offer to buy Milwaukee-based Midwest Airlines. Midwest said no, but AirTran is continuing the pursuit.

Of course, one of the other big players, which has not been mentioned recently, is Southwest Airlines. But Southwest tends to operate on the fringe of the mania that has traditionally engulfed U.S. airlines. Southwest's growth strategy has remained pretty much the same for years . . . slow and steady, a successful plan that recently rewarded shareholders with its 26th straight year of profits.

Generally, the marketplace views these dynamics as generally good news, which means it is definitely something to keep your eye on. The December 14, 2006 issue of the *Wall Street Journal* showed the stock at Continental sitting at $44.76, up 141.7 percent in the past year. Midwest Air Group, parent to Midwest Airlines, showed a stock price of $11.10 a share, up 154.6 percent in the past 52 weeks. Also, American's stock has been up 66.5 percent over the past 12 months. This solid economic news will be welcomed by pilots—7,000 of whom were still on furlough as of December 1, 2006.

From a United Airlines bulletin issued in late 2006: "On the recall front, to date we've announced 249 recalls in 2006 (generally evenly split between Airbus and 737 First Officer), and all have been filled, except a handful of slots in the last announced class (June 5). We continue to be on track to recall over 300 pilots in 2006, although the precise timing for future 2006 classes has yet to be determined.

"Interestingly, we recently passed the halfway point on the furlough list, as 1,175 pilots have been offered recall since November 2004. As we process the lower half of the list, the acceptance rate has declined to around 30 percent, with about 50 percent choosing to by-pass; the remaining 20 percent technically return to United, but immediately move to LOA status, typically military leave."

In a December 14, 2006 article, a *Chicago Tribune* staff writer, Julie Johnsson, tried to explain the major airline-merger fever to consumers. "Once the frenzy has subsided and all the airline industry consolidation has been digested, Chicago could find itself home to the country's biggest carrier, a super-size United Airlines, or home to none at all. Passengers could be looking at higher fares or more bargains . . . Tens of thousands of workers could be affected as airlines merge, with some getting bigger and others disappearing."

What you've read here is a glimpse of not the good or the bad news, but simply *the* news about the industry, period. Sometimes it will be as volatile, perhaps even more than now, and at other times it will seem a bit, well, boring by comparison. If you intend on flying airplanes for a living, this is the atmosphere you'll be living and working in.

For an interesting perspective from Europe, type in this URL—www.dft.gov.uk—and read the United Kingdom's Department for Transport's Air Transport White Paper Progress Report 2006.

Some might say the rules have obviously changed because of the industry's turmoil. Or, have they really? Twenty years ago, you considered yourself fairly safe if you managed employment with a United, an American, or a Delta Airlines, but there were layoffs years ago as well. A friend who is now a Captain for one of the top three major carriers was furloughed two different times in his first eight years with the airline and it first happened the day he left B-727 class.

Remaining Focused

Despite the earlier advice about sending paperwork off to a company before you have the minimum requirements, that rule has interpretations. Is it a waste of time to send your application to Southwest Airlines if you don't already hold a Boeing 737 type rating? I say yes. But if Southwest asked for 2,500 hours total time and you had logged only 2302, I'd absolutely send the paperwork in. Close to press time, I opened a web browser on the Southwest Airlines web site and pulled down the most current hiring info advertised for pilots. I did this to remind you that, despite all the economic struggling ahead for some of the airlines, a few are doing relatively well, and one of those is Southwest Airlines.

Another arena for job seekers that has emerged strong, especially since 9/11, has been the small-package freight carriers. Not only are large companies, such as FedEx and UPS, hiring pilots, but these companies—now considered major airlines—are reaping incredible annual profits, which have made them an anomaly in the U.S.: an airline that is making a solid profit. Visit their web sites at http://pilot.fedex.com/ and https://ups.managehr.com/screening/professional/.

Southwest Airlines New Pilot Requirements

Certificates/Ratings

U.S. FAA Airline Transport Pilot Certificate. Unrestricted U.S. Type Rating on a B-737 not required for interview, but required for employment.

AGE: Must be at least 23 years of age.

FLIGHT EXPERIENCE: 2,500 hours total or 1,500 hours turbine total. Additionally, a minimum of 1,000 hours in turbine aircraft as the Pilot in Command, as defined in the following, is required. Southwest considers only pilot time in fixed wing aircraft. This specifically excludes simulator, helicopter, WSO, RIO, FE, NAV, EWO, etc. No other time is counted.

CURRENCY: A minimum of 200 hours must be logged in the preceding 36 months.

MEDICAL: Must possess a current FAA Class 1 Medical Certificate. Must pass FAA mandated Drug Test.

AUTHORIZATION TO WORK IN THE UNITED STATES: Must have established authorization to work in the United States.

DRIVER'S LICENSE: Must possess a valid United States driver's license.

EDUCATION: Graduation from an accredited, four-year college preferred.

LETTERS OF RECOMMENDATION: At least three letters from any individuals who can attest to the pilot's flying skills, by having observed them over a sustained period of time.

The Hiring Process at United Airlines

The last time United hired pilots, the total time requirements were considerably less, about 350 hours total time. This carrier believes other means are available to determine whether you're the best possible candidate for their job. Here's a look at the hiring process at United from the perspective of those who ran it for many years.

An Interview with Bill Traub, VP of Flight Standards and Training, United Airlines (Retired)

"The personnel interview consists of two, or sometimes three, people meeting with the candidate. I must say that we would like every candidate invited to interview to be successful. Candidates need to do their part by displaying a positive attitude and enthusiasm.

"At United, flying the simulator is an extremely important part of the process. You can't be rusty on instruments and pass. The ride is all computer scored. We've also correlated that performance with a pilot's achievement level later on the job and that correlation is high.

"Essentially, the simulator check measures instrument skills and eye-hand coordination. The candidate should look at this as an instrument flight test and understand they must prove themselves. The test may use a Frasca simulator or, perhaps, a DC-10 simulator. Regardless of the simulator, the test is the same. Believe it or not, they both fly the same.

"You can make a mistake on the simulator check and still pass. If you make an error, we look at how you recover. If the rest of your ride goes to pot because of the error, you're probably not the person we want. It's like golf. Having played the course before doesn't make you a good golfer if you don't have the basic skills.

"You'll shoot several ILS (Instrument Landing System) approaches that get progressively harder. Getting off altitude is not a killer, but the longer you stay off, the more points you lose. You begin with 100 points, too: 90 to 100 is *A* work. 80 to 90 is *B*. We've never hired anyone who had less than a certain score on the simulator ride.

"We begin the interview with a little chitchat to break the ice and get to know you a little better. But our people will spend up to an hour, before they meet the candidate, going over the paperwork you've already sent us.

"The interview is structured with six or eight areas of concentration. For example, we want to know about your commitment to the profession, how you've prepared for achieving your career goals, and how well you work in a team environment. As a clue about these, we might look at participation in sports or other extracurricular activities in high school or college. We believe that past performance is the best predictor of future performance.

"We look at a lot of the nonverbal cues, as well. Do you have a firm handshake? What about your posture, your neatness? If you don't look us in the eye very often, well—that doesn't project much confidence. We also try and learn how well you accept critique. We want to know what you've done well and what you have not. We want to be assured that you will grow and not continue to make the same mistakes. Nervousness is a judgment call on our part. If a candidate's nervousness interferes with performance in the interview, it is unfortunate, but you must find a way to overcome it enough to present a positive and professional image.

"Candidates must also pass a first-class physical, but one that United doctors would give. There, we're trying to predict who these people are. Being overweight probably does not make a good first impression."

Traub also talked about some of the common "showstoppers" that candidates will want to avoid. "People who don't listen during the interview, who don't give us the answer to the question we asked. If there is a hint of dishonesty, we can tell. A history of check ride failures can be difficult to overcome, although one failure can be turned into a positive, by telling us what you learned, why you're now a better pilot. Logbooks must be accurate. Computerized versions are OK.

"Finally, a poor driving record is a tough one to beat, as well as not having a license in the state in which you reside. If you don't follow the rules while driving, or with respect to licensing, how do we know you will follow the rules when flying? Respect for authority is of the utmost importance.

"After the simulator ride and the interview, all information is sent to a board for evaluation," Traub added. "Pending a successful medical exam, we'll make the candidate a job offer if in the judgment of the Board of Review, the simulator evaluation and the personnel interview meet United Airlines standards. Remember, approach the process with commitment to be successful. United's goal is the same as yours; we want you to be successful. Make the most of this wonderful opportunity."

While a subscription to Air, Inc., FLTops.com, or any of the other information services is a valuable asset, they are not required to learn much of the basic information about an airline, although they can make the research work much easier.

Which flying job is right for you? You might be tempted to say "any" at this point, but I caution you against that. Take the time to read about the different airlines before you apply. Perhaps you should set your sights somewhere other than the majors or, perhaps, on some airline in addition to the big three or the big five. But, without an organized plan to approach these carriers or any other flying job, the work will become three times as difficult.

Where's Your Paperwork?

Let's talk about a few methods to keep your information organized. First, pick up a box of manila pocket folders, the ones that are open at the top, with closed sides. Label one folder for each of the airlines where you intend to apply. Find a location in your home where the dog won't chew on the folders or the kids won't try to use them for coloring books. From this moment on, every piece of information you collect about this airline should be stored in this folder. If it fills up, that's a good sign. Start another folder, a part two, to continue the process.

Record the date of every letter or update you send out, as well as the response, or lack of it. Look through all the folders at least once a month to make sure you haven't missed an update that was required or to learn whether a follow-up letter was ever answered. Then, if you're offered an interview, you'll have a veritable wealth of information on the airline to look over before you go in. Regularly scanning your files will also make life easier. You'll always have a copy of the letters and forms and you can more easily stay on top of your progress.

I haven't mentioned buying a computer because I'm assuming you already own one attached to high-speed Internet access as the basis for your efforts. Consider adding a piece of useful software to your computer—a scheduler and database—such as ACT or Maximizer. Originally designed for salespeople, either of these two programs can help more easily track all the contact information being sent out. Maximizer, for example, lets you develop an individual file for each company sent an application. In that file, you can keep records of all contacts—when you last updated, who you sent it to, and what your totals were at the time. It also enables you to schedule the next event, such as when to update. If United wants an update in six months, the scheduler can be set to alert you when the time is right. If you're applying to many companies at the same time, you'll welcome a reminder of exactly when to make the next contact.

Because so much information is available online, you also need both an organizer—MS Onfolio is one I like—that lets you not only bookmark web pages, but also to save actual copies of the HTML pages, including all the graphics, such as pictures, tables, and charts. Onfolio is a part of Windows Explorer. These pages can be extremely useful to jog your memory when the time comes to study for an interview.

Certainly, too, you need a good word processor. MS Word is the most common and is often the format in which an employer may expect to see your work, if they ask, although both Google and Yahoo are planning to offer online word processors they claim will rival Microsoft. Do not, I repeat, do not send a résumé out in a generic text format. They look so bad when they're printed that you'll make the reader squint at the first review.

And, if they cannot easily read your documents, they end up in the trash. Old story here, but you only have one opportunity to make a good first impression.

And, finally, don't forget to regularly back up your documents. Most everyone claims they back up regularly, but few do. And, of course, you know exactly when you'll remember the last backup date, don't you? That flash of brilliance usually occurs just when you realize your machine's hard drive has crashed and the computer won't start properly.

Another idea to consider is an online backup system. If you put everything on a CD that's filed next to the computer, you still run the risk of losing everything if there is a fire, or if the kids or the dog get too close. (My 90-pound Rhodesian Ridgeback recently grabbed a CD off the dock in my office and chewed it before I even realized what he was up to.) If you have a safety deposit box, take a backup copy of your data there at least once every few weeks. This might sound like work, but it's nothing compared to the agony you face when trying to re-create a month's or year's paper trail to dozens of companies, as well as related personal background material.

Another idea to consider is off-site online backup systems. These software-based programs operate in the background on your machine and automatically back up data according to any parameters you set up related to time or format. With high-speed Internet access, you must leave your machine on 24 hours a day, but it seems a small price to guarantee the safety of the data. Here are two for your consideration: www.carbonite.com and www.mozy.com. I'm evaluating each right now for my own use. But, don't forget to have antivirus, web-security, and antispam programs running, too, because that can always open your machine to a hoard of junk if you're not prepared.

The Psychological Exams

If you haven't heard of the Minnesota Multiphasic Personality Inventory (MMPI) hold on to your hats. This is one of the more common evaluations given to pilot applicants during major airline interviews. It's quite, well, fun doesn't quite describe it. This test contains about 500 multiple-choice and true-false questions. It's designed to tap into your personality, as well as your "truthiness," as Comedy Central's Stephen Colbert might call it. With this exam, there are so many questions that it's virtually impossible to beat it, so don't bother. You also can't study for the exam and you can't fail it.

Some people do try to outsmart the exam, however, and experts can tell you they only make themselves look worse in the end. The exam revisits the same question more than once to learn whether your answers are consistent. An example is "I never lie." True or False? If you answer true, most people would find that pretty hard to believe. If that's true, then say true, but realize that the question will appear again, worded somewhat differently. The number of conflicts to questions are part of your overall evaluation.

Seldom would any airline make a hire or not-to-hire decision based solely on the MMPI profile results. More good news is that not every airline requires these exams.

The Interview

So many variables exist in the interview process because of the differences in the carriers, that trying to handle each, on a case-by-case basis, would be nearly impossible unless you happen to be an aviation information service. One of the generalities that applies to most all of the carriers, however, is virtually none of them will hire you with simply one interview. One Midwestern airline brings people back for as many as four interviews, while United, for instance, now believes they have candidates pretty much figured out in two. There's no hard-and-fast rule.

Some airlines treat pilots as total professionals and won't ask for a written exam, while others can make candidates feel as if they're crawling from one phase to the next. One airline is known for asking extremely personal questions designed not just to gain information about you, but also to see how you react in stressful situations.

A comrade of mine was asked about how long he thought it might be before his wife became pregnant, as well as whether they were trying to have children in earnest. He told me it took every ounce of his strength not to get up and walk out of the interview. Luckily all interviews are not so unprofessional. The point is to be ready for anything, but also try to find someone who has been through the interview process ahead of you. That's where the information services and message boards can help. Personally, if I didn't know anyone, I'd hang out at the terminal and ask questions of the pilots I met who worked for the carrier I was interviewing with. FLTops.com and Air Inc.'s preinterview counseling could also help.

You've read it here already, but the most important part of the interview is your attitude. One Midwest carrier says there's not much an applicant can do to study for the interviews, so "We just tell the applicants to be themselves." The most important part (other than doing some homework on the airline itself—either through recent research or by using the information you've gained that is stored in your folder on the carrier) is to convey the idea that you're committed to working at that airline. No one is telling you to bounce off the walls when you arrive for the interview, but you must convince the airline you want the job. And don't forget the line at the end of the interview that so many people seem to assume is understood. "I'd really like to work for your company. Thank you."

Helping the Airline to Say Yes

by Judy Tarver

Author's Note. *I've known Judy Tarver (Figure 6-2) for ten years and every time I've ever had an interview question, she's the expert I ask. She's been around the airline-hiring business for a while—28 years to be precise. President of Pilot Counseling Services, Inc., Tarver is an aviation consultant, writer, speaker, and expert witness, specializing in the field of pilot selection and recruitment.*

Tarver has been responsible for hiring over 7,000 major airline pilots and has consulted with several major air carriers—including American Airlines—and other aviation associations, such as the Air Line Pilots Association and the University Aviation Association.

She is the author of Flight Plan to the Flight Deck: Strategies for a Pilot Career *and several magazine articles that have appeared in* Aviation for Women, Flight Training, Air Line Pilot, *and* Airline Pilot Careers. *Tarver is also a contributing writer at www.fltops.com. Her web site is www.pilotcounseling.com.*

Figure 6-2 Pilot interview counselor Judy Tarver.

Whether you have 100 hours or 10,000 hours, the job-hunt philosophy is the same. For low-time pilots, get those stepping-stone jobs quickly to expedite that trek to the majors. High-time experienced pilots may only have one shot at the company where they might spend the rest of their career.

Getting the interview is only one step in your goal to getting the job. Getting hired is the tough part. When the market is slow, the job hunt becomes more competitive. Here are some important tips to help clear the way for a successful journey to achieve your goal.

Do Your Homework

Make sure you do your research on the companies to which you are applying and consider the following:

Do you understand the corporate culture? There are subtle and, sometimes, not so subtle, differences in company philosophies. Talk to pilots. Make sure the companies to which you apply fall within your own expectations of the places you want to work. Many web resources today can provide any kind of information you need, including press releases, quarterly reports, and so forth.

Unless you are only trying to gain experience in this unstable economy, carefully evaluate the financial status of the companies where you apply. Nothing is a given but, remember, seniority is everything and you don't want to be forced to start over when a company goes out of business.

If you know a company has demonstrated unsafe practices or policies, avoid working there, if at all possible. When you're trying so hard to build hours, this may seem desirable, but it can impact the rest of your career if you end up pushing the limits, receiving an FAA violation, or if you leave because of unsafe or undesirable conditions. Employers who run companies like that are also likely to give negative feedback.

Keep Good Records

Not being prepared can cost you the job. These are things you can do along the way to avoid problems in the future. You should have an updated résumé available at all times, keeping the following factors in mind.

You will be asked to provide the following information.

- Certified copies of all your college transcripts.

- Some companies will ask for high school transcripts.

- Breakdown of all employment going back to college and, sometimes, even all the way back to high school.

- Sometimes, an application will ask for more information than you have space for on a résumé.

Make sure you always have the following information handy.

- Dates of employment.
- Names of all companies (aviation-related or not), addresses, supervisor's names, and phone numbers.
- If you had any aviation-related jobs in college or school, I recommend listing them.
- Keep a breakdown of aircraft flown at each company, the crew positions you held, and any other responsibilities you had.
- If your employment has any gaps, be prepared to explain and provide a contact who can verify the information.
- If a company is out of business, provide whatever information you have to verify employment, such as a trustee, or the name and number of a former supervisor or coworker.
- Military pilots should list each upgrade or move (transfer) within the military. Don't consolidate all military experience together.
- Keep a record of current and all previous addresses. Because pilots move frequently, it can be difficult to recall all your previous addresses.

LOGBOOKS:

- Make sure you have all your logbooks available from day one. Your logbooks tell a lot about you and are scrutinized carefully in the interview process.
- Make sure your logbooks are legible and accurate. Messy logbooks reflect poorly on your attention to detail, an important characteristic for a pilot. Do not put inappropriate comments in your logs—stick only to the facts.
- Don't ever stop maintaining logs. You never know what can happen to a company to put you back on the job market.
- Your hours and work experience should track.

IMPORTANT DOCUMENTS:

- Military DD214 papers.
- Originals, and copies of all licenses and certificates.
- FCC radio operator's permit.
- Make sure you always keep a current FAA 1st Class Medical Certificate. Even if you don't need a 1st Class Medical in your current job, the airlines will want to see one. And, if you are just starting out, you always want to make sure you can pass one anyway.

CURRENT PASSPORT:

- You'll need to complete an Employment Eligibility Verification I-9 to prove you have the right to work in the U.S. In addition to a U.S. passport, a certificate of U.S. Citizenship or an Alien registration Receipt Card will suffice.

Watch Out for the Pitfalls

While everyone admits these things could never happen to them, they often do. Consider the long-term consequences of short-term mistakes and don't burn your bridges.

- Always be forthright and honest in all your dealings. Don't be misleading in your information or documentation. If an employer ever finds out you lied or misled them, you will be dropped from the process or, if you are hired, terminated.

- Treat your coworkers and employers with dignity and respect. Don't abuse privileges and benefits. You never know when they can help or hurt you.

- Always give notice to your employer when you want to leave a job. A bad referral is difficult to overcome.

- Do not take unnecessary risks. So you thought buzzing the neighbor was fun. Not worth it. And please, don't put it in your logbook!

- You should know the airline is going to check your driving records and, often, it does a criminal background check. Follow the laws.

- Maintain healthy habits. Even though most airlines just want to see an FAA 1st Class Medical, some still go above and beyond. You want to keep this career until age 60.

- Even though it shouldn't be a factor, with the implementation of the Pilot Records Improvement Act (PRIA)—see the next section—don't ever take check rides or training lightly. Those records will be reviewed by potential employers. You are allowed to receive copies of these records. Do so. Review these records and make sure no discrepancies exist.

Pilot Records Improvement Act (PRIA)

Introduced in 1996, Title 49 United States Code (49 U.S.C.) § 44936(f)(1) mandates that all air carriers request and receive FAA records, air carrier and other records, and National Driver Register records before allowing an individual to begin service as a pilot.

By keeping good records and an exemplary employment history, you can avoid problems that could result in delaying your date of hire.

The Application Process: Getting the Interview   The airlines set minimum hiring criteria and rarely deviate from it. However, depending on industry trends, the criteria can change overnight. Therefore, it doesn't hurt to go ahead and apply, but don't expect to be called in unless you meet the specific criteria. Also, remember, the minimum criteria aren't necessarily the competitive criteria. Airlines are going to look at the best in the pool.

The process of applying varies from airline to airline. This can be as easy as submitting a résumé to a complex application-submission process. This is your first step to the interview. Use every tool you have to sell yourself on the application or résumé.

Follow the Directions Explicitly

- If you are applying through a database process, the first step the company typically takes is to download only those applicants who meet specific criteria. If you leave out one item that is required for the download, they will never see your application.

- The same applies to a scan type form. Missing information causes missing a job.

- It is often helpful, and sometimes required, to have a crewmember recommendation. Check carefully. Some companies want as many as you can give them, while others do not want to be inundated with referrals.

- In some cases, the referral will get you to the interview.

- In other cases, the referral is not reviewed until after you are invited for the interview and it plays a part in the selection process.

- Take advantage of all networking opportunities, such as aviation associations, which promote and encourage pilot development. Airline representatives regularly attend conferences and conventions given by a variety of organizations.

The Interview

Now that you have the interview scheduled, you want to enhance your opportunities for success. An airline doesn't invite you for the interview with the specific goal to turn you down. They would like to hire every candidate they interview. So, it is up to you to put your best foot forward and make it happen.

Preparation

- Take advantage of an interview skills preparation course. This can help lower your stress level during the interview and bring out your best side. This is especially helpful if you have an obstacle in your career path.

- Make sure you come prepared with every piece of documentation requested by the employer. Not being prepared can either eliminate you from consideration or delay the process.

- Arrive early and make sure you get plenty of rest before the interview. Make sure you have reliable transportation and a backup plan, just in case of a delay or cancellation.

- Make sure you are comfortable answering any questions that relate to your technical skills and employment history.

- Don't check any clothes or documents you need for the interview. Carry them onboard.

Appearance

- This covers not only what you wear, but also how you present yourself in the interview.

- Your goal is to have the interviewer focused on your credentials and your potential as an employee, not on what you are wearing or not wearing.

- This is not the time to express your individualism. Think about it. You will be required to conform to company uniform and grooming standards.

- Wear a conservative hairstyle and clothing. Don't wear funny ties, unless it is funny tie day at SunFun Airlines.

- Go light on the jewelry.

- Don't wear perfume or cologne, but do wear deodorant.

- Keep your attention on the interview. Let the interviewer remember you, your strengths, qualifications and qualities, not what you were wearing.

Behavior
- Your behavior will be under scrutiny from the moment you are invited to the interview until the final recommendation is delivered.
- Negative first impressions are hard to overcome, but it is possible. The best thing is to avoid it altogether.
- Treat each person with equal respect—from the person who calls you for the interview, to the gate agent, to the interviewer. If you don't, the word will get back quickly to the recruitment department.
- Don't ever assume you are safe from being seen or heard. You never know who is sitting next to you.
- Don't ever be condescending. Unless you are asked, always assume the interviewer knows as much or more than you do.
- Make sure your telephone message system is professional. A provocative or silly message can quickly turn off a recruiter.
- Don't try to be funny in the interview. It usually doesn't work.
- Your body language tells a lot about you. Try to stay poised and composed, yet not too stiff.

Interview Hints
- Do your homework, but don't use canned answers in an interview. The recruiters have a BS radar system that pops up quickly. Know the difference between being prepared and memorizing canned answers.
- Know yourself. If you want to know about yourself, ask your wife, husband, or mother.
- Know your business. You are going to be questioned about situations that relate to what you do. If you fly a Cessna Citation, they aren't going to ask you about the systems on a B777.
- Be positive in the interview. Avoid negative comments about former employers, coworkers, or anything else.
- Demonstrate your desire for the job.
- Maintain good eye contact, but don't get into a staring contest or look out the window.
- Be attentive. Listen carefully to the questions before answering. Don't cut off a question, assuming you know where the interviewers are going. They may be headed in a different direction.
- Taking some time to think about the question is OK. If you can't answer the question, don't dwell on it—move on. You may have a chance to go back later, but don't let it ruin the rest of the interview.

- If you know you have an obstacle to overcome, remember, it is not always so much the obstacle, but how you deal with it that counts. You definitely need to practice talking through it before the interview.

- Be factual.

- Don't get emotional.

- Give the details.

- Don't place the blame elsewhere.

- If appropriate, tell what you learned from the incident and how you turned a negative situation into a positive one.

- Stay away from discussing personal issues. This is not the time to talk about family, politics, or religion. Stay focused on the reason you are there, to obtain employment as a crewmember.

- Most questions require a narrative response. Understanding the qualities of a good pilot and employee can help you prepare your responses. These are characteristics airlines look for in the interview process. Read them carefully and use them to help you find your stories about times when you demonstrated those characteristics. Once you have a story, show how you demonstrated as many of the following dimensions as you can when telling it.

  - Technical Understanding

  - Professionalism/Operational Awareness

  - Teamwork/Crew Orientation

  - Leadership

  - Critical Thinking/Judgment/Decision Making

  - Initiative

  - Communication Skills

  - Enthusiasm

  - Stress Management

  - Motivation to Fly

  - Ability to Learn (Trainability)

In case you might believe interviewers only ask relatively easy personal questions, read on. Here are the first 25 of 201 question possibilities. The best part, though, is I don't have the answers listed here. You need to think about them yourself.

The 20-Minute Interview for Airline, Corporate, Charter, or Flight Instructor Candidates

It's that time again. Whip out that ol' video camera, put on your best duds, and strap yourself in. You might be in for a bumpy ride. Try answering some of these questions gathered from recent interviews at United, American, Delta, and UPS. Try answering the following questions with a video camera trained on yourself. Not only will this give you an opportunity to later review your results, but it will add a touch of realism to your responses (Figure 6-3). You'll need a partner to act as the off-camera interviewer.

And here's a tip someone offered that has saved me quite a few times. Interviewers are, at times, obsessed with learning how much a candidate understands, or does not, about instrument-approach procedures and approach-plate terminology. One of the most difficult issues is finding a good set of approach plates to practice with if you don't happen to own your own. This is where an Aircraft Owners and Pilots Association (AOPA) membership is also worth its weight. For that same $39 in dues each year, you have access to the AOPA Airport Directory, which offers you the opportunity to locate and print any instrument flight rules (IFR) approach plate in the United States.

Figure 6-3 Commercial charter operations also operate a wide range of aircraft.

You can spend a good afternoon of preparation with another pilot by downloading a half dozen plates and asking each other questions based on different scenarios. The scenarios are determined by you. Grab a handful of index cards and write words/terms, such as VFR, 100/ 1/2, Snow, Heavy Rain, Visibility 1/4, Slick runway, Missed approach at, and so on. When you begin your discussion, simply grab a card from the pile and that represents the weather at the airport. Learn how it changes your decision-making process.

- When do you have to declare an alternate?
- What are alternate minimums?
- Name all the elements for a holding clearance.
- If you do not receive an EFC time, is your holding clearance valid?
- What are lower-than-standard minimums for 1600 RVR, 1200 RVR, and 600 RVR?
- Can you file an IFR flight plan under Part 121 without listing an alternate? If so, when?
- What is a hydraulic accumulator?
- Name and explain the three types of hydroplaning.
- What is the maximum crosswind limitation for the aircraft you are flying now?
- Have you ever bent or broken a company policy?
- What was the last book you read? Why did you decide to read it?

An airplane enroute from Denver to Cheyenne has a crew of three: a pilot, a First Officer, and a flight attendant. Their names are Brown, Black, and White, but not necessarily in that order. Three passengers are also on the plane, named Mr. Brown, Mr. Black, and Mr. White. Mr. White's income is $9,999.99. Mr. Black lives in Cheyenne. The flight attendant lives just halfway between Denver and Cheyenne. His nearest neighbor, who is one of the passengers, has an income exactly twice his own. The flight attendant's namesake lives in Denver. Brown is a better chess player than the First Officer.

- Who is the Captain?
- Who is the Speaker of the House?
- Who is the Senate Majority Leader?
- How many members are in the House of Representatives?
- On an ILS approach, what is considered the final approach fix?

For even more mental exercise, look at a few hundred more possible interview questions (see Appendix B). Remember, the questions don't all necessarily need to make sense to you. Many are designed to gauge your reactions during a stressful time.

Let's listen in as an American Airlines pilot talks about what he experienced during the big interview in Dallas for the world's largest airline a few years ago.

American Airlines Interview Debrief

by David Manning

I was notified of my American Airlines interview on Monday, September 14th. I selected Wednesday, September 23rd as my day. I was offered either the 23rd, 24th, or some time later in October or November. I chose the earliest date, partly because of my excitement, partly out of a sense of reluctance to get any lower on the seniority list.

I was Federal Express'ed my additional paperwork two days after my notification. It detailed all the required documents. Most difficult for me were copies of all college transcripts, not just the university conferring the degree. After a flurry of faxes and credit card authorizations, I had my transcripts from three colleges and driving records from two states. I was missing one transcript on interview day, and they told me I'd need it before they could schedule a class date.

The Big Day
Arrive a bit early for the 0815 show time. Give all required documentation and logbooks to Barbara, the front office staff member. She's very friendly and professional, so don't be afraid to ask questions; however, she can add comments to the process, so treat her like gold.

The Applicant Profiles
1. Cont. Express EMB-145 RJ First Officer (former AA Flight Attendant), female
2. Mesa Airlines Beech 1900D Captain, female
3. Airborne Express DC-8 FE, female
4. Continental Express EMB-145 RJ Captain (Dad, two brothers pilots at AA), male
5. Chrysler Corp. Hawker 800/Gulfsteam G-IIIIIV Captain, male

6. Former U.S. Coast Guard Falcon Jet Aircraft Commander/civilian
   Learjet 60 Captain (Dad retired AA Captain/wife AA flight attendant), male (me)

Group welcome aboard and a brief discussion about the state of
affairs at AA hiring. One at a time, introduced ourselves with quick
rundown on past flying experience. Now is not the time to read your
doctoral dissertation—just a quick, "this is what I did." Leave to take
Psychological Inventory, which takes about 45 minutes. It asks questions like "Sometimes I feel so overwhelmed, I just want to hide" and
you mark a scale of 1–5, Strongly Agree to Strongly Disagree. No
advice on how to take it, except to be honest. I don't know if it'll say
that I'm deranged or whether it's just for a personality profile for my
future personnel file.

Next, one at a time, you'll be taken back into the depths of the
office to review paperwork and make sure everything is there, certified, and so forth. I had to add the other two colleges I attended to my
original application because I only put down my final university at
graduation.

Time to watch a quick ten-minute video presentation on the simulator profile. A quick break, then your group of six is broken down
into three pairs. Your partner will assist you during your simulator.
Important: you and your partner aren't competing! They want to see
you both work together! My partner was the ABEX DC-8 FE. We were
allowed 30 minutes to prebrief. We flipped a coin; she got the left seat.
Sim profile was very basic. In fact, more basic than I expected or was
briefed on. My partner's profile: one leg hop from CLE to ORD, fly
the airway, shoot a Localizer approach to a freebie landing. Switch
seats, and mine: I flew from ORD to CLE, although for me weather
went below minimums in Cleveland and I returned to 01W to shoot
the full NDB to mins (exactly what I would've done—yeah, right!).
While you're the Pilot Not Flying (PNF), expect a question about
holding, that is, if given this clearance, how would you enter, what
airspeed for a particular altitude. Whole session lasted from 1100 to
1240 or so. Upon return to the recruitment office, we were told no
time for lunch because our interview was at 1300. Quick cup of coffee, then into the interview. At the same time, the other pairs were
either coming out of their interviews or coming back from lunch. The
good thing about going into the sim first and missing lunch was that
we were done for the day first.

The Interview

Two Captains—a B-767 check-airman and a Fokker F100 line-Captain. First and foremost, they were the consummate professionals, but they did their best to make me feel comfortable. Offering a drink, asking if I wanted to take off my suit jacket, and so forth. They seemed very relaxed and wanted me to be relaxed, too.

When they interviewed me, they didn't just "ask questions," a la United Airlines. They actually crafted a set of questions into their own version of a short situation and asked me how I'd react or what I'd take into consideration when making my decision. It immediately put me at ease and let me speak "airplane," letting them know who I was as a person and an aviator. This is exactly what they want—to get to know you, not hear any gouge regurgitated. Questions/situations I remember (by the way, these are all on the gouge available somewhere on the Net):

- Initial—tell us about your flying experiences until this point.

- If it snowed a foot overnight, how would this impact your preparation for a flight?

- You're an FE. Your Captain arrives at the jet looking ill and smelling of alcohol. What do you do? If he's insistent on flying?

- What if the Captain is low and slow on an approach?

- How would my First Officers rate me as a Captain on a scale of 1–10? Why?

- What was your most pressing emergency? How did you handle it?

- What considerations would there be departing from or operating out of high-altitude airports? (Used my experience with Telluride, Aspen, Eagle Co.)

- What makes a good Captain?

- Have you ever had a conflict with a Captain? How did you resolve it?

- What is the worst mistake you've ever made in an airplane? (Then they said, "It's okay, all mics are off and the FAA isn't here.") Be careful.

- Do you have a mentor? (I understood "whom do you know at the company?")

- What would you do if the Captain was using nonstandard procedures, but not unsafe?

- Did I have any questions for them?

No systems or regulatory-related questions. In fact, when I got home and my wife asked me what technical questions they asked, my first reply was "none," but after thinking about it, I remembered that I detailed a few of the situational questions in a technical manner, as well as a procedural manner. Overall, very relaxed. They just wanted to see if they could stand flying with me for hours on end. It took about an hour and 15 minutes. Remember to smile and be enthusiastic!

After waiting for my logbooks to be returned, I was released to leave. I was told I'd hear something in ten days to two weeks.

I was called back for the physical in seven days. During the physical, they gave me a one-hour cognitive test that I had never heard anything about. It used a light pen and a computer, and seemed to be testing hand-eye coordination. I used the pen to touch a number—five, for instance. Then I'd move on to the next page and it would show me a series of new numbers and ask me to touch the number from the last screen. I had only a few seconds to decide. No results of that test were given to me before I left. They made it clear that you were graded for speed and accuracy. The attendant took the computer grade before I left, but I didn't see it.

My networking: I had plenty of friends who were pilots. I took a business card and kept a phone number from everyone I ever met. It was easy to meet people. Some companies want character references; others simply want people who have flown with you.

Thanks to David Manning, who is currently awaiting a class date with American Airlines.

The Top Ten Mistakes Made by Job-Seeking Pilots—and How to Avoid Them!

©1998 Bossard Publications Inc. d/b/a *Flying Careers* magazine. Used with permission.

1. Misrepresentation

Not all the skeletons in your personal closet can be overcome. But lying—about anything—on an application or during an interview can be a career killer. Practice answering sensitive questions—especially if you have a DUI, incident, and so forth—with an interview coach who specializes in airline interviews or another pilot who has recently been through the hiring process. The better prepared you are for the question, the less temptation there will be to misrepresent yourself.

2. Showing Up for an Interview Unprepared

Some airlines interview as many as five qualified applicants for each one they hire.

Preparing for an interview—regardless of whether it is a major or regional airline, or even a corporation—is not something that can be successfully accomplished in a few minutes. Time spent honing interviewing, nonverbal, and other skills is well spent. While most pilots are accepted or rejected for employment based on several criteria, the interview is often the first step, and the first point of elimination.

Prepare, prepare, prepare. Keep detailed, well-organized files on each company to which you've applied. Know everything about the company there is to know—not so you can dazzle them during an interview, but so you can speak with confidence and answer any questions intelligently. Taping your practice sessions and reviewing them with a good coach is a great idea.

3. Being Too Revealing

Just as it is important never to misrepresent yourself on an application or during a Human Resources (HR) interview, it is equally important to answer only the questions being asked or you risk sabotaging yourself. Answer completely, but keep your answers short and don't ramble. If the interviewer wants to dig a little deeper in some area, let that be their decision. Again, the more practice interviewing you do, the more confident you feel when the big day arrives.

4. Downplaying the Importance of Interpersonal Skills

Interviewers—both from HR and CRM/line representatives—want pilots they can see themselves working with for several days at a time. Communicating effectively in the recruitment process is imperative to getting the job; communicating effectively in the cockpit is pivotal thereafter.

One of the most effective ways of honing your interpersonal communications skills is by practicing the toughest of all questions—conflict resolution. Tackle each question individually, making plenty of eye contact while watching your nonverbal cues. Remember, too, an airline interviewer may ask a question with as much interest in the demeanor with which you answer as what you say.

5. Showing Up for the Simulator Evaluation Unprepared

In most cases, employers don't expect you to fly the simulator as if you regularly flew that particular aircraft. However, buying a little sim time—especially if you can learn what type you'll fly during the evaluation process—to sharpen your instrument scan and get a feel for the sensitivity of the simulator will be beneficial. Simulators don't fly precisely like an airplane, only close. The time to refresh your memory about that fact is before your evaluation, not during the session.

6. **Being Unwilling to Relocate**
   Some jobs require pilots to live in locations they might find unattractive at first. But applicants should remember such moves can be temporary and jobs that help build quality time are often difficult to locate. Look at each job opportunity carefully before rejecting any. Remember, reaching your career goal is a process of small steps—none of them permanent—until you reach your career position.

7. **Believing Your Qualifications Will Speak for Your Experience**
   Regardless of how many thousands of hours you have logged, airlines will ask you about your flying experiences. A common mistake is thinking you won't need to talk much if you already have 10,000 hours.

   Be assertive during an interview, but never cocky. Remember, despite your significant experience, the résumés of the next three pilots interviewing—as well as those of the last three—are probably similar to yours. You need to make an effort to convince a potential employer you are the best pilot for the job. Have at least one story in the back of your mind ready about how you successfully coped with some conflict in the cockpit or dealt with an inflight emergency.

8. **Not Understanding the Cyclical Nature of the Industry**
   Bailing out of the aviation industry is a personal decision. Many pilots have quit their flying jobs and ceased to log hours during downtimes, only to regret their decision once hiring resumed. Succeeding in this industry usually requires constant focus and unwavering dedication.

   Be persistent and patient. If a civilian pilot can't dedicate a significant period of time to building a career—time during which they may be poorly compensated—it may be time to look at another career.

9. **Lack of Correct or Current Career Information**
   Using gouge as a supplement is OK, but don't rely on hearsay as your only source of career information. The aviation industry is

unique. Knowing when application windows are open, where to send paperwork, how much of an application fee the company charges, minimum qualifications, and so forth are all an important part of a successful job search.

Purchase some form of career information service. But evaluate the service before you buy, to make certain they're not simply regurgitating old, incorrect, or bad information.

10. **Being Unprepared for a Written Exam During the Interview**
Most companies will ask pilot applicants to complete some form of written knowledge exam during the interview process. These run the range from a simple 10-question exam about regulations, ATC procedures, and weather to a complete 70-question Instrument, ATP or Commercial knowledge exam.

A number of books are available to help you review for the knowledge exams—if you know which one the company uses. There are also the standards every pilot should study before any interview: the FARs, AIM, Aviation Weather, and both NOS and Jeppesen IFR charts.

Honorable Mentions
1. Not Holding a Four-Year College Degree
Just looking at the statistics on the education level of pilots who get hired will tell you not to shortchange yourself here. Correspondence (external degrees) programs now make getting a legitimate degree easier for pilots of all ages and work schedules.

2. Being Unaware of the Competitive Qualifications of Pilots Who Are Getting Hired
Building on the education issue, pilots should know which certificates, ratings, and kinds of experience are required and which ones are preferred. Then, make sure you have both.

Simulator Checks

When you arrive for the interview at some airlines, you'll be expected to fly a simulator and demonstrate your skills as a pilot. Not all airlines require a simulator check, but many regional carriers do. Never arrive at a simulator check cold, with no recent IFR training. Unless you're a real ace, the chances of getting through the test are slim. Good instrument skills produce a good simulator session.

Preparing for a Prehire Simulator Check

There are different opinions about whether pilots should buy simulator
time before going to an airline interview. While there is no single
answer to the question "Should I buy some time?" some pilots have
found they needed it. Others have been successful landing a job without buying it. The bad news is, if you wrongly evaluate your need to
buy simulator time before a prehire check ride, it could cost you the job.

In the grand scheme of things, the prehire simulator check most
airlines put pilots through is only a portion of what these aviators are
evaluated on. Also considered is a pilot's aeronautical knowledge,
and their overall physical and mental fitness for the position. All too
often, unfortunately, pilot applicants treat the prehire simulator check
as an afterthought to the interview process, not worthy of nearly as
much planning as other portions. This can be a dangerous mistake,
according to some sources.

To Buy or Not to Buy

When the airline I was working for went bankrupt a few years back, the
closure threw hundreds of well-qualified and current pilots on the street in
search of new opportunities to ply their skills. Tom (not his real name) was
hired almost immediately by a USAir Express carrier on the East Coast, flying a SD3-360, notorious for a high-cockpit work level in bad weather
because the aircraft has no autopilot.

After a year of this kind of hard IFR flying, Tom received a letter from
United Airlines for the big interview. During a number of telephone conversations, I asked him whether he planned on getting some simulator time
under his belt before he flew the prehire simulator check at United. He said
"no," believing the many hours of hand-flown IFR during the past year
would suffice. Besides, he said, "the simulator at United was a Frasca. How
tough could that be?"

A few weeks later, Tom and I spoke after his simulator ride and interview. Apparently, the Frasca that United runs people through at Denver
was a bit more slippery than that lumbering Shorts Tom flew. His scan in
the unfamiliar Frasca was a little slow, as were some of his course corrections while he tried to get used to the simulator's sensitivity. When the letter arrived a few weeks later, Tom was surprised that the nation's largest
airline had decided "not to continue processing you at this time."

So what went wrong? After all, IFR currency should be enough to pass a check ride for a job, right? And isn't IFR proficiency what the airline is really looking for anyway? Well, yes and no. While the need to demonstrate your ability to fly an aircraft under IFR conditions is certainly the intent of the check ride, the spirit of the ride encompasses much more.

The ride will also—in most cases—measure your judgment, your knowledge of ATC procedures, as well as your overall use of cockpit resources, such as a copilot in many cases, according to airline sources. No airline is hiring you simply to manipulate the controls. They want a pilot who matches a set of their own special criteria, which tells them you are the person who will operate their aircraft safely and efficiently in all flight conditions.

Simulator Ride's Importance

Shane Losasso, president of Jet Tech, said, "The simulator ride is very important. Look at the number of [pilot applicants] who are turned away after a simulator ride. It is substantial." The strategy, then, for applicants is simple. Treat every part of the interview process—including the prehire simulator check—as if it were a make or break part of landing the job, because it is.

When John Shogren, now a First Officer on a Boeing 747 for United Parcel Service, arrived in Pittsburgh for a prehire simulator check with USAir, he said, "I did no training before the ride. In retrospect, however, if I'd taken some simulator time before the USAir ride, I might be flying for them now. I did that poorly. I had been hand flying a Brasilia and a Shorts 360 for nearly 1,000 hours and I thought my skills would transfer easily, but I was wrong." Shogren's logbook showed about 2,700 hours total time when he took the USAir ride.

Bill Mayhew, former vice president of flight operations at Wheeling, Illinois–based Airway Flight Service, said, "A lot of people who come to us for prehire simulator checks have flown aircraft, but [they] are not used to flying simulators. They are not used to the sensitivity a simulator might offer. Proficiency in an airplane does not necessarily equal proficiency in a simulator."

Process Viewed Differently

What makes simulator checks confounding to applicants is that pilots of various skill levels often view the process so differently. Todd Carpenter, a former C-141 instructor pilot, did not purchase any training either, although he had flown only about 500 hours during the previous 12 months prior to his prehire ride. "I felt pretty current before the rides with FedEx and Airborne," he said, "since I hand-flew the approaches and departures in the 141 most of the time."

Not All New Hires Buy Time

Carpenter believed the training was unnecessary before the Boeing 747 ride with FedEx, in particular, because "I did not believe they wanted me to fly a 747 perfectly. I think they wanted to see if I was someone who could learn as he went along, trying to get a good feel for the airplane."

Jenny Beatty, an MD-80 First Officer with Reno Air, added another spin to the subject by explaining that "most airlines do not expect someone with 1,000 hours to fly like someone who has 5,000." Beatty also did not purchase any simulator time prior to her prehire simulator checks because "I try really hard to maintain my skills . . . I felt I was prepared. But I also do some armchair flying by sitting in a chair and mentally going through both aircraft and IFR procedures. A large part of training your body is training your mind." But many pilots believe they need every point possible during a prehire ride and refuse to give up a single one easily.

"Worth the Money"

Tim Doreen, a Jetstream 32 First Officer for Atlantic Coast Airlines, said, "The entire simulator check and interview is a stressful time anyway, and you want to feel as comfortable as possible. If you don't pass the sim, you don't get to talk to anyone (at some airlines). If you don't get to talk to anyone, you don't get the job. I flew a Frasca in Denver before my ride with United, and it was definitely worth the money."

Citation pilot, David Hilsdon, said "The main reason I took some simulator training before my prehire check ride is that there is always a difference between the way simulators fly. Because I considered my check ride with FedEx to be a career-changing one, I wanted to put myself in the best light. I figured even an hour in the 747 would give me a leg up."

And, what does an airline really want anyway? According to Losasso, "Each company tends to look for basically the same thing, but goes about the process just a little differently. The end result is that the company can probably correlate how current someone is on instruments or whether they seem to understand instrument procedures well with their initial training washout rate."

Reno Air did not always use a prehire simulator ride for its applicants, according to Jim Hubbard, the carrier's director of training. "We were hiring people with lots of experience in our aircraft. Now we're hiring less-experienced pilots. We're really looking for basic instrument flying skills and airmanship . . . anticipating power changes, that sort of thing."

Good Instrument Skills Crucial

Hubbard also said that Reno tries to keep the ride as objective as possible so "a person's lack of experience in the MD-80 does not count against them. We have a series of standard profiles and choose them at random. In a one-hour flight, they are given a briefing ahead of time, as well as written instructions without having to know anything about the MD-80. It wouldn't hurt someone to have some MD-80 simulator experience, but it's designed so that any ATP-rated pilot with good instrument skills should perform very well."

The other side of the prehire coin is Southwest Airlines, which requires no simulator check before hiring a pilot. Paul Sterbenz, vice president of flight operations at Southwest, said, "We do not use prehire simulator checks because of the time and energy it takes to do them in the first place. But, also, the objectivity of what you find out is somewhat suspect. It tells you whether or not a person can fly a simulator, but it doesn't tell you that much about what their general airmanship is in terms of situational awareness and getting around in the real world. Secondly, our requirements are pretty high. We require an applicant to possess a type rating in the Boeing 737 before we will interview them. With those considerations, we are having pretty good luck with a very low-failure rate in training. We don't have any indications that another screening device like a prehire simulator check is really necessary."

What can you expect to see on a prehire simulator ride? Nothing outrageous, really—nothing more than a profile to perform some specific flight maneuvers. Here's a look at a sample profile from one airline's prehire simulator check. Flights often begin with the aircraft already airborne, as this one does—flying straight and level at 250 knots. First, decelerate to 200 knots and accelerate to 250 knots again in level flight. Next, climb at 250 knots, descend at 200 knots, and accelerate to 250 in the descent, and then, a turn. Next, comes steep turns in both directions, and then an entry to a hold. In this profile, flown in a Boeing 747, no instrument approaches were made.

If you've decided to make the extra effort (and incur the expense) to buy some training time before your simulator ride, some problems you'll face are what simulator to use, as well as how to locate the company that owns a simulator you can rent. Opinions about which simulator to use vary.

What Type of Simulator?

Don Buchanan, president of Allied Services, a simulator training company, said, "Fly the heaviest simulator and the one that is the toughest to fly." Todd Carpenter advised that "If you had no heavy time at all, I would buy some time in a simulator like the [B-]747." Mayhew added, "Try and find out what simulator will be used on your ride, and try and get something as close as possible."

Local networking with other pilots, as well as reading aviation publications, seems to be the way most pilots learn about the right simulator company. But one relatively new method is with a personal computer (PC), by searching any of the dozens of World Wide Web (the Web) aviation sites available on the Internet. Other sources of information are the online services' aviation forums, where hundreds of airline and corporate pilots—as well as the wannabes—hang out. A word of caution if you decide to seek advice through your computer: Take the information you read with a grain of salt because you never know for certain who you are corresponding with unless you verify it in another way.

Can you afford simulator time before a prehire ride? A better question is whether you can afford not to buy simulator time before your ride? How important is the job you're interviewing for? How will you feel in five years if you don't get that job at United or FedEx or Northwest that could eventually pay you more than $175,000 per year because you didn't want to spend $500 to $1,000 for some additional training? Hilsdon said, "If you have to beg or borrow the money, do it."

One caveat, however: Not all pilots hired by airlines, of course, buy prehire simulator time. Some pilots "ace" the simulator evaluation portion of the interview process without the expense or time in a full-motion or other simulator. Some pilots who are flying often for an airline decide the expense is worth the peace of mind it gives them going into the interview. It's an individual decision and not something that should be taken lightly.

At Jet Tech, the prehire simulator package costs $375 and includes an hour of pre-briefing about the profile, an hour in the Boeing 737 simulator, and half an hour of debriefing. At Airway Flight Service, $225 buys two-and-a-half hours in a Frasca 142P, similar to the one used at United Airlines. "Tom" spent nearly $1,200 for the Boeing 747 training time he purchased. And, also, don't let too much time elapse between your training and the ride. More than two weeks and you're probably wasting your money because you'll have forgotten so much of what you learned. Remember, too, prehire simulator training is not the place to relearn IFR procedures and regain your IFR currency. You can do that for less at your local fixed-base operator in a smaller, less-expensive simulator, sources said.

Tom received his interview letter from a major airline a few weeks back and called me to talk about it. The first thing he mentioned was the need to buy some time in a simulator before he headed out to Los Angeles for the ride. From Tom's contacts in the industry, he learned the simulator this airline uses is also for rent and he bought three hours. Tom wanted every advantage he could muster during the interview process this time. Tom told me he was ready this time for the ride like he'd never been for the one at United.

There was one more call from Tom last week, although it was hard to understand what he was saying because he was yelling and screaming so loudly. Tom had just been awarded a class date with that major airline. I asked him whether the money he spent on the training was worth it. "Best money I ever spent," Tom replied.

The Schedules

Major airline schedules, as well as the pay, are one of the reasons why so many people want these jobs. Schedules can offer as few as 8 days off per month or as many as 20, depending again on the routes, the aircraft flown and a pilot's company seniority. Most major airlines operate under FAR Part 121 in the United States, which prescribes a maximum of 100 flying hours per calendar month for a pilot. But before you run off to tell your friends how little you'll have to work each month, understand that 100 flying hours relates only to the flying portion. Another important consideration is how many hours of duty, which includes preflight and post flight chores, and sitting around between legs of a flight or around a hotel on an overnight.

In general, a pilot who flies ORD-HNL—Chicago, Honolulu (about a 16-hour round-trip flight) might only make the trip about four or five times a month before they have flown up to that 80-hour guarantee. The flight probably required a minimal amount of sitting time because, once off the ground, you fly until you land in HNL and go to the hotel for your rest period. On the return, it's basically the same thing back to ORD. Pretty cushy job, most likely flown in a Boeing 747 (Figure 6-4) or a Boeing 777, so the pay is good. Obviously, these routes are highly sought after because the time off is tops. This is where a high-seniority number will be worth its weight in gold. Without it, anyone more senior can outbid you.

A more typical domestic route might be ORD-HOU-MEM-DCA—Chicago, Houston, Memphis, Washington, D.C.—and then overnight and, possibly, DCA-ORD the next day. In a Boeing 737, for instance, the pilot will most likely have flown about seven hours in two days. What you

Figure 6-4 Cathay 747. (Courtesy Dean Heald: www.jetphotos.net.)

might not notice at first glance, though, is in Houston, the crew sat for two hours before the flight to MEM, where they might sit for another hour-and-a-half before heading to DCA. This can make for a pretty long day, especially if you run into any weather or traffic delays along the way. You have various methods of choosing schedules each month, based on days off, maximum flying time, minimum flying, certain days off, certain bases, and so on. Today, the work of choosing your airline schedule, called *bidding*, has been significantly reduced because much of the tedious sorting work required years ago can now be accomplished with a PC at home (more in Chapter 8).

Rather than printing reams of paper with individual lines of flying to consider, pilots now normally bid their monthly flying online using software that easily considers all the variables to the schedules that reflect that pilot's own values—and dozens and dozens exist. One airline system contains over 100 variables, in fact.

For example, if pilots want to fly weekdays and have weekends off, they can make that a high priority. If they want trips that begin midday—great for commuting pilots—they can make that a priority, as well. If they want maximum takeoffs and landings each month or minimum approaches, they can make their bid reflect this.

The airline then builds the lines of flying for each pilot based on their desires and their seniority. The line the pilot eventually receives is as close as the computer can organize it.

Duty and Trip Rigs

The major airlines, unlike the regionals, provide their pilots with pay from the moment they leave their base until the moment they return. Some regionals only pay their pilots when they fly, although this is beginning to change. If the aircraft breaks down or gets weathered in somewhere, they don't get paid. Major airlines use two systems to assure proper pay to the crews: the duty rig and the trip rig.

The *duty rig* assures a pilot of a specific amount of money for a certain amount of time on duty, regardless of how much actual flying was involved. For example, a 1:2 duty rig means one hour of pay for each two hours on duty. If a pilot signs in for duty at 11 A.M. and flies one hour to the destination, and then sits for two hours before flying back home, they only flew two hours total. But, if the pilot signed off duty at 5 P.M., they were on duty a total of six hours. But, the pilot will be paid on a two-hours-duty-equals-one-hour-pay duty rig, so they are paid for three hours.

A *trip rig* is similar in that it pays pilots a minimum amount for a given time away from their home base, regardless of how much flying was involved. If a trip rig were, say, 1:3, a pilot would be paid one hour of pay for every three hours away from base, regardless of the time flown.

Additionally, pilots are paid an hourly per diem allowance when they're away from base, and they are also eligible for various free-space-available passes on their own or other airlines.

The Big Bucks

This is what everyone calls it. An airline job is where pilots believe they'll find their fortune, so let's take a quick look at some of the money involved in this end of the industry, with figures from the FLTops.com pilot salary survey.

This is pay information for the 15 major airlines, soon to be 14, after America West and US Airways complete their integration. The top three airlines related to pay are freight carriers.

- Average Starting Pay—$36,883
- Average Sixth Year First Officer Pay—$88,092
- Average Junior Captain Pay (Sixth Year) —$136,128
- Average Senior Captain Pay (12+ Years and largest aircraft)—$170,920

At the regionals, retirement plans—other than a 401K—are virtually unknown. At most of the majors, retirement is a serious affair. The two major plans are the A-Fund, a defined-benefit retirement plan, and the

B-Fund, a defined-contribution retirement plan. The *A-Fund* pays a defined amount each year to the employee beginning at retirement, based on the pilot's earnings. The *B-Fund* gives the pilot a monthly retirement sum based on money added to the account over the years by the company, the employee, and accrued interest. One major change to the industry since 9/11 and the rash of airline bankruptcies is the elimination of many of the A & B Fund defined-benefit plans in favor of 401Ks, which are less costly to the airline.

The choice of airlines is still substantial, although many of the old industry players—such as Midway, Braniff, Eastern, and Pan Am—have departed. While all the jobs those carriers sucked from the aviation industry have not returned, there's hope on the horizon with the new start-up carriers. Where some people were astonished at the thought of regional pilots paying for their own training, some start-up airlines may also require various sorts of financial concessions from pilots just to pick up the job. Pilot pay at most of the start-ups is also considerably less than most major airlines.

I spoke with Peter Larratt, a Boeing 737–200 and 737–400 training Captain for British Airways. I asked if he would recommend this career. "Yes," Larratt said. "There are really several aspects of flying professionally that appeal to me. One is the initial challenge of flying large aircraft. Then, the challenge of managing the entire operation. Each day, a new job is started and you see the job through to a finish. Generally, too, you don't take the work home with you. You're essentially your own boss, even when you're a junior member of the crew." When I asked if he had any tips to pass on to future generations of pilots, Larratt said, "For the first few years, maintain a low profile and learn your trade well. You might be the best aircraft handler in the world, but that's a very small part of the job of a professional pilot. You can't teach experience" (Figure 6-5).

Labor Organizations

The airline industry is probably one of the most unionized of all industries. During the course of your career, expect to confront the decision of whether or not to join a union at most U.S. and many international airlines. Corporations and their pilots tend not to be union, however, leaving negotiations for pay and benefits up to the individual pilots themselves, just as you'd expect in a regular job. Also, flight instructors are seldom unionized and you won't find unions in pipeline patrol or banner towing, because those operations are just too small—often just a few pilots and aircraft. Regional airlines are mostly represented by unions as well. At the majors and many of the U.S. regional airlines, pilots are represented by the Air

Figure 6-5 Ryanair Boeing 737. (Courtesy The Boeing Company.)

Line Pilots Association (ALPA), while some carriers, such as Southwest Airlines or United Parcel Service pilots, have their own in-house unions. One noteworthy exception to ALPA representation is at American Airlines, where pilots are represented by the Allied Pilots Association (APA). However you might feel about unions, you're probably going to have to deal with one if you remain in the aviation industry long enough, and that can bring on some interesting choices.

A list of most of the airline industry unions' web sites can be found in Chapter 8.

Unions: Can Pilots Exist Without Them?

Gregg Watts developed his love for flying a little later than some—age 23—when his older brother Glenn, an agriculture pilot, started letting his younger sibling hang out with him at the airport. Glenn took Gregg along once when he sprayed for mosquitoes in an ancient Aztec. One ride at 200 feet at 160 mph and the younger Watts was hooked. Gregg won his private pilot's certificate in six months and worked two jobs to pay for the rest of his ratings. It was no surprise to the rest of the family when Gregg's first paying job turned out to be as a spray plane pilot, too.

A few years later, Gregg's career took him to the left seat of a pewter gray Beech 18, hauling night freight along the Gulf Coast for $1,200 a month—before taxes. Gregg upgraded to a Cessna Caravan for turbine time, and then to Lockheed Electras for Part 121 experience, each step a part of his plan to someday command a jet aircraft under Part 121 himself. In January 1991, Gregg Watts began his current professional sojourn with Southern Air Transport, a worldwide freight carrier, where he now flies in the left seat of a DC-8. It's been a dream come true—almost.

"When we decide to become pilots, we know up front that we are going to miss some things at home—our children's birthdays, our anniversaries, graduations, and even weddings sometimes," Watts said. "It's just one of those realities of the job. But there also has to be some balance between giving your company what is fair and seeing your family." An aviator's lifestyle can be rugged and often takes its toll on even the strongest families.

"On one trip, I was told I would be gone 30 days straight. I knew that, and I was prepared for it. When the first 30 days ended, the company told me I'd be out another 30 days because there was no one to replace me. When I protested, the Chief Pilot told me he had 400 résumés on his desk and asked me what I wanted to do. I flew the trip."

Strained Relationships

The aviation business is tough—on pilots and on managers. Both are often forced to push the other in directions that wreak havoc on professional and personal relationships. But, when one group possesses authoritarian powers over the other, the system can swing out of balance quickly. Pilots often feel they can only maintain their equilibrium in one way. What Watts termed a "change in management's approach to getting the job done," brought Southern Air Transport pilots to the doorstep of the International Brotherhood of Teamsters (IBT). Now the bargaining agent for Southern pilots, the IBT is currently negotiating that group's first labor agreement.

Trade unionism has been a part of the American work culture since the late nineteenth century when Samuel Gompers, an early president of the American Federation of Labor and the Council of Industrial Organizations, fought the early oratory battles with the titans of the new industrial revolution over higher wages, shorter hours, and more freedom for workers. Much of what Gompers accomplished has been carried down through the years by thousands of other workers, who have been building the union movement piece by piece.

Union Numbers Declining

At its peak in 1945, approximately 35.5 percent of the American work force was unionized. Today, union membership—still locked in a two-decade tailslide—has declined to 14.5 percent of working men and women. In 1995 alone, unions lost 400,000 members. Despite the total decline in union membership nationally, the U.S. airline industry remains a stronghold, one of the most unionized of all professions.

According to ALPA spokesman John Mazor, "at least 80 percent or more" of eligible pilots join that union. And, in the grand tradition of past union members, the aviation industry is still a platform from which thousands of pilots speak out each year about issues that help create today's pilot working environment, such as safety, duty schedules, and merger considerations. The recent standoff between American Airlines' pilots represented by the APA and the management of AMR Corp., the parent company of American, clearly demonstrates the passion of the movement.

Frustrated by stagnant wages in recent years, as well as conflicts over who will fly the new regional jets—mainline or commuter pilots at American's four regional carriers—American Airlines' pilots walked off the job at midnight on February 14, 1997. The highly charged job action lasted only a few moments, however, until President Clinton—citing that significant economic loss would reign—stepped in and ordered the pilots back to work. Clinton called for a presidential commission to study the problem and asked both sides to return to the bargaining table. The two sides eventually signed a new contract.

But American Airlines isn't the only airline mired in labor troubles. United pilots were initially unhappy about the pay raises they were recently offered. This is despite the fact that they, along with the airline's mechanics, possess a controlling interest in the carrier. In early March, however, pilots agreed to a new pay structure and turned down the pressure cooker that had begun boiling in January. This piece was written in the late 1990s, but is applicable today.

Additionally, a strike by one union can create difficulties for another union as the American strike did. Because the American pilots want to fly all jets and protect their own jobs, they could be putting the jobs of regional pilots at risk—pilots who are also union members, albeit a different union.

Also, these commuter pilots seemingly stand to earn higher wages if American Eagle—like many other large regional or national carriers—gets the nod to fly the jets. Continental Airlines pilots are also beginning contract talks, only their second in 12 years. Continental's pilots got their first contract in 1995. Northwest pilots' recent contract negotiations ended in a three-week strike. At USAirways (formerly USAir) intense negotiations continue to gain pilot concessions that will reduce the carrier's costs, still higher than those of competitors like low-cost Southwest Airlines.

Unions Offer "Protection"

One USAirways First Officer said, "Union membership is the only protection we have against the [Frank] Lorenzos, [Carl] Icahns, [Robert] Crandalls, and [Stephen] Wolfs of the airline industry," naming some of the more well-known airline chief executives of past and present. In a defeat for ALPA last fall, that union was ousted as the bargaining agent for the pilots of all-cargo FedEx, to be replaced by the in-house union called FedEx Pilots Association, led by newly appointed president, Captain Mike Akin. ALPA has since been reinstated at FedEx.

In the aviation industry, unions are pretty much a fact of life. But Bret Henry, a pilot for Horizon Airlines believes, "Unions help at some companies that have poor management, but when you have a leadership that values safety as a No. 1 priority, you don't need the outside help."

Becky Howell, a Boeing 737 Captain at Southwest Airlines and a member of Southwest Airlines Pilots' Association, an in-house union, said even though that airline is unionized, "We are fortunate enough to have a union that is very close to management. We still realize that we're in this together. Cooperation with one another gets the most benefits."

Continental Airlines Boeing 747 First Officer, Ed Neffinger, said, "Unions are not always necessary, but management generally forces them on to the property by their actions. Once a union is in place, the relationship is necessarily adversarial, but need not be confrontational."

If you join a union airline, you may hear the term "agency shop agreement," where pilots are not forced to join the union, but they must pay for the maintenance of the labor agreement in place at a particular carrier. This occurs because, even though a union's membership might not include all pilots, the union is still the sole bargaining unit for all the pilots.

Nonmember pilots have no vote and are not required to take part in union activities. At American Airlines, APA spokesman Tom Kashmar said, "There is no agency shop arrangement [present]. Pilots may choose not to be members of the union and pay no maintenance fee. However, only about 50 pilots—[about one-half of 1 percent]—of American's list of about 9,100 have chosen not to join."

What happens when you don't join the union at a union carrier? Legally, there should be no pressure on nonmembers. But, sometimes, that union member-nonmember relationship can become strained. Many months after the 1985 United Airlines' strike, nonmember as well as union member pilots who crossed the picket line to fly aircraft reportedly were ostracized in the crew room at most United bases. Small clicker devices were handed out to union pilots to make their feelings known. Anytime a pilot who crossed the 1985 picket line entered the lounge, dozens of clickers went off until all heads turned to acknowledge the offender. Twenty years after the strike, some animosity remains between the two groups of pilots. Some

Captains simply limit conversations in the cockpit to essential business when flying with pilots who crossed the picket lines in 1985.

The probationary period of one year—even at an airline—is based on the Railway Labor Act. During this time at American Airlines, pilots do not pay dues and do not enjoy the protection of the contract in force at the time. A pilot may be represented by a union officer—usually another regular line pilot—in meetings with the carrier's management on request, however. At the end of one year and provided the pilot passes their 12-month check ride, they become full dues-paying members of APA.

Dues at carriers vary considerably, too, but ALPA carriers charge about 2 percent of a pilot's base pay per year, while at other newer union carriers, such as Southern Air Transport, dues are expected to cost 1.25 percent of base pay per year. Dues at FedEx's in-house union, for example, are one flight hour's pay per month.

Many pilots believe the primary benefit of union representation is better wages. It would be misleading, however, to categorically state that unions always mean higher wages. Pilots at American and United are concerned that, despite representation, their salaries have not kept pace with the changes in cost of living. In May 2007, American Airlines pilots demanded a 30 percent wage and benefit increase to cover past losses.

Role in Safety

Most unions also display a formidable role in aviation safety. ALPA's Safety Department numbers 15 full-time staff people, while ALPA field workers—line pilots volunteering their time—add almost 600 more workers to the cause.

Unions offer pilots protection in disputes between a cockpit crewmember and management. The union representative stands in on such meetings as a witness to what is said and also to ensure all provisions of the current labor agreement are followed closely. If there should be an FAR violation, union pilots enjoy the availability of legal assistance to mediate between themselves and the FAA. Unions usually offer pilots access to various insurance programs as well, such as loss-of-license insurance.

But one of the major advantages to flying for a unionized carrier is the labor agreement that outlines exactly what a pilot can expect from management and what kinds of behavior management may expect from pilots. An example might be the limiting of a duty day to 14 hours. The pilots will not remain on duty longer than that, and management should not ask them to do so.

An advantage of most contracts is they are hammered out locally, by locally appointed negotiator pilots working with management, not members of some union hierarchy simply flown in for the occasion. This also aids in contract enforcement later, because many of the same parties will come together during this phase, as well. "First contracts are often modest and can take a year or more to negotiate," said ALPA spokesman Mazor. "But they become something to build on as both pilots and management get more comfortable with the process."

If a downside exists to pilot unions, it is that they must represent all members of the bargaining group, even those who might not be superb pilots. One Midwestern regional carrier Captain's poor flying habits were so well known among the pilot group that many First Officers preferred not to fly with him. When his employment was terminated for another allegedly near-FAR violation, the union was forced to negotiate the rehiring of this pilot, much to the dismay of his coworkers.

Another potentially unpleasant aspect of unionization is "that some pilots tend to want someone else (the union) to manage their career," said one pilot at a major carrier. One Airborne Express DC-8 pilot said, "Nonunion pilots with little knowledge of unions and their role in the industry [are the biggest threat]."

There are other threats to unionized pilots as well, according to Shem Malmquist, a Boeing 727 Captain at FedEx. These include code-sharing passengers and freight on foreign airlines. "It's incredible to me that a passenger can buy a ticket on most any major U.S. airline, climb aboard a jet operated by a foreign flag carrier, and go someplace and return without ever setting foot on an aircraft operated by the airline named on the passenger's ticket. It's a very ominous trend."

Another practical example of a union's power can be demonstrated over the issue of pilot pushing, such as Watts being forced to fly 60 days as a good example. In another, a Part 135 nonunion carrier allegedly attempted to force one pilot, currently a First Officer with USAirways, to fly a Cessna 402 with an engine that would not develop full power. When the pilot complained, the Chief Pilot allegedly told him to either fly the airplane or find new employment.

Unions Can Change

Sometimes, one union is ousted as the bargaining agent in place of another, such as the case at FedEx late last year. ALPA was out and the FedEx Pilots Association was in. Some of the union animosity began at FedEx when it merged with Flying Tigers in 1988. Tigers was represented by ALPA, and FedEx was nonunion. Eventually, however, ALPA did win at FedEx, much to the dismay of many pilots there who were staunchly nonunion. As with many merged carriers, the split between the two pilot groups widened as ALPA attempted to negotiate its first contract with FedEx management.

Eventually, the pilots began arguing more among themselves than with management, putting FedEx management in an enviable position.

ALPA membership never climbed above 65 percent during its tenure at FedEx. In late summer 1995, the organization of the FedEx Pilots Association began as pilots decided they wanted representation at FedEx, just not ALPA. ALPA eventually won the election, certified on Oct. 29, 1996, but much animosity reportedly remains in FedEx cockpits today over the issues stirred up during ALPA's tenure. Not long before their contract expired in 2001, the FedEx Pilots Association decided to merge with ALPA, once again returning over 4,000 pilots to the union's membership roster.

Despite a show of unity during contract negotiations in 1998, 98 percent of FedEx pilots joined the union—when both sides arrived at the stare-down point in those negotiations, the pilots backed down under FedEx CEO Fred Smith's threat to restructure the entire company and leave the pilots out in the cold.

Logically, career decisions about whether to join a union shouldn't be made with only the heart, but also with the head. An essential part of being well informed is reading publications that are not prounion, which offer distinctly different perspectives of both management and labor. In *Air Transport World*, a magazine aimed primarily at airline executives, the publication's editor, Perry Flint, called United Airlines' pilots greedy, believing their request for hefty pay raises abrogated their agreement signed during the leveraged buyout of the airline in 1994.

Is a union in your future? If you remain in the aviation industry, that answer is probably yes. But whether you take an active role in a union can only be answered with a maybe. A union's strength depends on the active help of many pilot volunteers at the local level, many who are not used to directing their own destiny to the degree that a union membership can thrust on them. But, without that help, without that involvement on an individual level, there is no union.

The Upgrade Decision

Captain, oh my Captain! Crew scheduling says the airline will choose ten First Officers for upgrade to Captain in the next few weeks and your seniority number puts you in the middle of the potential group of candidates. You finally have the opportunity to make your own decisions using the knowledge you've gained over the years of sitting in the right seat. You're going to be in charge.

"The ego can be a big deal for some people (in upgrading), also the pleasure of being the boss and running the cockpit the way you like it," said Malmquist. The decision to upgrade is an individual one, a decision made by most pilots based on a variety of factors. It also turns out that upgrading to the left seat, as well as that fourth stripe, is an opportunity some pilots choose not to accept if their airline employer allows them that choice.

Years ago, when the upgrade opportunity first appeared, the conclusion was that a First Officer who had been sitting in the right seat for 10 or 15 years would jump at the chance. Bill Traub (now retired) reports, "Since the average upgrade time is about seven years here, it is not very common for a pilot to turn an upgrade down. In fact, we delve into this issue during the initial interview. Those who think of their profession first are the kind of person we look for." Ken Krueger, director of flight operations at Midwest Express—a single-hub airline—said, "We've not seen a time when someone would not upgrade when offered the opportunity. But a new hub (the company recently added Omaha, NE) may change all that here," he added.

Today, however, the upgrading pilot faces a "Pandora's box" of questions and decisions that could force that pilot to choose to remain in the right seat. If money were the only issue, one B-747 pilot summed it up nicely. "The object of the game (whether to upgrade or not) is to maximize pay without working. Period! If you can make more money sitting shotgun, then it's a better deal."

The chance to command an aircraft is expected to be a part of most First and Second Officer's career plans. Indeed, most First Officer's job descriptions portray them as a Captain in training. Traub said, "We hire pilots to be Captains at United and we like to see them upgrade at the earliest possibility. But, we won't argue if a pilot's family or possible commuting problems would prevent them from upgrading." American Airlines also hires its pilots to be Captains, not permanent First or Second Officers.

Although airlines like American, United, and Southwest hire only potential Captains, each airline expects its pilots to upgrade at a different time. Some companies also are willing to allow a First or Second Officer to remain permanently in a noncommand position, although somewhat reluctantly. At United, for instance, if a pilot chooses to remain in the right seat of a B-767 for the remainder of their career, the airline allows that as an option. Some pilots change their mind about upgrading later, too. Traub said a problem with pilots who remain too long as a First or Second Officer is "people that stay an extended period in one seat often find the upgrade process difficult. They tend to stagnate. Their command skills just don't develop as a First Officer. Some Second Officers have almost forgotten how to fly, as well as how to command."

At American Airlines, a pilot's right-seat time is limited. "You must upgrade within one year of the time when a pilot junior to you chooses the left seat," said American B-767 First Officer Gary Mendenhall. Southwest

Airlines' Chicago Chief Pilot Lou Freeman said a pilot cannot choose to remain indefinitely in the right seat. "Everyone here is Captain-qualified when they walk in the door. We hire you to be a Captain." At Southwest, a pilot can delay the upgrade process if, as a Junior Captain, he would have to hold a reserve line. But once a Southwest pilot can hold a regular line schedule, they must upgrade. The only other exception: Southwest will not force a pilot out of their domicile to accept an upgrade in another city.

Continental Airlines does not force its pilots to upgrade. Many of the airline's Los Angeles–based, DC-10 First Officers could hold a Captain's schedule in another city, but choose to remain in L.A. because they enjoy the southern California lifestyle. USAir pilots also won't be pressured to move to the left seat until they choose to do so.

Many pilots believe the decision to upgrade is one they wanted to make on their own terms, or at least as much of their own as a seniority system would allow. Few said they did not want to become a Captain at some point in their career. One of the first items most pilots consider in the upgrade decision is the salary increase a move to the left seat brings them. At Midwest Express, First Officer's pay skyrockets nearly 70 percent after an upgrade to Captain. The pay raise for a United pilot to upgrade to the left seat initially can range from about $35,000 for a Senior First Officer to as much as $75,000 for a junior one. Continental pilot Kaye Riggs said his position as a Los Angeles–based, DC-10 Second Officer pays "about $50 per flight hour." A junior Captain at Continental earns more than $80 per flight hour. At USAir, B-737 Captain Bob Gaudioso said, "a senior B-767 FO could get paid more than an F-28 Captain. The incentive to upgrade, then, is very small."

The high-end pay rates at the regional airline level are considerably less than the majors, making the difference in pay after upgrade considerably smaller, as well. The importance of the upgrade at the regional level, then, might not seem as important. First Officers flying a sophisticated aircraft, such as an ATR-72—a new generation glass-cockpit turboprop—may qualify for food stamps. Turboprop pilots seldom pass an opportunity to upgrade because of financial considerations. Pilot Robert Lastrup confirmed: "The regionals are not known for paying their pilots very well, therefore, without the money, there is no lifestyle choice." According to BE-1900 pilot Dave Oberlander, another factor in the upgrade decision for a regional pilot is the "need to build valuable [pilot-in-command] time. I also think that once a regional pilot makes over $40,000 per year, they will also choose lifestyle over money."

The pay disparity between left seat and right in corporate flight departments is not as vast as in the airline industry, because beginning pay rates in an aircraft, such as a Lear or Falcon 10, are substantially more than their first-year airline counterparts—in the upper-$30,000 range to the low-$50,000 range.

In the corporate world, however, the upgrade process operates some-what differently. Many companies hire First Officers who are either already type-rated in the aircraft the corporation flies or pilots who they believe will easily achieve that goal. Because many corporations fly with a Co-Captain arrangement, achieving left-seat status in a corporate flight depart-ment also does not seem to carry the same sort of need to command as the airline industry, especially because most Co-Captains usually exchange seats on different legs or days of a trip—something unheard of in the air-line industry.

A corporate flight department upgrades its right-seaters after they gain a certain total time in the aircraft, which varies greatly from company to company. But, a corporate flight department normally makes upgrade choices based more on pure talent, with an eye toward the bottom line, and not so much on company seniority. When a pilot is deemed ready for upgrade training, they go. That plan does have its natural checks and bal-ances, too, however. Falcon 50 pilot Valerie Dunbar said, "I needed the complete concurrence of all eight of the other pilots in my flight depart-ment before upgrade training was approved."

At the major airlines, the upgrade decision usually is based on lifestyle. But lifestyle means different things to different pilots. Most want to earn the largest salary possible with the least amount of time away from home. And while more money can often buy a better lifestyle, if pilots must commute to another city to hold a Captain line, the extra money may not be worth it. For instance, Riggs is based in Los Angeles by choice. A resident of San Luis Obispo, Calif., he commutes about an hour to Los Angeles for his trips. He could hold a DC-10 First Officer position in Newark, NJ, or Cleveland, but chooses not to. "To me, the lifestyle of California is more important than the money," he said. A Newark line would offer more than the 12 days off a month he now has as a reserve pilot in Los Angeles, but "If I were based in Newark, I could spend as much as five hours just getting to work. And commuting to Newark is difficult because the flights are always so full." He also said he would not be interested in living in either New Jersey or Ohio, so com-muting would be a must. Riggs estimates 70 percent of Continental's pilots commute to work.

L.A.-based regional pilot Andrew G. Weingram said, "I already fly banker's hours, 110 hours per month, based in [Los Angeles]. If I become a Las Vegas Captain, I'll be getting up at 4 A.M. to work 16-hour days to fly 60 hours a month. No thank you!"

Besides the personal time lost, commuting to grab that left seat costs money. A Junior Captain and a reserve line holder usually is required to live within a two-hour call of the airport. If they happen to reside in anoth-er city, this means, "a pilot could possibly spend five nights a week in a hotel waiting for the phone to ring," Freeman said. That is not going to be a cheap. Some pilots choose to share the expenses of a small apartment in their domicile city with several other commuter pilots. These "crash pads" help defray the cost of living part-time in another city, but on a night when most of the pilots who share the apartment all need a place to stay, the quarters can become quite cramped. Sleeping bags and cots are common fixtures in crash pads.

Robert Lastrup said, "Commuting is a necessary evil. With so many changes in the airline industry . . . it's not really practical to move with every job or domicile change. Commuting is very stressful, too. For instance, you just get home from a trip and you're already planning your strategy for getting back to work. What about the weather? Should I leave a day early? You may even lose a whole day commuting on both sides of your trip. Commuting, as I've explained it, will never be worth the money."

Robyn Sclair, a Detroit-based DHC-8 Captain, said she passed on her first upgrade opportunity in the SA-227 Metroliner because "I just didn't want to be a Metroliner Captain. In general, I was just tired of the airplane." Sclair waited an extra six months until her seniority allowed her to move from the right seat on the Metroliner to the left seat on the DHC-8. She saw her pay "just about double" when she passed her check ride. She also said, "moving to another domicile for the Captain job would not have been a problem . . . but maximum days off is also very important." Her Northwest Airlink airline, Mesaba, has no policy to force a pilot from the right seat to the left. Dunbar flew as Co-Captain on a Falcon 50 and a Hawker 800 for Phillip Morris until some recent cutbacks in its flight operations.

"In our department, the upgrade process was not that competitive. Everyone upgrades eventually," Sclair said. Although the firm had bases in other cities, "each one was autonomous. Seniority at one base did not necessarily allow you to take a slot at another if you wanted to." Sclair felt airline seniority is predictable, while corporate seniority and its outcome on the upgrade process were just not as well defined. Most corporate pilots, like Dunbar, are salaried employees, so their pay is the same if they fly 10 hours per month or 110.

Southwest's Freeman said he could not recall a pilot who did not choose to upgrade when offered the opportunity. "A Captain at Southwest is the greatest job in the world. We let you take an airplane out and just do your job. We pay you well, you make your own decisions, and there's no big brother looking over your shoulder." But, the upgrade decision is not always so easy to see. At American, Mendenhall said, "As a First Officer on an international wide body, I have a good schedule with 18 days off per month. If I accepted a position as a Junior Captain, I'd have 10 days off per month [flying] an F-100. I make as much money now as a Junior Captain. To me, the schedule is more important."

And, last, some pilots won't upgrade because of a fear of failure during the training process itself. America West B-737 First Officer Ray C. Phillips said, "We have First Officers who won't even attempt upgrade. Some have tried and failed, and simply won't try again." Freeman said Southwest had few failures in the upgrade process. However, "If you fail the first upgrade, you'd go back on the line for 90 days. If you fail the second, you'd be terminated from the company." Dunbar also remembers corporate pilots whose attempted upgrades ended in failure and termination from their job. A United pilot who fails upgrade training returns to the right seat for 18 months before getting another chance to become a Captain. Although Traub also reports second failures to be a rarity, "we'd also be much less inclined to offer a pilot a third opportunity to try."

When pilots get the opportunity to sew that fourth stripe on, they have more decisions than just how to spend the money once the training is completed. The decision to upgrade or not involves many factors about lifestyle—money, days off, commuting, and so forth. A command position is the ultimate goal for most pilots, but pilots should be certain the first decision they make as a potential Captain is whether becoming a Captain is right for them (before they begin the process to upgrade). But, at some airlines, such as American, once pilots prove their ability to upgrade, they have the opportunity to bid back down to a First Officer where schedules most probably will be better—if pilots' egos allow them to go back to being a First Officer, that is.

Keeping an Eye on the Ball

To be certain readers have a feel for what the airlines are telling the world about their pilot requirements, here is the latest pilot-hiring information from JetBlue, the Jamaica, NY–based, low-cost carrier that competes strongly with Southwest Airlines.

JetBlue Airways

REQUIREMENTS: First Officer (posted 05/02/2006 Jamaica [JFK], NY)

POSITION SUMMARY: The First Officer is responsible for the safe, consistent outcome of a flight. The First Officer assists the Captain in ensuring a safe outcome and a positive JetBlue experience. In addition, the First Officer maintains compliance with applicable Federal Aviation Regulations (FARs), and JetBlue Airways policies and procedures.

ESSENTIAL FUNCTIONS:

- Assists the Captain in ensuring the safe outcome of a flight in accordance with all Federal Aviation Regulations, and Company policies and procedures
- Safely operates the aircraft in accordance with all Federal Aviation Regulations, and Company policies and procedures
- Provides the JetBlue experience to all customers
- Maintains skills, training, and qualifications in accordance with all Federal Aviation Regulations, and Company policies and procedures

MINIMUM QUALIFICATIONS:

- 1,500 hours total time in airplanes (excluded: helicopter, simulator, Flight Engineer time)
- 1,000 hours turbine in airplanes
- 1,000 hours Pilot in Command Time*
- 1,000 hours in airplanes at or above 20,000 pounds (maximum takeoff weight) or 1,000 hours in large, jet-powered airplanes at or above 12,500 pounds (maximum takeoff weight)
- Recency of flight experience will be considered
- Federal Aviation Administration (FAA) Airline Transport Pilot (ATP) Certification
- Current FAA Class 1 Medical Certificate
- Federal Communication Commission (FCC) radio license
- Valid passport with the ability to travel in and out of the United States
- Three reference letters from pilots who can personally attest to the candidate's flying skills (must bring the originals to the interview)
- Vision corrected to 20/20
- High school diploma or General Education Development (GED) diploma
- College degree preferred, but not required
- Regular attendance and punctuality
- Organizational fit for the JetBlue culture, that is, exhibit the JetBlue values of Safety, Caring, Integrity, Fun, and Passion
- Well groomed and able to maintain a professional appearance
- Pass a ten (10)-year background check and preemployment drug test
- Legally eligible to work in the country in which the position is located

- When working or traveling on JetBlue flights, and if time permits, all capable Crewmembers are asked to assist with light cleaning of the aircraft

*JetBlue will only consider PIC time when the Pilot has signed for the aircraft. Please use only this time when completing the areas of the application asking for PIC time to be entered.

Candidates who have previously interviewed will be ineligible for another opportunity to do so for a period of six (6) months from the date of their last interview.

Competitive Qualifications

Our computers surface the most competitive applicants in the database. These files are then reviewed by a team of line Pilots. Because of the high volume of applications we receive, competitive qualifications can be significantly higher than the minimum requirements. On average, we are currently seeing the following background in candidates selected to interview:

- Between 3,000 and 10,000 hours total time in airplanes
- Greater than 2,000 hours turbine PIC in jets
- Greater than 2,000 hours in airplanes at or above 20,000 pounds maximum takeoff weight, or in jet-powered airplanes at or above 12,500 pounds maximum takeoff weight
- Experience with more sophisticated aircraft utilizing Electronic Flight Instrument Systems (EFIS), Flight Management Systems (FMS)
- Bachelor's degree

Knowledge, Skills, and Abilities

- Knowledge of Microsoft Office, Acrobat Reader, and the ability to use browsers effectively
- Ability to speak, read, and write English, and to possess excellent communication skills
- Ability to think creatively
- Excellent interpersonal skills

7

There's More
to Flying Than
Simply the Airlines

Corporate Flying

Many corporations have lost valued pilots to the airlines in the past few decades, but today many pilots are thinking twice about their job choices right from the beginning. Corporate flying can be not simply professionally rewarding, but even more financially advantageous than working for the airlines (Figure 7-1).

Corporate aviation operates quite differently from the airline business, however. Some of those differences are why some corporate operators are not always keen to hire former airline pilots into their flight departments. For starters, corporations don't want to spend valuable training budgets on an airline pilot who they believe might leave at any moment to return to an airline cockpit if a new opportunity presents itself.

Another issue for pilots is business airplanes are often treated as a travel expense in private enterprise while the airlines make money—hopefully, enough to turn a profit—from shuttling around as many people as possible. Business aviation is simply an efficient means to shuttle company executives to the places they need to be when they need to be there.

An airline pilot seldom handles more of the flight workload than stepping up the stairs, turning left into the cockpit, shutting the door, and taking off. A business aviation pilot is responsible for making certain the aircraft is fueled and ready before departure, and they also handle all the flight-planning tasks, whether the trip is 500 miles or 5,000. They'll often be the people responsible for ordering the catering and keeping the airplane

Figure 7-1 Gulfstream G-450. (Courtesy Gulfstream Aerospace Corporation.)

stocked with refreshments. If the boss likes Pepsi, best not bring Coke on board. The pilots will usually load the bags, as well, so pilots with weak backs had better head for the airlines. Most of all, business aviation pilots are expected to interact with the people in the passenger compartment, so an engaging personality is a necessity.

Louis Smith, FLTops.com president, said, "Business aviation, and especially the fractional sector, are growing rapidly and [will] offer nearly as many pilot jobs as the airlines in the next 12 years. Job security is always an issue for pilots working for corporate flight departments, because they are always one merger away from a shutdown of their flight department. The fractional sector has successfully attracted large numbers of high-yield passengers from the airlines, which is one of the reasons the fractionals are able to provide competitive pay and benefits."

Corporate Pilot Profile: David Ball

Falcon 2000 Captain
A Chicagoland Fortune 50 company

David Ball grew up in the northwest suburbs of Chicago, watching airplanes sail overhead toward nearby O'Hare and Palwaukee Municipal Airport (now Chicago Executive Airport) (Figure 7-2).

"I was one of those kids of eight or nine and standing outside the fence at Palwaukee watching airplanes," Ball recalled. "My first goal was that I'd be a Navy pilot, then fly with the Blue Angels demonstration team, then spend the rest of my life flying for Delta Airlines." His career took a few different turns, however, as do most in aviation.

"I never joined the Navy and I never flew for Delta Airlines," Ball admits. The 40-year-old pilot now flies a Falcon 2000 for a Fortune 50 company. Ball says he almost missed business aviation as an option. "No one ever spoke much about flying cor-

Figure 7-2 Corporate pilot David Ball.

porate at school. This option used to be just an afterthought. Today, though, flying corporate is a major direction young pilots should be thinking about. You can make a good living and have a nice lifestyle as a corporate pilot. I'm very proud of my career."

Friends are surprised when Ball tells them that, despite an intense interest in all things aviation—"I read every library book I could find about airplanes"—David didn't take his first airplane flight until he was on his way to begin the search for the right college. "Although I was a little nervous on that first DC-10 flight from Chicago, I knew I wanted to fly. I never thought it might *not* work out, but it would have been helpful to have a mentor in my life." Ball said his high-school guidance counselor was a big help in staying focused. And, he also wanted to enhance his lifelong love of playing soccer. Ball eventually chose Parks College of St. Louis University for his academic training.

Looking back at his flight training days, Ball's interest in competitive sports helped him soar past a few learning plateaus. "I remember holding patterns were tough for me. Some of the instructors I had were not very supportive. I thought about playing soccer and said 'if they can do it, so can I.' I remember my school on the Falcon 2000 EZ, as well. So many people said it was going to be incredibly tough. Same thing. If others could pass the type rating ride, so could I. And I did."

Flight education anywhere is not cheap. Ball revealed that he made it through college with "no magical inheritance." He did accept some support from his parents, but much of the rest was in the form of "loans that I didn't pay off until after I was married."

When it came to that first flying job, Ball tried a technique few others consider: the Yellow Pages. "I found a directory that covered northern Illinois and Wisconsin, and sent résumés to every flight school that had a good size fleet and a charter department. A school in Milwaukee called back and I spent the next year instructing and flying around traffic reporters." He also added 500–600 hours to his logbook.

His next job evolved from pure networking when he connected with a friend from Parks who had a flight school at Palwaukee. "Three or four students per day, plus more traffic reporters built my hours quickly." Then, Ball moved on to his first real charter job at Milwaukee Mitchell. It only happened "because I bothered the Chief Pilot all the time," Ball recalls. He spent the next six years of his life with Scott Air Charter learning the flow of a Part 135 operation. Ball flew the King Airs and the Learjets, as well as picking up a type rating in the Challenger and Citation 650.

Ball tried a few airline interviews during that first ten years. "I knew the first one would never work out when the interviewer showed up two hours late." Then, there was the potential lifestyle change, which most likely meant uprooting the family and a huge salary cut. The guidance his Chief Pilot at Scott had given him came to mind again. "He told me he would make it difficult financially, in a good way, for me to leave. And he was right. I was probably making $60,000 at Scott. Some of the airlines I interviewed with would have started me at $25,000." It didn't happen.

Ball realized his future was truly in business aviation, but recalled a frustrating interview that confirmed he had yet to arrive. "I'd sent résumés to one corporate department for quite awhile before they called me in for an interview. I went to lunch with a few of their pilots, which I thought went well. When we got back to the hangar, I was suddenly faced with another interview with five or six more pilots, who threw all kinds of questions at me. I knew they were trying to rattle me about my talents as a pilot for their own internal political issues." Ball spent 4 hours on that interview. "Six months later I'd not heard one word and I called back. I didn't get the job."

After a few more years, Ball connected with a Chicagoland flight department, the old-fashioned way, "I used the National Business

Aviation Association (NBAA) member directory," a resource he highly recommends. The NBAA directory is available only to members, so plan on joining the association. Access to NBAA's online job board could be worth the price of membership alone.

"I sent my current Chief Pilot a résumé and a letter. One day, he called out of the blue and said, 'What are you doing these days?'" Ball's job flying a Hawker 800 was then on the verge of being eliminated. "The Chief Pilot and I seemed to click on the phone and he invited me for an interview. I've been with the company for six years." Ball said having a good résumé, being in the right place at the right time, and possessing good credibility as a pilot helped.

The company flight department includes two Falcon 2000 EX Easy and five other pilots. Ball says most trips are scheduled ahead of time and "99 percent of my flying is domestic." He might also fly to any city where the company has a major office. "Although our schedule has peaks and valleys during the year, we fly pretty much day trips only. I show up at the hangar about 5:30 A.M. and we leave for Boston, New York, or Washington by 7 or 7:30 A.M. We're usually home by 6 P.M. that same night."

By the time the aircraft is put away, the work day has stretched to about 14 hours. "This job has given me all the pension plans and stock options I could have ever wanted in a quality company." It does take him away from home, which pretty much leaves his wife Jodie to fend for herself with their three kids. "I try to take care of as much as I can before I leave, like cutting the lawn, filling the cars with gas, and even making waffles on Sunday morning, so the kids will have some during the week while I'm traveling."

Like many other young pilots, Ball had no relatives or friends in aviation who could act as mentors. "I've acted as a mentor for a would-be pilot, too. I showed him around the hangar and explained what my work day was like. I was never the kind of person who thought 'I'll only be a real pilot if I fly a particular kind of airplane.' I wanted a good living with a stable company. I also wanted to be home for little league with my family. I told him that even though I don't fly international on a 747, I have a career I'm very proud of. A good mentor doesn't make the final decisions, though. We only present people with the best information we can" (Figure 7-3).

Figure 7-3 Dassault Falcon 2000 on takeoff. (Courtesy Dassault Falcon.)

More Corporate Flying: An Interview with Janice K. Barden, President Aviation Personnel International

New Orleans, LA. API also has an office in San Francisco, CA— www.apiaviation.com.

Janice K. Barden has been telling aviation companies who to hire for 25 years. A veteran industrial psychologist, Barden has worked with many Fortune 100 corporations, in addition to airlines like Pan Am, American, Braniff, United, and National. She remembers that "in the late '60s, you couldn't get on if you were over 32."

"Right now," Barden says, "airline hiring is stimulating everything. A good corporate flight department is still one of the best kept secrets, because pilots keep these jobs for a long time. While I think corporate hiring will remain close to what it is right now, we'll be retiring many of these pilots and the corporations don't know yet where the replacements will come from."

Barden's company, Aviation Personnel International, is a research organization dealing in personnel. There is no charge to the applicant for API's services. "At some companies, we handle everything for a flight department, from hiring of pilots to hiring of mechanics and dispatch personnel." Barden said that "the difference between corporate flying and other types is that these pilots are going so many different places each day and doing all the work with the people in the back as well. Today, a corporate pilot needs to think more than just flying.

"Remember that a pilot is in a key slot when they are responsible for the top management people at a company. They should pick a good school to attend to get their basic education. The corporate world recognizes a degree from a good university." Barden added that most corporations today are also looking for pilots who see themselves in a flight department management slot in the future, as well, so advanced degrees are also becoming more valuable.

API has an extensive screening program for applicants that they only offer to "about three people out of a hundred who walk through the door," said Barden. "To find a good pilot, I believe we must study people—their weaknesses and their strengths. For a corporate pilot, attitude is very important. In fact, it is the number one factor. Is the pilot good with people, a good team player? Do they have good values and are they willing to work? Corporate flight managers look for résumés that have some depth to them."

But Barden also knows physical fitness is important. "We haven't placed a smoker, for instance, in many years. We'll ask, 'Do you work out? What is your cholesterol level?'

What else does it take to begin a corporate piloting career? "For an entry-level position at the corporate level, a pilot should have a degree," Barden says. "A business degree will be an asset, too, as well as good technical and interpersonal skills. We test them and learn about their leadership potential, too. They have to believe in themselves and be self-directed. They need to be easy to get along with. As far as their flying skills, they need an ATP and should have been exposed to some great flying experience at this point. That's where charter or regionals can really be some great training grounds. A total time of 2,000 hours—at least—in turboprops and jets is the best."

Many pilots may not be considering a corporate piloting career for a number of reasons, which Barden believes she already understands. "I think that years ago, there were war stories about flying in the middle of the night, but that seldom happens anymore. Most of these companies operate under Part 91, but fly to Part 121 standards. This is a good businessman career. I think there is also a great pride of ownership in being a part of a corporate operation. They are a very close-knit group. They are only as strong as every one of the links."

But don't expect the road to hiring on at a corporation will be easy. "When I do career counseling, I tell pilots this is one of the hardest industries to break into," Barden said. "They've got to pound the pavement and make their own opportunities. They must develop their own network."

Corporate-owned and operated aircraft come in all shapes and sizes, from the low end of a Piper Navajo or Beech Baron on up through King Airs, Cessna Citations, Falcons, Learjets, Gulfstreams, Challengers, to even airline equipment. Schedules in corporate flying are sometimes a bit more unsettled than at the airlines and somewhat similar to charter flying. That means days off are not always cut in stone, but they are much more firm than a charter pilot experiences.

Some large corporations employ a crew scheduler and, often, the crews know a number of days in advance who is scheduled in what airplane to go where. Sometimes, though, the flight could be a last-minute trip scheduled because of some unforeseen business situation. Corporate flying can also involve a great deal of sitting around while you wait for the boss to complete their business. But, because most corporate pilots are salaried, you're not losing money. The extra time could provide you the opportunity to catch up on your reading or even do some work for a business of your own. I never travel anywhere without my trusty laptop computer with its word-processing software stored inside. Part of this book was written on a two-day layover I had during a Cessna Citation trip.

Corporate flying salaries have always trended higher than charter flying. And, with the growth of business aviation today, more people are after an ever-increasing supply of business flying. In my career as a corporate pilot, I managed to fly some pretty nice airplanes—Cessna Citation 500s and 650s, as well as the Hawker 800A. As a magazine journalist, I've also had the opportunity to fly aircraft I would most likely never have had the chance to fly if I'd only worked for a single company, such as the Boeing BBJ, the Embraer Legacy, the Sino Swearingen SJ-30, the Cessna Citation Mustang, the Pilatus PC-12, and more (Figure 7-4). I've added a few Pilot Reports in Chapter 9 to give you a better idea of what you might experience when it's your turn.

Figure 7-4 Sino Swearingen's SJ-30 offers one of the fastest long-range aircraft in service. (Courtesy Sino Swearingen Corp.)

Education and the Future of Business Aviation

Reprinted courtesy *Aviation International News*—www.ainonline.com

An important part of the never-ending debate between airlines and business aviation has often been heavily focused on the requirement for a college degree until the past few years when the supply of university-educated applicants began to dry up. Because supply-and-demand dictated hiring more people without a formal education, the industry looked toward people who have worked their way up without college. But a degree can, and should, be much more than simply something a pilot uses to check off a box on a job application.

What makes a degree worth more than simply the paper it is printed on is the value it delivers to both the pilot and their employer. The need for people with more formal education to lead this segment of the industry, as well as the long-term changes in the job market for people employed in business aviation, may soon cause aviation department managers—and their bosses—to reconsider educational qualifications in a way they never have before. The future for corporate aviation may soon witness an unbending requirement for a college degree in aviation.

According to Dr. Tom Carney, professor and associate chairman in the Department of Aviation Technology, Purdue University, "One could argue that, sure, if you're technically qualified, you can be perfectly safe. But with safety being the spectrum it is today, the right education will add dimensions to a person's capability that are beyond those needed to fly an aircraft, but are harder to define. The ability to think more clearly, express one's self in a professional manner, orally and in writing, is critical to success today."

In what may already be a cultural trend toward better-educated managers and pilots, the 2000 NBAA Operator Profile and Benchmarking Survey showed nearly half the companies with sales of more than $2.5 billion required managers at the Chief Pilot level and above to hold a four-year degree. In companies under $100 million in sales that number drops to 22 percent. In the pilot ranks, 25 percent of the largest companies in the nation required their Captains to hold four-year degrees. At companies under $100 million in sales, the figure drops to 11 percent. For copilots, the number requiring a degree ranges between 6 percent at the smallest companies to 18 percent that require a degree for the largest companies. Those numbers are expected to rise.

Traditionally, the most senior pilot in a flight department was eventually promoted to the Chief Pilot or flight department manager position, under the assumption that the best pilot would make the best manager. This often proved to be a fatal mistake for some flight departments, as results have shown. Jeff Lee, IBM's director of flight operations and the former chair of the NBAA Corporate Aviation Management Committee, said, "The people who lead an aviation department in the future may not necessarily be pilots. They could certainly be mechanics, as well. The skills to lead have very little to do with flying an airplane or turning a wrench." Kevin Harkin, aviation department manager at Corporate Air Service, added, "If you want to go beyond flying, you need the manager's skill set, as well."

Today and in the future, pilots face the choice of whether to continue only flying airplanes for the rest of their careers or to move into areas that may offer new challenges and additional responsibilities. But, as Lee added, "Like mechanics, pilots often don't get a lot of training in how a flight department actually works. They're those who are pushed out of their comfort zones in a department and are totally lost. What I'm a bit frustrated by are that few pilots really want to manage and right now have only a pretty narrow look at how a department operates."

Part of the problem stems from confusion between the rationale of a degree and an education. The two can be mutually exclusive because a pilot might pick up a degree from any of the hundreds of degree-granting organizations around the country, academic or not, that simply deliver a piece of paper. But those degrees do little more again, than allow the pilot to check a box that says they possess a degree. The difference between an education and a degree is, with the right education, a person's mind is trained to view life and the solutions to life's problems differently than those with a less-formal education. They think differently.

While the precise purpose of a university degree varies with the individual, experts pretty much agree a correlation exists between a good formal education and a pilot's ability to successfully complete ground school, but little connection has ever been proven between a college degree and a pilot's ability to fly an airplane. Many career counselors tell a pilot to get the degree and to get a degree in some field of study they are interested in. The underlying meaning is, when the cyclical wave furloughs a cockpit crewmember or manager, they'll have some skills to fall back on to earn a living with while they await their return to the cockpit. By the time you read this some 8,000 major airline pilots will be on furlough. Numbers on corporate pilot furloughs are considerably less, but the exact figures are unavailable.

Some pilots are already seeing the handwriting on the wall. Robert Neve flies helicopters in Canada. "In the Canadian helicopter industry, there have never been any requirements for post secondary education for a pilot position. And, to the best of my knowledge, all advertisements for aviation management positions require a specific type of experience. Not many request a particular level of education. Having said that, I'm finding that a greater percentage of entry-level pilots are college or university educated." Neve has no college degree, but he is working toward a Corporate Aviation Management certificate through Embry-Riddle University.

"There's also a very pragmatic requirement for a college degree in a corporate flight department," said Janice Barden, president of Aviation Professionals International. "One of the first things the CEO says is 'I'm not paying someone this kind of money who does not hold a degree.' Many discerning company officials want to know where a pilot received their degree from, as well." Carney agreed. "In a corporate setting, a degree from a good university holds more value." For a cockpit crewmember or mechanic, value often translates into continued employment, while a less-qualified individual is furloughed.

Susan Anderson is president of Factory Pilots Plus, in West Palm Beach, Florida, a company for contract pilots. Anderson does not have a degree, although she does have college experience. "Even if I did have a bachelor's degree, how would that help me now?" Anderson wondered. "I could have a degree in basket weaving and I could say yes to the degree question, but again, what value does it really have? Certainly getting a degree shows you have staying power and a commitment. But what does an ATP certificate show? But, seriously, I also know I have gone as far as I can without a degree."

"I sacrificed college to be an Army helicopter pilot," said Kevin Baker, a professional jet Captain currently between jobs. "Even after the Army, I had no time to work on a degree because I was working seven days a week running my own charter business. I've been working with Embry-Riddle Aeronautical University online, though, since 1985, taking a class here and there to finish my degree. It is a personal thing to finish my degree. But, I think a degree makes for a well-rounded individual. It gives you background and depth to converse with people on another level."

Baker added, "I have a short fuse with pilots who want to do nothing but drive and believe everything else is someone else's job. I think you need to be as robust as possible and help your employer get the job done. If you won't go the extra mile, someone else will." That desire to go above and beyond, including the education necessary to accomplish a new set of tasks, is the precursor for future managers. Lee agreed, "A degree gives you a hint of a person's abilities, but it only shows when you assign them a task." Carney said, "You need pilots today who understand they provide travel solutions, not just fly airplanes." Miles East, a pilot who holds a four-year degree and currently flies with Textron, said, "I think that piloting portion of my job is a very small part. I'm a data manager, safety manager, navigator, and a customer service manager." Today's pilots need to decide. Do they only want to fly airplanes or is there enough insight into this career field that they can see a time when they either may not want to fly or may be unable to fly because of a medical condition?

Iris Critchell, former chair of Harvey Mudd College's Bates Aeronautical Program and an aviation educator for 55 years, said, "The managers who feel pilots lack a business side to maintain connections with management are speaking of people who are primarily pilots. They are trained largely by the companies in how things work and to support company goals. The typical pilot applicant in recent years with the aviation management degree is actually not expected to go into management. Those pilots would be giving up flying, and that isn't what they came for."

The degree prerequisite is not new. "A college degree became a default requirement (for pilots) in the '70s with the military pumping so many pilots into the civil system," Lee said. "It could simply have been a discriminator that became the standard until the past couple of years when the military hasn't been delivering as many aviators as it used to. People available to business aviation recently have often been those who worked their way through the system as a flight instructor and charter in lieu of college. Now we have a lot of technicians, instead of college grads. And, while the degree is a reflection of the marketplace, we must ask what a flight department needs in the future. We have to grow these people within our companies. We haven't planned the succession issue very well." Lee believes business aviation needs to begin hiring people with the right skills to take a flight department into the future or needs to begin developing these skills with people already employed within a department, who have shown the aptitude *and* the desire.

Although being overtaken by someone smarter than they are is a threat almost everyone in corporate American has been dealing with for years, the same issue is certainly headed for flight operations. Despite that threat, "Some managers and pilots just don't want to change," said one flight department manager in Atlanta. Harkin said, "Some managers are always very intimidated by other people's knowledge. A manager of today must recognize who is around them and find something of value to the company outside of their own position." Jay Evans, NBAA, senior manager of airman and commercial services, said, "It takes a very astute manager to realize they have been successful and not to feel threatened by this better-educated pilot."

But, in the end, there is no choice for management and pilots if they want to remain competitive. "Pedal faster and learn what you need to compete or be left behind," said Lee. "I'm actually a bit disappointed that we as an industry are a bit threatened at change." Change means better educated employees. "Everyone actually benefits when employees broaden their skills," Lee added. "They just don't always know it." Baker questioned whether managers might be intimidated to bring someone with their previous management experience on board. "They shouldn't be, though. They can leverage my background because I can make them look great."

Change will not occur easily, whether it is convincing pilots of the need for more formal education or managers of the need to hire better-educated pilots. One Seattle-based pilot, who did not want her name used, said one manager she worked for asked her point blank, "What does a pilot need with a master's degree?" and "What good is a pilot who doesn't fly?" "This sort of manager likes feeling superior," she said. "They don't like people around who have the capacity, much less the inclination, to question their decision making. Their ideal candidate is someone who can fly well, read, and write adequately to fill out the manifests and write up squawks, but [who] does not identify problems, process them, and generate solutions. And, heaven forbid, they memorialize anything in a memo or letter. You wind up with no one within the department who can be promoted though, and the next manager probably comes from outside the company. I think there is a real disconnect between the folks most business aviation managers are genuinely comfortable with and the type of candidate the universities are trying to prepare."

Tom Carney acknowledges that some managers might feel vulnerable by bringing on people who are better educated than they are, but he also sees the light at the end of the tunnel. "Whether you have a degree or not, no current manager should feel threatened. Managers today shouldn't be put at risk if they are already successful because they don't have a piece of paper. We all lift each other up by more knowledge. I just can't see a downside to a better-educated workforce. We've raised the bar and that helps everyone. The goal is not to alienate people, but to bring everyone along. The goal is also to develop a successful flight department. But it won't happen overnight."

Miles East is dual-rated in the CE-650 and the Bell 430 helicopter. Despite the fact that he holds a degree in English Literature from Brighton College in England, East said, "No one at Textron asked me if I had a degree when I interviewed. But I think they assumed I did. But I think persistence is what got me this job." Surprisingly, East added, "Some pilots won't even apply at Textron without a degree because they think they need one. But the company has no formal degree requirement."

Aviation professionals who believe in the need for more formal education are also scratching their heads about what sort of degree they should pursue, especially because most corporate and airline-degree requirements seem to care little about the subject. Iris Critchell said, "A bachelor's degrees in aviation management, for instance, prepares students to be airline pilots. It is limiting because there are not as many academic course as for other kinds of degrees. They fill the need of the college time but, practically, these students have an associate's degree. Without a four-year degree, a pilot with no education is completely dependent on the medical. If that goes, they are on the street."

The earlier NBAA chart also highlighted an important insight . . . the size of the flight department does reflect the company's overall interest in education. "Promotion possibilities are seldom even considered in a small flight department," said one pilot. "In a small flight department, the model is that you'll be a copilot or a Co-Captain for a long time. In small departments, you're stuck," said Harkin. "74 percent of NBAA is small flight departments." Critchell said, "Some companies don't have any interest in pilots who want to move ahead." Evans said, "Sometimes small departments don't want a degree because they are just looking for a pilot. But, the larger departments do care about a formal education. I think small departments will be changing the way they look at pilots, but the change will be industry-driven."

The NBAA is in the final development stages of a new education program leading to a Flight Department Manager certification through testing by 2003. While the new qualification may appear on the surface to be only a way for pilots without a formal education to climb the ladder, Evans believes even pilots with degrees will want to participate. Lee said, "If you're certified, you have the skills, the knowledge, the motivation, and the aptitude for management. This is the equivalent of a degree."

Essentially, the standards for the program are set by NBAA. The actual classes are offered though eight major academic institutions—many through online distance-learning programs—in conjunction with the association's Professional Development Program (PDP) classes.

Evans said, "The rationale for the certification is to make sure the flight department is properly mangers and provide an industry-wide standard for management in corporate aviation. Right now, there is no formal training to become a flight department manager. This is a career path for future managers. We're still working on the details for the certification test, but it should include a written component, a practical exam, and some additional prerequisites to be worked out."

Miles East has taken 22 of the NBAA-approved classes since the summer of 1999 and will take the certification test as soon as it is available. "I didn't see anything else other than the PDP (and certification) that would differentiate me from someone else when it came time for promotion." East added, however, "My main criticism of the program, though, is that no one has marketed the concept of certification to the aviation department managers, so they'll see the value. I also wish I had more guidance in getting this organized before I started. There were no written prerequisites." Harkin said, "I would like to see all my pilots go through the NBAA classes, even if they don't go into management, so they can understand what managers do." Baker added, "I think the NBAA certificate will help add some value to a flight department by demonstrating that a pilot understands the functions necessary to run the department. This credential also says I'm not afraid to invest in myself."

Many people interviewed for this story agreed that no matter what a candidate's education level, it should not be the tiebreaker for a job. Harkin said. "The degree gets you in the door for the interview and, after that, it is what you can do. We're trying to find people who can add value to what our companies are doing today." Baker added, "People need to be interviewed, not their paperwork. The attitude might be better with a lesser qualified candidate." East said, "All the degree does is say you're capable of learning and manipulating information. I think it is generally hard to find the right individual to employ and you should look beyond a college degree as a discriminator. You will pass over some really good people. If you feel that strongly about a degree, give a person a chance and pay for their education." Anderson questioned, "How many good pilots have been turned down for a position because they didn't have a degree?"

Costs associated with furthering an education can be an issue. At many of the larger companies, such as Textron, East said, "We have 100 percent tuition reimbursement." But time is also a concern. Anderson said, "I want something that will expand my thinking, I want to be more knowledgeable in my field. One of the problems with going back to college is starting from English 101. Full-time college right now would ruin my flying career. That is why the online program looks so good. And what is the rush? If I finish my degree today, what is it going to get me?" Neve said, "Management positions are few and far between. And any senior manager will fill those positions with people they know and trust, and will pay to educate them." East believes, however, that despite holding a degree, "If I had taken the four years and just started flying, I would have been further ahead."

Carney asked, "Who is going to take over as senior corporate people retire? Who will take a leadership role? Watching what the airlines have done, I'd say, let's get the best blend of people to manage. If a pilot exhibits a lack of knowledge or skills, those are fixable. You can train to standards and get rid of those rough edges. But, we can't change who a person is. Isn't all that interaction important in a corporate flight department? We should want the best blend of interpersonal skills, as well as a person who is a technologist."

"People with an aviation qualification added to that bachelor's degree are in great demand on the aeronautics employment market, earn higher starting salaries and, above all, have more flexibility to meet the wide-ranging changes that occur in the industry," said Critchell. "They are the ones who become the senior managers and marketing leaders. Industry wants the best-prepared and most flexible employees it can get. The people with the most and best formal education have more choices in aviation and provide the vital ingredient—flexibility."

Do you need a college degree to be hired and succeed as a pilot today? Not really. But will a college degree be of value to a pilot a few years down the road? Absolutely. Barden countered, "I think a person with only a high school education sends a message that they're not a part of the world today." But, Lee added, "Aptitude and interest are also important. It will help employees develop if they have the skills and aptitude. Leaders are made more than born."

Carney spoke to the future. "We are headed toward a degree requirement for most everyone in this industry, pilot or manager. You're so limited without one." Louis Smith's web site—FLTops.com—publishes decision information for active pilots and applicants. "There is simply no reason not to get a four-year degree. It opens a lot of doors once you're hired. A university degree will often get you more contacts within the industry, as well." According to Anderson, "There seems to be a real camaraderie between alums." Smith also said, "The degree can still be a strong tie breaker."

"Times are changing," Neve concluded. "A post secondary education is now what a high school education was 20 years ago. And, more and more colleges and universities are offering aviation degree programs. So, the competition increases. This is good for the industry as a whole. But, don't forget, intelligence and ability aren't just based on education, and flying experience isn't just based on hours in the logbook. It's what you can do with them."

Fractionals

This category of pilot career did not even exist ten years ago when an earlier version of this book was being written. Now, fractional jet operations have evolved into positions significant enough that many pilots don't want to leave for any other job (Figure 7-5).

The fractional concept was developed by Rich Santulli, the Chairman of NetJets, the first real fractional ownership company. NetJets is now owned by Warren Buffet, Berkshire Hathaway Inc. The concept was simple. Many companies were wealthy enough to own their own business aviation aircraft and run their own flight department. But a significant number of companies were not, although they would have liked to enjoy the benefits of private aircraft ownership The barrier was often cost. Santulli developed a mathematical model that would allow a company to purchase a portion of a business aircraft, much like the concept of time-sharing on vacation real

Figure 7-5 NetJets operates the G-550 as part of its fractional jet fleet. (Courtesy Gulfstream Aerospace Corporation.)

estate. Fractional ownership has opened the business-aviation door to many thousands more companies for considerably less up-front cost and none of the hassles of running a flight department and managing the aircraft.

The idea evolved into what is today NetJets—www.netjets.com—a company comprised of over 2,500 pilots flying over 650 business aircraft around the world. These range in size from the smallest light-cabin aircraft, such as the Cessna Citation Bravo and the Hawker 400A, to mid-size aircraft, such as the Hawker 800 XP and the Citation X, to the largest cabin, international aircraft, which include the Dassault Falcon 2000 EX Easy, the Gulfstream 550, and the Boeing BBJ, the business version of the Boeing 737–700 airliner.

Fractional flying is a blend of corporate flying and charter. It offers top-notch equipment with the same variety of destinations known to pure charter companies and some corporate operators. Training is normally handled by only the best factory-affiliated companies, such as FlightSafety and Simuflite.

Netjets is not the only fractional company. Other pure jet operators include Flight Options, FlexJet, Delta AirElite, and Citation Shares. Avantair operates a fleet of Piaggio Avanti P-180 twin turboprop aircraft. Alpha Flying, in business as Plane Sense, operates a fleet of Pilatus PC-12 turboprop aircraft. Web addresses for the other fractional ownership companies can be found in Chapter 8.

Many pilots were initially unsure about the strength of this kind of flying. Fractionals looked like a good stopping-off point until something better came along. The following pay figures for NetJets pilots show why the fractionals have emerged as a great career. Most of the fractional companies offer traditional benefit packages to pilots—including health, dental, and retirement—specifically because they want to be certain they remain with the company. NetJets offers an established presence in the United States, as well as in Europe and the Middle East.

NetJets breaks down its pilot payroll into a number of different categories, based on both the size of the aircraft and the position occupied, for example, left-seat Captain on a Hawker 400A vs. Captain on a BBJ or First Officer on a G-550.

A first-year Officer on the smallest aircraft can expect to earn $39,000. By year five, a First Officer in this kind of aircraft will earn about $48,000. The name of the game in aviation is the same at the fractionals—flying a larger aircraft while accepting additional responsibilities is the key to a larger paycheck. A first-year captain on the smallest aircraft NetJets flies will earn $52,000 and watch that figure rise to about $90,000 by year five. In the largest aircraft, such as the international Gulfstreams and Boeings, a First Officer can expect to earn $61,000, and watch that figure rise to $73,000 at year five, and nearly $90,000 by year 10 of being in the right seat. Again, Captains on international airplanes are the best paid—and gone from home the most as well—at $139,000, increasing to $161,000 in year five and $194,000 by year ten.

Charter Flying

Some people believe charter flying is only a time-builder until a real job appears. If you believe this, you could be missing out on an important piece of the job market. A friend of mine has been flying for a charter operation at a fixed-base operator here in Chicago for 12 years, and he loves it. The operator has grown from a combined piston/turbine fleet 10 or 15 years ago to a pure turbine operator today. My buddy is typed in a Falcon 20, a Learjet, and is soon headed to FlightSafety for training in the G-III. The hours are varied, so while he knows just what days off he will have, there's always the possibility that he could be called in on his day off. But, the destinations are almost always different, as are the passengers.

While this pilot is flying jet equipment, hundreds of charter operators at smaller fixed-base operators (FBOs) are flying piston equipment (Navajos, Cessna 414s) and light-to-medium turbine equipment (Beech King Air &

Piaggio Avantis). Those FBOs also offer significant possibilities. A call to the National Air Transport Association, the FBO trade association in Washington (703-845-9000), can put you in touch with a listing of hundreds of FBOs. Often, some of the more successful FBOs and their charter departments aren't necessarily located in major U.S. cities. Recently, I visited an FBO in a small Midwestern city and was surprised to see the size of its flight department: 23 pilots flying King Airs, Learjets, and Falcons on mostly a five-day-per-week schedule. Normally, however, charter flying tends to be an on-demand kind of service. The salary of charter pilots runs all over the scale: higher in the big cities and lower in smaller towns. The rate also varies with the type of aircraft flown but, as with life, everything is negotiable. And, one new company, SATS Air, based in Greenville, SC, is starting an air-taxi charter service using a fleet of Cirrus Design SR-22s. By the way, they also need pilots (www.satsair.com).

Charter Flying: The Good, the Bad, and the Ugly

Historically, if you wanted a career as a professional pilot, there were only three routes—the airlines, corporate aviation, or the military. But, the recent boom in airline hiring has thrown open alternatives some pilots may not have once considered. One such possibility is Part 135 charter flying. Thanks to a robust economy, the charter business is also expanding as never before. Andy Cebula, vice president of the National Air Transportation Association, the trade organization for charter companies and FBOs, reports that "we're seeing 18 to 20 percent increases in charter business over last year."

Unpredictable schedules have always made it tough for charter companies to hold on to pilots, but the shortage is becoming worse. "Attrition has been terrible," said Monty Lilley, president of Congressional Air, based in Gaithersburg, Maryland. To counter that, though, Cebula added "In a tight employment market, you'll always do whatever you can to hold on to employees." In some parts of the country, this is already translating itself into limited improvements in wages and schedules, the two areas that, traditionally, sent pilots scurrying from charter as soon as any airline called.

Gil Wolin, then president of Denver's Mayo Aviation and now with TAG Aviation, added that "It's just the nature of the Part 135 charter business that to maintain your position in the marketplace, you need to control costs—so charter just can't pay what corporate flight departments or airlines can. We've just resigned ourselves to a high turnover rate." Fred Gevalt, publisher of the Air Charter Guide, admits operators "are having a tough time holding on to pilots because charter always has been just a stepping stone. But, you can credit charter with providing a real-life training ground for pilots."

Part 135 companies typically fly unscheduled, on-demand charter flights carrying passengers, cargo, and sometimes both. The overnight delivery of checks is a typical 135 operation. One of the great business benefits of aircraft charter is they'll fly into airports not normally served by the airlines. Some charter departments even deliver enough service to a customer in a year that they begin to look like a corporate flight department to regular company passengers. The cost to charter an aircraft per mile is normally much higher than that of a corporate aircraft, but for a company that does not want the responsibility or the work of managing their own flight department, "Aircraft charter is the most productive use of air transportation," said Gevalt. "You simply buy it when you need it."

The key word in charter flying is "on demand," a term that has proven a great boom to the *pager/cell phone industry*, which is the typical way a charter operator communicates with cockpit crews. *On-demand* means flying on the weekends, in the middle of the night, or even on Christmas. An airline, for example, can't make money when its $100 million aircraft is sitting on the ground in Des Moines AOG awaiting parts. They'll often charter an aircraft to carry mechanics and parts from a maintenance hub to the crippled machine. Hang out around any FBO and look for the people asleep in the snooze rooms. They're probably charter pilots who've been up since 2 A.M.

Gevalt said that in the United States, "508 charter operators are currently using some 1585 turbine-powered aircraft in Part 135 service (figures for piston-powered aircraft under 135 are not available)." The total number of aircraft operating under Part 135 was also considerably larger until just a few years ago when many, that were operated as commuter carriers, were mandated to upgrade to the higher training and operating standards of Part 121. Currently, Part 135 flying applies to nonscheduled operations in turboprop aircraft of less than ten seats. On-demand service in jet aircraft of less than 30 seats may also be operated Part 135, as well as single-engine aircraft that meet certain equipment regulations. Some scheduled 135 service—a tiny percentage—still exists in small aircraft that do not meet Part 121 requirements.

One Part 91–135 regulation maverick is fractional ownership, an area of corporate flying—and pilot hiring—that has flourished under the leadership of Rich Santulli and is being copied by other companies, such as Raytheon and Bombardier. Many of the larger charter services provide backup lift to Santulli's NetJets on a regular basis. While operating under Part 91, fractionals provide on-demand service to their owners while following Part 135 standards, such as crew duty times. One of the main benefits to fractionals remaining Part 91 is this: Part 135 requires weather reporting service or VFR conditions at the destination before landing. Part 91 pilot can shoot an approach to any airport, in any weather. But, the FAA is currently taking a long, hard look at fractionals to determine whether they should be brought under the Part 135 umbrella, remain Part 91, or possibly have a new set of regulations developed specifically for them.

One necessary key to a successful charter operation is a good charter pilot, someone with a pleasant personality and the ability to easily converse with customers during the flight. Mayo's Wolin adds, "We establish a profile (for pilots). We don't just fly airplanes. We hire people to fly people. Until the customer is delivered to the other end, our crew's job is not done. Pilots have to be Boy Scouts—total technicians are not going to succeed in the long run. Pilots have to think about the little things, like whether a customer likes Pepsi or Coke. They have to listen to a customer's concerns. Charter pilots also develop a rapport with people that you'll never have in a 737 after you punch on the autopilot."

Although the range of aircraft flying Part 135 is vast—everything from a Cessna 172 to a Piper Navajo to a Citation to a Gulfstream G-5—most major charter-flight departments are finding turbine aircraft—both turboprops and pure jets—to be their main sources of bread and butter. Charter companies also do not ordinarily own their aircraft. Most are leased back to the 135 operator by a local aircraft owner, typically a corporation. The benefit to the aircraft owner is additional income to offset fixed costs, as well as a significant tax break.

Piloting charter aircraft is more demanding than many other facets of flying, simply because Part 135 crews don't have the benefit of a licensed dispatcher to help out with many of the chores, such as checking weather and filing flight plans. Charter crew check their own weather, make their own go, no-go decisions, and normally load the aircraft with coffee, ice, snacks, catering, and even passenger bags when they arrive. A charter crew also seldom sees the same city twice in a week, or even a month. Brian Ward, a pilot for Mayo Aviation, said, "We buy some food for early morning flights and tidy up the aircraft interior between legs. We also clean the aircraft when we get home at night," but Ward adds that "we don't clean the exteriors."

Like their brethren at the Part 121 majors, the FARs limit the number of hours Part 135 pilots may fly. By contrast, Part 91 pilots have no duty- or flight-time restrictions to get in the way of an overzealous boss. In a one- or two-pilot cockpit, no Part 135 pilot may fly more than 1,400 hours in any calendar year. By contrast, a Part 121 pilot is limited to 1,000 hours (most pilots fly considerably less). In a quarter, the figures are 800 hours and 300 hours, respectively. In a single day, the Part 135 pilot can fly a maximum of ten hours if two pilots are aboard, while under Part 121, the limit is essentially eight.

While Part 135 flight time limitations are vastly different from those of Part 121, they are a significantly better option than none at all . . . sometimes. Tom Deutsch, who now flies for a major U.S. airline, remembers his recent job flying bank checks for Airpac in Seattle, WA. "I started in light singles, like the Cessna 172RG and 177RG, and worked up through the Senecas, Navajos, and Beech 99. I'd often average 14-hour days. Sometimes, I'd only log two hours of flight time. The rest was sitting around." Part 135 also has duty limits, while Part 91 has none. Part 135 pilots can be scheduled for no more than a 14-hour day, normally followed by a ten-hour rest period. By contrast, some Part 121 operators still fly both continuous-duty and reduced-rest schedules. In a *continuous-duty schedule*, a crew takes the last flight to an outstation, often arriving about 10 P.M. They head for a hotel for a few hours, and then take out the first flight in the morning, often garnering no more than three or fours of sleep for the evening. Some airlines make crews fly four of those in four days. *Reduced-rest schedules* call for a crew to arrive at the outstation—often after a full day of flying—and return for duty exactly eight hours after they shut down the engines. The following day is often comprised of another full day of flying. Even charter-flying schedules are not as tough on a person's body as some of these.

Part 135 operations require pilot training that, on the surface, appears similar to that of the major carriers. Training is where you first begin to discover the variations on Part 135 regulations: in this case, the content and the delivery of training material. For example, FAR135.245 says a second-in-command must "hold at least a commercial pilot certificate with appropriate category and class ratings, and an instrument rating. For flight under IFR, that person must meet the recent instrument experience requirements of Part 61 of this chapter." Essentially, anyone who has had three takeoffs and landings and some ground instruction could find themselves in the right seat of a turbine-powered aircraft.

Some operators train in the aircraft at their home base, due to cost of training issues and, to fly a Piper Navajo, a few hours of in-flight training may make sense. But that kind of training in a fast, turbine-powered aircraft flying at high altitude—while it may often meet the letter of the regulations—would hardly be safe. More and more operators are opting for a more traditional FlightSafety or Simuflite training environment, when possible. For Kathy Julien, now a Falcon 2000 pilot, her new position as a pilot for ACM Aviation in San Jose, CA, included a trip to FlightSafety as soon as she was hired into the right seat of the company Citation 1. Pilot training should definitely be a subject for discussion during the first interview.

With the numbers of pilots joining the major airlines on the increase—nearly 13,000 expected in 1998 alone—some Part 135 operators are investigating methods to assure themselves of a return on their investment in pilot training. Monty Lilley says, "There is no loyalty to any employer in this business." Lilley is strongly considering a training contract for pilots on any of the six piston and turbine aircraft his company flies. Mayo Aviation's Chief Pilot Cody DieKroeger said, "We have a training contract (at FlightSafety) that we value at $6,000. If the pilot leaves prior to the end of the first year, they reimburse us on a prorated basis." Pilots at Air Castle, based in the Northeast, pay for their first type rating at FlightSafety before they begin flying. Stuart Peterson, an Air Castle pilot, adds, "Once you pay for your first rating, the company pays for the rest."

On the other side of that argument, Jack Stockmann, director of operations at Wayfarer Aviation, White Plains, NY, said "We believe that safety can only be obtained through training, but we don't ask pilots for training contracts or agreements. We believe the fundamental of a good relationship is trust that works both ways." Wayfarer's fleet consists of 26 aircraft, the smallest a Beech 400A and the largest a Gulfstream 4.

Schedules are a subject certain to raise the blood pressure of some would-be Part 135 pilots, as well as many who currently fly charter. Part and parcel to on-demand flying is that scheduling, the when and where of charter, is often difficult, if not impossible, to predict. Many married pilots avoid charter flying for this reason alone. For that same reason, many managers like hiring single pilots with few responsibilities. The Catch 22, however, is that single pilots tend to be young and focused on airline jobs down the road.

What might a Part 135 schedule look like? Wayfarer's Stockmann said, "The crews are off two hard days a month. They have to be available to fly the rest of the time. But, since we do high-end charter, we have more notice and more time to plan. Pilots here feel we do recognize them and not treat them like a machine." At Air Castle, Peterson reports they work "20 days on and 10 days off, and average about 500 flight hours per year." In the Denver operation of Mayo Aviation—it also has a few Arizona bases—DieKroeger reports "pilots are on five days and off two. They fly about four to four-and-a-half days a week, which translates into about 55 hours a month." Mayo tried a six days on and three days off schedule, but found pilots were simply too exhausted. In its Arizona medical evacuation bases, Mayo's crews live in a trailer at the airport for a 12-hour shift. Kitty Hawk Charters' operations manager, Mike Joseph, said from their Detroit base that "our pilots are on call 24 hours a day, seven days a week." Kathy Julien, has "no hard days off. In August, I was flying almost 21 days. Then, there is always pop-up stuff." The regs state that Part 135 pilots need have only 13 days off in a quarter.

In the Midwest, one operator—who declined to be named—felt they had a solution that kept pilots happy by providing them with "a life." The upside for this operator is they hadn't lost a single pilot since they instituted the new schedule. In their scenario, there are three dedicated crews—all salaried—for each of the company's jets. Each day, one crew is scheduled as the first to be called, the other the second, and the third is off. When one trip is scheduled, the first crew knows they'll get it, and so on. While the crews don't know their destinations, they can plan their lives. Pilots can also trade trips with other equally qualified pilots. But, as in all charter operations, crews can still be called when a pop-up trip occurs. One element of charter flying that bears repeating is this: while pilots may be on call for many days during the month, they will not normally fly all of those days, resulting in a day off. It just isn't a scheduled day off for planning purposes. Most charter pilots carry their uniform and an overnight bag with them everywhere they go while they're on call.

Another over discussed element of charter flying is the need for pilots to carry a pager. Some hate it, while others see it as a part of the territory, even when it goes off at 1 A.M. DieKroeger asks potential Mayo pilots how they feel about carrying a pager during the initial interview. But, no matter what a pilot's feelings are about pagers, without them, a pilot on call would need to remain near the telephone—always! The response time, once paged, varies widely and is dependent on the type of operation. Mayo Aviation's air ambulance pilots report they can sometimes "be in the air within ten minutes." They're required to be within 30 minutes for regular charter flights. Stuart Peterson said, "We have a half-hour call out to be airborne within an hour at Air Castle." Kitty Hawk Charters requires its pilots to "be off the ground within 45 minutes," according to Mike Joseph. At ACM Aviation, Julien says they need to "be at the airport within two hours of the call."

The pay range for Part 135 pilots also varies significantly by the type of aircraft and the part of the country where they fly. Some companies offer pilots a weekly salary, others pay per hour, and some a combination of both. Many pilots receive a per diem on the road, others simply cover actual costs. Rebecca Traudt, a charter pilot and flight instructor with Mount Hood Aviation in Troutdale, OR, is paid per trip on some segments and per hour on others to pilot the C-172 the company flies under Part 135. On one regular 100-mile freight run, Traudt is paid $28 per trip. If she's flying on a per-hourly basis, she's compensated at $14 per flight hour. At Congressional Air, Lilley said "some pilots get paid $150 per day and others receive $25 per hour."

Top-end salary at Mayo Aviation is $50,000, according to DieKroeger. Wayfarer's Stockmann reports "First Officers begin at $50–60,000. Captains on small jets begin at about $68,000." The top pay at Wayfarer—typically in the largest aircraft it operates, such as the G4 and the Challengers—hovers at "about $112,000." At Kitty Hawk Charters, Joseph said, "We pay $250 per week for a pilot to fly the turbine Beech (Beech 18 turboprop conversion)." Peterson reports that Air Castle pays $24,000 the first year in the Lear, with about a $6,000 raise the second year. Lear Captains earn about $40,000 and Challenger pilots about $48,000." Twenty full-time pilots work for Air Castle.

So, do you have the necessary experience to consider Part 135 flying? That depends on the company and the aircraft you might fly. While some operators don't require experience in the aircraft they fly, previous Citation time would seem to be a step up during the initial interview for a company operating that aircraft. At Mayo Aviation, you need 3,000 hours total time—1,000 multiengine and 2,700 PIC—to be considered. Wayfarer wants 2,500 total time, an ATP, and a four-year degree. Kitty Hawk wants to see a minimum of 1,500 hours, of which 500 is turbine time.

And, just when you thought there couldn't be too much more variation within this segment of flying, you'll find hiring procedures are vastly different, as well. Most charter operators reported receiving dozens of unsolicited résumés per week, many of which did not meet their minimum qualifications. DieKroeger said, "Because our hiring requirements are high, we conduct an initial phone interview with only about one in four applicants. We talk about their mountain- and previous charter-flying experience." Applicants selected pay their own way to visit Mayo in Denver for the one-day interview. This includes "a written exam about ATP procedures." This is followed by a one-on-one interview with the Chief Pilot to see how conversant a person might be. "Then they'll fly an AST-300 simulator for about an hour." DieKroeger interviewed 30 pilots to fill their recent class of 7. The company currently employs 40 full-time pilots.

At Wayfarer, the process is even more intensive. But Stockmann believes that "if you put in the time to find a quality employee, you save much more time and possible problems later on. We have a responsibility to our clients." At Wayfarer, the Assistant Chief Pilot conducts all the hiring and begins the selection process with a telephone call to assess the applicant's experience and goals. They even ask an applicant why they'd want to work for Wayfarer rather than an airline. "Some applicants will say they are looking (at the airlines), but we ask them for two years with us," said Stockmann. He said a successful candidate will visit the White Plains HQ at Wayfarer's expense for a more in-depth interview. "We listen to their view on things, how they think, and assess how they might fit into the company. We look for a good demeanor, a team attitude, good interpersonal and listening skills, and a strong knowledge of cockpit resource management (CRM). They also meet with the company president.

"If that is promising, we schedule a simulator ride in a Falcon at FlightSafety." Stockmann says it's more advantageous for a pilot not to have flown the aircraft before. "We're not looking so much at their ability to make the airplane fly, as to how they ask the questions about an aircraft they don't know. Do they use CRM skills? You can understand their knowledge base and skill level (that way). We ask about a lot of situational issues, too." Next, come background checks, from driver's records to a security and even a credit check. All listed references are called, as well.

The final piece in the hiring pie at Wayfarer is to meet with the CEO of the company that owns the aircraft. If that is successful, an offer is made. The entire process often takes just a few weeks to complete. Wayfarer has 87 pilots on board.

At Congressional Air, applicant processing is rather simple because the president does all the hiring. "I like to see how someone dresses. I like to hear what they've been flying and what they think they can do for me," said Lilley. "If I like what I see, the applicant will take a check ride in one of our aircraft." Lilley adds that "the FAA check ride is not as hard as the one I give an applicant. We know them pretty well before they get hired." Congressional Air employs four full-time and four part-time pilots.

Because charter flying often appears to be a moving target, why would anyone want to live with a pager and never know when they were leaving home or when they were coming back? Tom Deutsch said, "When you fly for the airlines, it is a long time before you are a PIC. The nice thing about (some) 135 flying is that you're on your own." George Dunn said, after flying 135 part-time for three summers in Maine, "You really learn how to handle yourself in stressful situations. You may be sweating on the inside, but with passengers so close to you, you learn how to be professional. That's what these people are paying you for." Dunn added, "A lot of people don't give a Part 135 operation credit for what they do. It's often just a few people with a few airplanes putting their heart into the business."

Brian Ward flies charter because "the people are really great. I enjoy that I get to see different places and meet different people. I also think this is a good stepping point. I'm just not a fan of commuters, and the amount of work and responsibility for what they get paid." Oregon's Rebecca Traudt said, "I don't think there is a downside (to 135 flying). I've heard other pilots say they would never fly a single-engine aircraft at night in this rugged (Oregon) terrain. If the flight is planned well and the airplane is in good shape, you'll know how to handle emergencies. You also learn to keep you wits about you and keep your options open." Stockmann said, "Pilots think that charter is only high demand for low pay with lots of standby, a minimal benefits package, and minimal security. At Wayfarer, compensation and quality of life are two major issues."

When asked what he liked about being on call almost constantly to fly a Lear 55 on Part 135, one disgruntled pilot had a simple answer: "Nothing!" Kathy Julien countered that. "If you're a pilot who loves to fly, this will be more satisfying. I fly really great equipment to some really great places. A great day is going somewhere fun and finding out you're staying for a few days. I also think Part 135 is a great training ground. If you work hard, you can work your way up to almost anything you want—if you can stick it out."

The differences between Part 91 (corporate) and Part 135 (charter) flying are:

	Part 91	Part 135
Duty day restrictions	None	14-hour maximum per day
Flight time restrictions	None	10-hour maximum per crew-member
Rest period requirements	None	10 consecutive hours in 24-hour period
Annual training	Not required	Yes, in type
Management personnel required	None	Director of Operations Chief Pilot & Director of Maintenance
Additional equipment requirements	None	Cockpit voice recorder, digital flight data recorder in aircraft of ten seats or more. TCAS, ground proximity warning system
FAA-approved operations manual required	No	Yes
Drug and alcohol testing	No	Yes

(Courtesy of Wayfarer Aviation)

Profile: Diane Powell, Charter Pilot, Lear 35

Diane Powell was luckier than some people. She had a great mentor for her adventure into a professional pilot career—her father, a man who just happened to be a certified flight instructor, as well. But, while her father was a catalyst for her career, Powell still had to overcome the same cash-flow problems as other aspiring pilots while she learned to fly. She worked full-time while learning to fly and got a break on her lessons from the FBO that employed her.

Powell also picked up a degree in business along the way to her other ratings. "Having a degree is more important today," she said. "But I wanted one for myself as a fallback, just in case I couldn't fly someday." She won her commercial and multiengine rating in 1989 and added an instrument rating in 1990.

Powell flew long, grueling days on traffic patrol around Fort Lauderdale in a Cessna 172 and logged 600 hours in one year. "I took off at 5:15 each morning and flew until about 9 A.M." She'd return to the airport in the afternoon and fly another few hours for the evening rush. "There were also no breaks during the flying," she said. "If you had to land for any reason, the bosses were not happy." Powell built time flying photographers around Florida, too, when she landed an opportunity to fly newspapers in an Aztec to Nassau from Opa Locka, just north of Miami. She also picked up her CFI rating along the way, and she even managed another full-time job with Miami's Metro Dade Port Authority as a dispatcher to pay the bills.

Now, with a little over 1,500 hours total time, Powell said, "I was up surfing around on an America Online (AOL) aviation message board one day and connected with a Lear operator who said he'd be in Palm Beach and asked if we could meet. They were concerned with my low time, but said they liked to give new pilots the chance that someone once gave them. Unfortunately, I didn't get the job. But they kept my résumé while I returned to flight instructing."

Then Powell had another opportunity to turn pro when she was hired by St. Louis–based Trans States Airlines. "The two-day interview for the job was pretty stressful," she said. "They really wanted to see that you already knew the airline-style procedures. I went through three simulator sessions and quit. I never did get the simulator thing down pat," Powell recalled. "Coming from general aviation, the transition from little airplanes to turbine flying was tough. They gave us no time at all to learn. They didn't seem to care how tough it was."

Powell recalled leaving St. Louis felling pretty despondent. Just after her return to Florida, the phone rang. "It was the Lear operator. He said he needed someone in the right seat—right now. I took the job. After two solid days of training, I was flying 135 trips in a Lear 35. This is something I always wanted to do and couldn't be happier—despite the fact that I must carry a pager. One of the best things here is that they let me fly left seat on the empty legs."

Powell—who says she'd eventually like to fly for United—remembers some of the words of wisdom her father passed along. "He said you should help someone else when you have the chance because it always comes back to you. Sometimes the pilots I've met always seem to be out for themselves and no one else. They're worried about only their job."

Powell has already begun the process of giving back to the industry that has given her so much. Her connections online are making that process even easier for her, too. "I just helped a guy get a job in a Merlin because he was online. Networking really is the key to finding a good job. I find that online, I can send an e-mail to people and ask lots of questions, and they send me plenty of information back. I find the message boards there are very helpful."

As you begin the search for a new job, you want every conceivable extra point on your side, such as a type rating. But buying a type rating has become quite a controversial subject when cost versus value received is considered. Some airlines, such as Southwest, require a B-737 type rating just to apply. Others, like Midwest Express, basically say don't bother because they're going to put you through their own training program anyway. Then, airlines like United say, sure, a type rating is worth something, but only if you also have a fair amount (about 500 hours) of PIC time to go along with the rating. In the corporate world, a type rating can be an asset, but if you're typed in a Learjet and the job is for a First Officer on a Falcon 50, the type rating might not count for much, other than the fact that you're trainable.

Contract Flying

Contract flying, or flying as a temporary pilot, translates essentially into running a small business with you the pilot as the chief product, service, and company CEO. The upside is pilots who are between jobs or who are the kind of people who want more control over their lives than traditional employment offers, can pick and choose which contract jobs to pick up. The benefit to the aircraft owner is a ready pool of pilots who don't need to be brought on full-time (Figure 7-6).

The key to success as a *contract pilot* is experience on the kind of airplane that a company needs at any given time and the skills to market that availability to potential customers. Obviously, no one can be trained on everything. What contract pilots tend to do is gather experience in a few types of aircraft during some portion of their careers, and then stick with those as they transition to contract flying. Take another look at the contract opportunities I showed from the Rishworth list at the end of Chapter 4.

Make no mistake, while a contract pilot can decide which contracts to accept and which to decline, they are running a business. That means making certain the service—themselves—is always current. They must also advertise in the right places to land customers willing to use their services. Contract pilots are traditionally paid on a daily, weekly, or sometimes a monthly basis, depending on the job at hand. Out of their fees, pilots must

Figure 7-6 Contract flying means being ready to fly a variety of aircraft.

pay for their own training expenses, as well as taxes and insurance, so this style of life is not for everyone. A few places to begin looking at the life of a contract pilot, even if it might be only for the future, is www.avcrew.com. Richard Harris is the owner and a business aviation pilot himself. Another is International Pilot Services, at www.intl-pilot.com, where you can find a considerable amount of information about the life of a contract pilot. Roger Rose runs the place.

Will a Type Rating Get You Hired?

In a profession where tens of thousands of dollars can easily be spent just to land a bottom-of-the-ladder job, is an additional $10,000 to $30,000 for a type rating worth the gamble?

If a pilot's goal in picking up the type rating is to hire on with an airline, they should look at how the potential employer evaluates a type rating in the interview. During an airline interview, a pilot applicant is evaluated in many different areas. How pilots present themselves and how they perform on airline written exams often can be the deciding factor. If the decision is between two equally qualified pilots, one with a type rating and one without, you'd imagine the pilot with the type rating would win out. But, because most pilots never learn why they were or were not hired, that premise is pretty tough to verify.

The only major or national airline that requires a type rating is Southwest. The company will not interview applicants who don't have a B-

737 type rating. Southwest also operates a B-737 type rating school, and about 35 percent of the school's students get hired at Southwest after the type rating is earned.

Viewpoints from other major airlines include that of USAir spokesman, Jim Popp, who said, "Our minimum requirements don't ask for a type rating. A type rating is just reviewed along with all other qualifications a pilot might have." America West, which operates a type-rating school, also doesn't require a type rating. At America West's Contract Pilot Training, Dee Rush said, "Of all the people who have gone through our type-rating school, only a small portion have been hired. There are many factors that are taken into consideration when a pilot is hired."

United Airlines' spokesman, Joseph Hopkins, said, "United is seeing an overall increase in the qualifications of its applicants, but the type rating really is evaluated along with all of a pilot's other strengths." At regional carriers, company officials are singing the same tune as the major airlines. Glen Bergman, Chief Pilot at Business Express Airlines in Windsor Lock, Conn., said, "To have a type rating is important because it shows us that the pilot can pass a type-rating course. We do, however, begin all new applicants now in the BE-1900, regardless of experience in other aircraft . . . I would much rather see previous turbine experience."

Drew Bedson, assistant Chief Pilot at Pan Am Express (now out of business), said, "Possession of a type rating doesn't really affect anything in terms of being pulled for an interview. What's more important to those who do hold a type rating would be whether or not they really have used the rating as part of their job. If someone walked through the door with a type rating in a Jetstream he picked up at a school, but had no practical experience flying the airplane, then he has no more experience in that machine than he would have on completion of our approved training course, so the rating wouldn't change anything."

Paul Rogers, formerly with FAPA's Aviation Job Bank, said this of his corporate clients looking for pilots: "The type rating at the corporate level makes a pilot applicant look a bit better than the competition, but there are plenty of people with lots of experience who just don't present themselves well during an interview. Corporations are more interested in a well-rounded professional applicant who works well with passengers and can be an asset to the company in other ways with their education or skills."

Airline statistics show that, in 1998, nearly 61 percent of the pilots hired by the majors were type rated. In 1988, two out of ten pilots hired at the nationals already were typed. These numbers have been on a downward slide from past surveys. It's not known how many of these type-rated pilots bought their rating and how many got the type rating from a previous employer. Also, it should be noted that all types of type ratings are included in these percentages. The type rating could be in any turbojet-powered aircraft or in any aircraft that has a maximum gross takeoff weight of more than 12,500 pounds.

Some pilot applicants believe that when their application is screened, the extra points from a type rating are what they need to get an interview or land a job. Pilot Mike Roebke said, "I bought the 737 type rating because most of my time in the last five years was in crop dusters and I didn't think anyone would look at me. I would definitely do it again if I had to." Roebke recently was hired as a B-737 First Officer by America West.

Although pilot Henry Schettini was planning for the future when he paid for his type rating, he still can't find work. "I really felt my opportunities would be better if I had the type rating. I still think it's good that I have the rating, even if I don't have the job right now." It could be just a hunch on the part of the applicant that says the type rating is worth the expense. For some it's not. As one regional airline Brasilia Captain said, "I've spent enough money on this career. Another $10,000 to $15,000 just to possibly add one point in my interview score is just too much to ask."

The schools that sell air-carrier type ratings believe the rating is the pragmatic way to approach the interview process. Nancy Wilson Smith, manager of customer service at Dalfort Aviation, said, "Having the type rating makes a pilot more marketable. It proves you're trainable." America West's Rush said, "If you bought your own type rating, it shows you're pretty serious about your career." Aero Service's chief instructor, Steve Saunders, explained, "Hiring on with a major is extremely competitive . . . A type rating will give you that competitive edge." Ray Brendle, owner of Kingwood, Texas-based Crew Pilot Training, said, "The airlines know that, with a type rating, the pilot has made the transition from civilian general aviation or military flying to the air carrier side."

If a pilot decides on a type rating, he should be prepared for a whopping bill, somewhere between $5,000 and $20,000. Rates vary by aircraft type and previous pilot experience. And, a pilot shouldn't worry if they only have 2,000 hours total time. As Saunders said, "We're seeing considerably more lower-time pilots entering the type rating program . . . many from commuter operations." Pilots who never have been type rated in a turbojet aircraft will be considered an initial student by most schools. If the pilot were already typed in a turbojet, they would be a transition student. Pilots moving from the right seat of a B-737 to the left seat are upgrades.

An important decision for a pilot is in which aircraft to type. Currently, you can choose from the B-707, B-727, B-737, B-747, DC-8, or DC-9. A few of the schools also offer the B-757 and A320. The most prudent choice should be the aircraft that will do the pilot the most good all around. The B-707, B-727, and DC-8 were fine aircraft in their time, but they're part of an ever-decreasing portion of the aircraft fleet. Certainly, a DC-9 type rating is similar to the newer MD-80 series but, for overall usefulness, most critics agree that the B-737 is the aircraft to fly.

Crew Pilot Training's Brendle said, "The 737 is the most popular type rating in the industry because about 50 percent of all major and national airlines use the aircraft." Does this mean that a pilot's money is wasted if they type in a B-727? Perhaps. As Joe Marott, manager of Southwest Airlines Training Center, said, "It (a B-737 type rating) is still a requirement here at Southwest to apply for a position. Only the Boeing 737 type rating would meet our requirements for employment." So, a pilot should choose the aircraft carefully by first deciding just who they want to work for. Many smaller aircraft type ratings also are offered, but they aren't nearly as useful or valuable.

Smart type-rating students should choose a school like a good shopper buys a new car. Students should be wise, knowledgeable, and ask questions until they're satisfied. Price is only one aspect. Dalfort Aviation's manager of flight standards, Ben Williams, said, "A pilot should consider where the school does its training, how long the course will take to complete, and how long the school has been in business. Some schools will organize the ground school in one location, and then ask the student to travel to a second location for the simulator training, and perhaps a third for the aircraft training. In this case, the cheaper school would not be a bargain."

Some apparent benefits might be intangibles, too. Southwest's and America West's schools are part of those corporate structures, but both schools make it abundantly clear at the start that attending their school will not guarantee a pilot an interview with that company. On the other hand, a few private schools reportedly tell students indirectly that they have an affiliation with a particular airline. In most cases, this is untrue. Remember, all schools aren't created equal simply because they're FAA-approved. Consider the staff of that school, too.

Advanced Aviation Training's president, Robert Mencel, said, "Each instructor here is currently serving in a training capacity with a U.S. airline." At other schools, this might not be the case. Currently, if a school is designed to accomplish at least 90 percent of the training in a simulator (most are), the school must be FAA-approved. The enormous cost of aircraft training virtually assures FAA approval, too, but students should ask before signing up. In most cases, an initial type-rating student need only have a commercial, multiengine, and instrument rating to begin. In other phases of training, such as transition, the requirements can vary, so check with the school prior to enrollment.

What can a pilot expect from a type-rating program? First, most schools require some form of a deposit in advance to hold a position in the class, ranging from 20 percent to the entire cost of the course. None of the schools interviewed provided any financing, so be prepared to find a loan, if necessary. Most schools agreed to a specific amount of training for a specific price, but no school offered a guarantee of the rating. Additional training required involves additional funds, which the student should ask about first. At one school, the simulator costs $375 per extra hour, while time in the B-737 goes for $45 per minute.

One bright spot on the horizon for potential type-rating candidates is the new federal Veterans Training Act, the GI Bill of the 1990s. Pilots who served in the U.S. armed forces might qualify for this professional training cost-assistance of up to 60 percent of the type rating bill. Pilots should contact the local Veterans Administration office for details.

According to the U.S. Master Tax Guide, a type rating may also be deductible on a pilot's personal income tax return. The guide says that "education expenses are generally deductible if the education undertaken maintains or improves a skill required by the individual in his employment. . ." To be certain, though, pilots should check with an accountant about their specific situation before signing up.

While the price of lodging was included in the package price at only one school, all had some sort of deal with a local hotel to provide accommodations and transportation to their facilities during the student's stay. The length of that stay for the course varied considerably, from two-and-one-half weeks, to the longest at six weeks. How a pilot can carve a hole in their schedule for that kind of training is another problem a student must solve before committing to the training.

The schools run by Southwest and America West place their contract type-rating students into open slots along with their regular company pilot training classes, so students are able to talk and learn from the instructors and students already working for an airline. Southwest's Marott said, "The best thing in our school for type ratings is that we use the same manuals, the same procedures, the same simulators, airplanes, and instructors that are used to train our regular line pilots." Currently, Southwest's school teaches only the B-737–200 course to contract students. Marott said the school opened in the spring of 1987. "Since then, we've graduated about 175 B-737 type-rated pilots. Of that number, approximately 60 were ultimately hired by Southwest Airlines," said Marott.

Saunders says Aero Service's B-727 type-rating course includes 120 hours of actual classroom ground school. The student then moves on for 16 to 18 hours of cockpit procedures training in a nonmotion aircraft simulator. Then, they'll spend 18 to 20 hours in the full-motion simulator, usually in teams, with about ten in the left seat as Pilot in Command (PIC) and ten in the right seat performing First Officer duties and observing. Finally, the check ride is performed both in the simulator and in an aircraft the company leases. In most cases, an initial student can't complete all the training, plus the ride in the simulator. There can be some reductions in training time for the rating, based on previous experience. These changes are made on a case-by-case basis by the FAA's Principal Operations Inspector, who oversees the school.

America West's Rush said, "The ground school lasts 12 days. This is followed by five days of cockpit procedures training, and then the FAA oral. The student next moves to about 12 hours in the simulator and about 1 hour of aircraft time to finish off the takeoff and landings. America West currently is under contract with the FAA to provide all that agency's initial type training to its air carrier, as well as maintenance inspectors." Crew Pilot Training's Brendle says that, although the total amount of ground school there is similar to other schools, his "is approved for 80 hours of home study. The student must show by a test on arrival that he has completed the required work." Students receive an additional 40 hours of ground school when they reach the classroom. Brendle remarked that "some schools send you for the FAA oral right after ground school, without ever having seen the inside of the CPT or the simulator. When a student learns about an aircraft from a book only, without ever having had the chance to move a knob or switch, they give a weak oral."

Information, such as ground school time, simulator time, cost, and reputation of the school, along with other considerations listed previously, should be kept in mind if the pilot feels they need the type rating because of qualifications required by a specific airline.

In the end, a pilot must decide if the type rating is worth the money. That decision should be based on experience as a pilot and whether the pilot feels their credentials match up to those of other candidates applying for the job, as well as whether the company to which they are applying requires a type rating. It's a decision that must be made at the right time under the right circumstances.

Very Light Jets (VLJs)

Also new to the scene since the first edition of this book are the Very Light Jets (VLJs). The idea for VLJs evolved from the Small Aircraft Transportation System funded by NASA 15 year ago. Industry visionaries were already able to see the direction air traffic numbers were headed for the hub airports around the United States and in other parts of the world. The solution seemed to focus on how to better use reliever airports, as both the launching and retrieval ports for more business aviation airplanes. To make these operations efficient, the aircraft would need to be capable of operating from runways as short as 2,500–3,000 feet. Enter aircraft such as the Eclipse, the Adam, and the Cessna Mustang.

VLJs don't fly far because the majority of the trips their designers envisioned were typical business trips, with only one or two passengers on routes of under 750 miles. While the performance of each airplane is slightly different, the slowest of these jets is the Cessna Mustang at around 350 knots and the fastest, the Eclipse, is only about 20 knots faster.

In addition to the idea of using nonhub airports for increased business jet flying, one company—www.dayjet.com—is attempting to use a fleet of Eclipse aircraft in the first true air taxi service. And its web site asks the question . . . *Do You Have What It Takes to Become a DayJet Pilot?*

"We're looking for motivated leaders who are committed to a culture of safety, integrity, and customer service to build on the first truly new idea in regional business transportation in more than 50 years—and to have fun doing it! We're also looking for individuals who possess the ability and commitment to help mentor and develop a rapidly growing pilot rank."

The Face of DayJet

Pilots will represent the 'face' of DayJet™ to our customers. In addition to uncompromising flight safety, DayJet pilots will be responsible for delivering a consistent, efficient, and enjoyable customer-flight experience . . . every time. As such, DayJet pilots need to be professional, friendly, caring people, who interact well with customers socially.

Minimum Qualifications
To be able to deliver the best customer experience in regional travel, DayJet pilots must meet the following minimum qualifications:

- 3,000 hours flight time (excluding helicopter, simulator, and flight engineer time)
- 1,000 hours as Pilot in Command (PIC)
- 1,000 hours multiengine
- 500 hours turbo-jet PIC
- Federal Aviation Administration (FAA) Airline Transport Pilot (ATP) Certification
- Current FAA First-Class Medical Certificate
- Federal Communication Commission (FCC) Radio License
- Valid passport with the ability to travel in and out of the U.S.
- Valid U.S. driver's license

- Must pass a ten (10) year background check and preemployment drug test

Additional consideration will be given to pilots with:

- Check Airmen and/or Instructor experience
- Experience in Electronic Flight Indicating System (EFIS) equipped aircraft
- Recent flight experience

DayJet is also leading the way to lifestyle changes that are bound to transcend the industry. DayJet understands that one of the biggest hassles for pilots are the nights away from their families. They intend to make a dent in that problem by building pilot schedules with *no* night away. Each DayJet airplane will return to its base at the end of the day.

Flying for the Federal Government

An often-overlooked area of flying is with the federal or state government. Positions often exist for both fixed-wing and helicopter pilots, but the search can take time because there are so many agencies to check into. The FAA, for example, uses pilots to fly its fleet of flight check aircraft, while the U.S. Customs service monitors border traffic with its aircraft. Some of the publications we've already discussed might carry ads for government pilots and, certainly, a number of flying employment services offer publications that list flying jobs for the government. An alternative to searching the various publications for government flying jobs would be to contact the various agencies directly. A call to the Federal Information Center at (1-800) 333-4636 should get you started with phone numbers and addresses of federal agencies.

Some of the U.S. agencies that might need pilots are

- Department of Defense
- Department of Homeland Security
- Federal Aviation Administration
- National Oceanic and Atmospheric Administration (NOAA)
- Department of Transportation
- National Aeronautics and Space Administration (NASA)
- Defense Logistics Agency

In addition, state and local governments might use pilots. Your state's Department of Aviation would be a good place to begin the search.

Finding a Flying Job After Age 50

Finding a flying job as you near, or pass, retirement age is more difficult than it would be if you were 25. But, as the United B-747 cabin decompression near Honolulu and the DC-10 crash at Sioux City both proved, there simply is no substitute for experience. The Captains of both aircraft were near 60 years of age. Some companies recognize the experience factor in their hiring practices—but not all.

If you're an older pilot in search of a flying position, you may have to work some to locate them. According to some projections, 15 percent of new hires will be older than 40, some possibly as old as 57 or 58. Even if you retired from the airlines when FAR 121.383 suddenly made you totally ineligible to fly revenue trips anymore, there is hope in jobs other than the airlines. But, the road to work contains a number of potential hurdles (Figure 7-7).

If you're still young enough to work for a Part 121 carrier, you'll compete with younger pilots, which a company looks at as long-term investments. Even though you already may be type rated in a B-737 with 10,000 hours total time, you may not get the job. Attitude plays a big role. For example, some carriers find that crews coming out of 10 or 15 years of flying with another airline "do not mold themselves very well into the ways

Figure 7-7 Finding a flying job past age 50 is not as difficult as it was once. (Courtesy Embraer Aircraft.)

of the new carrier," as retired DC-10 Captain Jim Minning said. "These pilots have very strong ties to the old airline." Today, Minning ferries transport category aircraft for corporations.

Another problem facing Second Officer rehires is the cockpit intimidation factor, real or imagined. A man or woman who commanded a B-727 for 14 years may have a difficult time transitioning into a subservient role in the cockpit. A relatively new Captain, on the other hand, may well have difficulty flying with a much more experienced pilot constantly looking over their shoulder.

Captain H. McNicol of Flight Crews International, a crew-leasing company, said, "It's the persistent pilot who finds a job . . . the man or woman who does not give up and is willing to be aggressive in his or her search."

As Minning said, "Sometimes, getting work is like being a racing car driver who needs a sponsor. You must go out and look for them." He also said that "a contact network is very important." To find a job after retirement, Minning said he "kept very involved in our industry and made a great many contacts that I kept over the years. Many of these people remembered me."

If attitude and aggressiveness are important in a job search, so is a realistic view of the world, especially for the pilot who no longer can fly for a Part 121 carrier. In hundreds of other professions, thousands of good people are put out of work for one reason or another, and they must face the fact that the comfortable, well-paid job they once had may not be replaceable. So, too, with the airline industry. Sixty-five-year-old Hal Ross, currently flying a Citation III for a corporation in Southern California, believes, "You have to be pretty open about what you accept. I just don't think there are that many good jobs out there for pilots my age." Ross also alerted pilots to the fact that, in many jobs today, "you'll get paid what the market can bear, but they certainly will not be the airline salaries some of these people are used to."

What are some jobs for older pilots and how did pilots find them? What are the companies looking for and where are the majority of jobs coming from? For pilots who have reached 60 years of age with a Part 121 carrier, the easiest solution to keep flying, if the airline allows it (most do not), may be to accept a position as a Second Officer and move back to the flight engineer's panel, where the regulations do not stipulate a maximum age. Lee Lipski, age 62, successfully moved from the left seat to the back seat at Continental Airlines. "I took a tremendous pay cut to move from Captain to Second Officer . . . but a lot of us age 60 pilots are just not ready to retire." Certainly, though, the future will see the eventual elimination of these jobs as more and more airlines receive aircraft designed for two-member crews. Also, some pilots may find it difficult to deal with the politics involved in such a move. Younger flight engineers don't like such reversals because those more senior engineers may block their own progress to larger aircraft with higher pay, or they may cause them to be the ones who get furloughed because they have less seniority than the older pilot. And, of course, pilots on the job search don't like the idea because it reduces the number of positions available for new hires. For now, however, the option exists.

Another possibility is to work as a Professional Flight Engineer (PFE), although the number of aircraft in the Western Hemisphere using PFEs is steadily decreasing. These jobs are mostly at nonscheduled charter and cargo carriers because most of the major airlines use only Second Officers who are expected to upgrade. A PFE must hold a flight engineer's certificate in the particular aircraft the employer flies and often is required to be a licensed A&P mechanic. Most companies that use PFEs do not have a maintenance facility or a company representative in all the cities they serve. The PFE is the company mechanic there to make repairs, and sign logbook entries and maintenance releases that, otherwise, would ground a valuable aircraft. While the number of PFEs in use is not staggering, companies that use them include Airborne Express on DC-8s; American Trans Air on B-727s and L-1011s; Evergreen on B-727s, B-747s, and DC-8s; Key on B-727s; Reeve Aleutian on L-188s and B-727s; Southern Air Transport on L-388s and 747s; and Tower Air on the B-747.

Another option is crew leasing. You don't need to be a type-rated crew member to take advantage of companies that lease crews to airlines around the world on short-term contracts. California-based Airmark Corp.'s President Ron Hansen remarked that he receives only about half of his résumés and applications from Americans. The remainder arrive from around the world. Hansen said he has "many leased crew members over the age of 50," such as a number of ex-Eastern Airlines pilots in Japan on the L-1011. While Hansen admitted "the pay is good," his comments on American crews is food for thought during your job search. "Americans are generally difficult to work with on these foreign assignments. They want more money, they complain about the housing, and, in general, they consider many things inferior to the United States. I can't say I really blame them, but this is where some of the jobs are. The alternatives might be flying freight out of Detroit or mail for the U.S. Postal Service where you get paid less than a truck driver. Many of the choices are not very good for pilots today."

FAR Part 135 contains no age restrictions— many former Part 135 commuters now operate under Part 121—so it is a viable alternative for some pilots. Paul Cassel, age 65, flies a BE-1900 for a small commuter in California. Paul knew he wanted to continue flying and made his plans accordingly. "I just don't think that after flying as a Captain for so many years, I would have enjoyed being out of the actual flying end of things, so staying on as a Second Officer with my previous company wouldn't have worked." At age 65, landing any kind of job can be tough, but Cassel reports, "it was not difficult to find this job on the 1900 at all. This regional was glad to have me and my experience." Several regionals interviewed admitted that even though the law restricts them from discriminating on the basis of age anyway, they would welcome the high-time older pilots if they could get them to apply. After flying with a Part 121 carrier for many years, some pilots don't want to fly for a commuter because they are used to having someone else do so many jobs they must now do by themselves, such as manifests and flight plans. Also, the salary at a commuter is not as good as the majors and neither are the duty limits of Part 135. Cassel said, though, "There are some really good things about flying a commuter. I'm usually home every night and I don't have the time-zone changes anymore."

Another aspect of Part 135 flying could be with a charter company like Columbia, S.C.–based BankAir, although Chief Pilot Jeanne Cook admitted she doesn't get many résumés from airline types because "they tend to look for more glamour than we can offer." *BankAir* is an all-freight operator of approximately 28 aircraft, such as Lears, Jet Commanders, and MU-2s. Former airline pilots "tend not to like this kind of flying, but it certainly isn't because we don't want them. Personally, I think some pilots like flying for a 135 company more than a 121, since the pilots seem to have more input here."

An older pilot also may want to consider Part 91 corporate aviation as a viable option to the airlines. In a recent poll conducted by the National Business Aircraft Association (NBAA), only 25 percent of its members have a mandatory retirement age of 60 for pilots. (Whether this is legal is another issue because the regulation that mandates age 60 retirement applies only to Part 121 carriers. As this book goes to press, the FAA is planning to raise the mandatory retirement age to 65.)

Corporate flying is a new kind of flying for some. With virtually no duty restrictions, a corporation is free to fly a great deal in a short amount of time. Pilots may have regular schedules with assigned days off or, possibly, they could be on call and carry a pager to call them for flight duties. Corporate Citation pilot Hal Ross had only good things to say about his employer who he found through his own networking efforts. "I did not really meet any resistance to my age in my search for work. The most important thing to the companies I've spoken to is the state of my health. Once they find out it's OK, they listen to my qualifications. My experience and my safety record are what got me this corporate position. I'm on call seven days a week, but I probably only fly about two or three [days] each week."

Government flying is yet another possibility. While FAA regulations prohibit pilots older than 60 from flying for Part 121 carriers, there is no such age restriction for pilots who want to fly for the FAA. Lee Lipski said, "It's very frustrating to be a check airman in a Boeing 747 at 59 years, 11 months, and 29 days, and then find yourself totally disqualified three days later (when nothing else has changed). I can't fly a revenue trip because I'm now 60, but I could hire on with the FAA itself as an air carrier inspector and even fly their B-727 after 60."

The same is true for most state governments. Richard Wray, of the State of Illinois flight department, said, "There are no age restrictions here. One of our 13 pilots is 64, in fact. Age plays absolutely no factor in our deliberations on whether or not to hire someone. It's the person's qualifications we're most interested in." Wray went on to say that many state fleets are larger than those in Illinois, so more opportunity may be available in those other states. Most states give a preference to state residents and veterans. The largest aircraft in the Illinois fleet is a King Air BE-200.

A pilot looking for work must also consider the possibility of a nonflying job, and they are available if you hold the proper qualifications. A spokesman for FlightSatety International said they recruit their instructors from the ranks of retired military and airline pilots, as well as those who have found themselves between airlines. FlightSafety wants its instructors experienced in the specific aircraft they will run the simulator for. FlightSafety emphasizes, too, that it's a pilot's experience, not their age, that decides who gets the job. FlightSafety currently runs 36 training centers around the country and employs approximately 800 instructors. Many other flight schools may need instructors and it also may be possible to hire on as an instructor in a major airline's simulator training facility, such as that of America West.

Space limitations prevent covering all the kinds of flying jobs around, so don't forget agricultural flying, forest-fire fighting, or even basic or advanced flight instruction at a local Part 141 school. No doubt finding work as you get older is tough, but the successful pilots seem to be those who use their networking skills, a good list of possible employers, and plain persistence.

Air Inc.'s Kit Darby had this to say about pilots who believe they did not get hired, or will not be hired, because they are too old—past 45 years of age. "Get over it! Age is not a disadvantage. It used to be, but numerous lawsuits have made the airline very gun-shy because the federal law against age discrimination applies. Sure, fewer people are hired over age 45 or so, but few actually apply. These older pilots are their own worst enemies. It's now a much better environment to hire older pilots than in years past."

Landing a Flying Job Outside the United States

If searching for a flying job leaves you somewhere between angry and distraught, take heart. There's another source of potential work waiting to be tapped . . . flying overseas. But, don't expect flying work outside the United States to be easier to find than in the States. You might find it tougher to get hired by an international carrier. You also might find yourself traveling further to work each week, sometimes thousands of miles. Residency outside the United States also accentuates the enormous differences in living conditions and customs from those you're accustomed to.

Before you begin shipping résumés to Bahrain and Taipei, check one important item of your personality . . . your attitude. Carefully consider the changes you might put yourself through to keep flying, especially outside the United States. Some pilots look at the possibilities and turn in their Jepp bags forever, rather than endure weeks or, possibly, months away from home. One former Pan Am pilot, a type-rated Captain, found initial employment in Alaska, but complained about having to commute from Fairbanks back to his New York home each week. Then, just after he completed training, his new airline furloughed the entire class and he found himself working in the Middle East on a short-term contract with no benefits. When he returns home now, he hopes for a jump seat ride or buys a ticket, and he said he thinks the Fairbanks-Kennedy Airport trip was not such a bad ride after all.

If you're thinking of commuting 10,000 miles to work, ask yourself if this routine is practical for you. It's not difficult to pack everything you own into a few suitcases and move to Europe or the Pacific Rim if you're single, but if you have a house and family here in the United States, a cockpit job based in Hong Kong could make commuting next to impossible. If you don't take your family with you, a consideration might be how long you're willing to be away from them. How long will your marriage survive with you out of the country? One pilot, who requested anonymity, left the United States to fly freight in the Middle East with his marriage intact. After four months away from his bride, the letter came to tell him she wanted a husband who resided at least in the same country. He left Bahrain hoping to rescue the relationship, but found it was too little, too late.

Maybe a short-term overseas contract might be a better idea. You'll stay current and probably keep your family intact while you wait for times to improve in North America. Dublin, Ireland–based Parc Aviation, the pilot-leasing arm of Aer Lingus, regularly uses pilots on six- to nine-month overseas contracts on aircraft as small as an EMB-120. Another option could be to take your family with you to Saudi Arabia or Indonesia. But know how your family feels about this before you accept a job. How will your teenagers enjoy living in a land that doesn't have MTV or a Blockbuster Video, or where it might be impossible for them to stop with their pals at the McDonald's down the street? Some overseas positions might neither provide for nor encourage you to bring your spouse or family with you because, for example, some Middle Eastern countries don't allow their women the freedoms Western women have.

And, then, you have the security considerations. Americans often are in great danger in other countries merely because they're Americans. Is it a good idea to expose yourself to such conditions? If you decide the job is worth such a risk, then the job search begins.

A recent aviation magazine editorial said the time to network is before a U.S. pilot needs to look for work. Many pilots found international work through a tip from someone else—usually another pilot—who saw an ad or heard about someone looking for crews. In 1992, a number of former Midway Commuter EMB-120 pilots found contract employment in Belgium from a tip passed on through the local Air Line Pilots Association (ALPA) office. You'll have to spend time on the phone calling airlines, leasing firms, and old pilot pals for leads. As Parc Aviation's Tim Shattock said, "The more experience you have, though, the better your chances are."

Whether you get a tip from a friend or connect directly with a leasing firm or airline outside the United States, expect a market that doesn't give Americans preferential treatment. And, you'll have to meet international requirements that vary widely among countries and employers. Cargolux's Graham Hurst, vice president of flight operations, said the quickest way to find employment with a European Union (EU) airline today is to "get a European passport and a European license. We don't have any restrictions to hiring U.S. pilots except that we're supposed to try to find Europeans first." France and Germany, too, are notoriously tough places for U.S. pilots to find permanent work. British Airways is quite open about not hiring any pilot who's not either a British citizen or the holder of an European Community (EU) passport. U.S. pilot Larry Schweitz, now a Boeing 737 Captain for an Egyptian charter airline, said, "I think it's as difficult to find work overseas as it is in the United States, especially with the new EU. It gives European pilots a leg up on American pilots." he EU has united Europe into a single economic and monetary unit having a powerful central bank and a single currency by 1999, common approaches to foreign policy and defense, and centralized authority, in such areas as the environment and labor relations. The stance with labor will be for European employers to give preference in hiring to Europeans.

Besides the roadblocks the EU might create for U.S. pilots, a potential stumbling block for some could be the language requirements. While English is still the international language of air traffic control, many foreign airlines would like to hire pilots versed in another language. Former Northwest pilot Mike Henderson, now flying for KLM, said part of the requirement to fly as a permanent crew member at KLM is to "learn to speak Dutch to be able to make the normal and emergency cabin announcements." Henderson said he takes Dutch language classes at his Amsterdam crew base. Similarly, EVA Air encourages, although it doesn't require, pilot applicants to learn Mandarin Chinese. Shattock said, "Language other than English is not normally a problem (for hiring by Parc Aviation)." Cargolux's Graham Hurst said, "If you live in Luxembourg, being able to speak French and German is an advantage." A former Pan Am pilot said that, while flying for All Nippon, "I learned the Japanese numbers, just so I could do the weight and balance their way. I think the other Japanese pilots appreciated the efforts I made to learn their language, too."

The pay and benefits on international jobs vary greatly. When the above-mentioned pilot began flying a B-747 on contract for All Nippon from Tokyo, the pay and benefits of more than $100,000 per year were a significant increase over what he earned at Pan Am, where he was at the bottom of the B-747 pay scale. (The contract initially was between Pan Am and All Nippon. When Pan Am ceased operations, another outfit picked up the contract.) Henderson said his wages were about the same as what he'd been paid at Northwest.

Major medical care is available, although some contracts might not provide this benefit. Schweitz's individual short-term contract in Egypt (he heard about the job via word-of-mouth, instead of a crew-leasing firm) keeps him current on the 737, but, he said, "There are no benefits. Either party can cancel the contract with 30 days notice. I also have to pay my own way up and back whenever I return to the United States because I receive no pass privileges." At All Nippon, one pilot said, "The pass policy was fairly restrictive . . . but they gave us two positive space international tickets each year." As part of the basic compensation package, some international companies (such as All Nippon) provide living quarters for Americans living abroad, but many expect you to find your own lodging at your own expense.

Because all countries require some form of work permit for noncitizens (besides a U.S. passport), Americans find it helpful that many of the international airlines and corporations assist the employees in getting their paperwork in order. Most wait until the pilot arrives at the new-duty station to complete the paperwork, although Henderson remembered KLM sending him a complete packet of material before he ever left the United States. Another pilot, however, said that when he began flying for All Nippon, "We did all the work for visas and permits on our own."

How do pilots find overseas flying work if they don't hear about it from a friend? It can be difficult. Australia's Qantas said it "only advertise[s] for pilots in the Australian press." Some airlines, like Taiwan-based EVA Air, search for pilots with ads in "internationally distributed aviation magazines." Pilots should also consider word-of-mouth as a supplemental source.

Search all the aviation publications, both U.S. and international. These are usually available at the library. And, don't overlook the World Aviation Directory (WAD), also available at most libraries. Robert Orr, who flew freight for DHL from Bahrain, said, "I think the World Aviation Directory would really help because it lists addresses to international carriers and leasing agencies throughout the world." Finding out whether that company is hiring or even accepting résumés is where the pilot's work really begins, however. One pilot said he believes a pilot seeking work outside the United States "must spend the money to call people directly who are working for an airline or company you're considering."

Parc Aviation's Shattock said, "Our (contract) placement of a pilot can depend on the season of the year, a pilot's qualifications, and their experience level. A 737–400 Captain, for example, needs about 500 hours PIC to be hired. Although we hire mostly Captains, the First Officers we do use would also need at least 500 hours, plus a type rating." For an overseas flying job, U.S. pilots usually need some type of flying certificate issued by the host country. However, it's common for the host employer to arrange and pay for the certificate. Henderson said, "In Holland, they have a B-3 license issued by the Dutch equivalent of the FAA that allows me to fly here. It requires me to maintain my FAA physical and U.S. flying currency, however." Canadian Airlines pilot Brian Rasmussen, also a former Eastern Airlines pilot, took no chances with his future. "I became dual-rated with an ATP in both Canada and the United States while I was still flying for Eastern. When Canadian agreed to hire me, I was already current in Canada." To fly in Japan, one pilot said, "We had to start from scratch to qualify for a Japanese license, and the training was as tough as any I've ever been through." Because Luxembourg doesn't issue a pilot certificate higher than a private, potential Cargolux crews must obtain a Luxembourg validation to their U.S. license. Hurst says, "This means a U.S. pilot can fly a Luxembourg-registered aircraft as long as he maintains his American license and medical" (Figure 7-8).

Figure 7-8 Many pilots find flying outside the U.S. provides new career options. (Courtesy Embraer Aircraft.)

It's difficult to say whether the scale tips more toward a pilot finding international employment through a contractor or through the airline itself as a permanent employee. Cargolux's Hurst said, "We have about 30 nationalities (of pilots) working here, quite a lot of American pilots and flight engineers, about 20 out of 120 pilots." Most are permanent employees, but some are on short-term contracts.

Cargolux hopefuls must have 1,500 hours minimum time, with at least 500 hours of pure turbine time. If you don't have 500 jet, 1,000 heavy turboprop time will do. The airline intends to lease a few short-term aircraft, but there's a chance it could take on more permanent crew members. Cargolux will be one of the first carriers to fly the freighter version of the Boeing 747–400. Parc Aviation or Sam Sita in Monte Carlo handles crew leasing for Cargolux. Those seeking permanent positions apply directly to Cargolux.

While most jobs outside the United States are for air-carrier-rated pilots, some agencies and companies also hire corporate crews. Houston, Texas–based Aramco, for example, hires fixed-wing and rotorcraft pilots to fly in Saudi Arabia. One crew-leasing firm manager, who requested anonymity, said he also recruited regularly for rotorcraft and fixed-wing pilots in all parts of the world. His last round of contracts were for Learjet, G-2, and G-3 rated pilots to fly in the Middle East. His firm advertised the openings in newspapers of major cities, such as Houston, Dallas, and Miami, but said they try to stay away from interviewing pilots from places like California because "those pilots just want more money than we can possibly offer." His last Lear pilot was paid approximately $50,000 per year, with a 30-day vacation and a one-bedroom apartment provided in Saudi Arabia. He warned pilots, however, that in places like Saudi Arabia, even though the U.S. pilots work for an American company through their contract, "over there, the Saudis call the shots."

Both you and your family must prepare for culture shock if you fly outside the United States. An All Nippon pilot said, "At least half the American pilots over here (Tokyo) wouldn't touch Japanese food . . . it just tasted different. The Japanese are a very structured society too, so in the cockpit, there's always a boss, a worker, and an elder kind of attitude. These companies also have a very different attitude towards their employees and what their employees' responsibilities are to the company. They seem to expect the employee to be grateful for having a job, so they're really not very interested in providing great travel benefits or things like that. Even the scheduling is hard set by the company. The pilots have no input."

Schweitz said, "There's a tremendous culture difference (in Egypt), and if you're not prepared to keep an open mind, it can become a very difficult situation. There's a big difference in the way they treat foreigners in Egypt, too. It's a lot more than just going up in an airplane. Cairo is not Phoenix."

Certainly one of the brightest highlights of flying overseas is the tax break available to a U.S. pilot. If you maintain a permanent residence outside the United States for more than 330 days in a year, the IRS might allow you to exclude your first $70,000 a year from U.S. taxes. That tax break can afford you a significant pay raise for a one-year contract, depending on your salary. Any money you earn will be free of U.S. taxes. However, Mike Henderson said that, while his money was free of U.S. taxes, "We'll soon be paying Dutch taxes of about 12 percent for a U.S. pilot." Pilots who are not sure if their pay is subject to taxation by the host country can get that information through the host country's U.S. embassy in Washington D.C. or through the U.S. Embassy in the host country.

A flying job outside the United States is not easy to find or, sometimes, maintain, but if you're the adventurous type, it can be quite profitable. If there's any advice for a U.S. pilot looking for work outside of North America, Shattock probably said it best from Dublin. "Aviation is a very cyclical business with many good and bad times. I hope we're starting to climb out of this aviation recession and that will give us a need for more pilots. Don't ever give up . . . keep on looking."

Surviving the Loss of a Job

While it may at first sound a bit contradictory to offer an opinion on how to survive being put out of work in a guide designed primarily to help you find a job, it isn't really. Although job loss is a subject many pilots have never dealt with, losing a job in this business can be just as much a part of the industry as finding a new one.

As you'll read in the accompanying piece, most companies seldom disintegrate in a heartbeat. Most start with a slow, downward spiral that they never pull out of. The question is this: how do you prepare yourself for a job loss? Easy. By always taking the pulse of the company you work for. That means learning all you can about the finances of the company, their customers, when they add a base or a new city, and when they lose one. Read the *Wall Street Journal* and the *New York Times* online to hear what the analysts are saying.

If you fly for XYZ Corp. and are planning to upgrade in their new aircraft a few months after it arrives, and then you read that the company has had its second worst quarter in history, a bell should go off—one that says you may be lucky to keep your job, let alone worry about upgrading. If your airline begins retreating from cities, you should start asking questions about how that might affect your career. Plenty of other methods are available to learn what's going on within your own company, such as talking to the Chief Pilot, but why not also talk to people in finance? The flight department often hears about things after the decisions are made—not before.

But, if the inevitable occurs and you find yourself either furloughed or simply canned because the flight department or the airline shuts down, first, take the time to grieve. No rational person wants to be out of work and no pilot I know feels good about hearing their services are no longer needed. (One caveat is I'm assuming you did not do something to cause the company to let you go. If that's the case, you don't need to grieve; you need a new line of work.)

Assuming, again, that the company fell out from under you, the most important thing you can do is to believe there is hope and, although things did not work out quite the way you'd expected this time, they will improve. You will find a new job. But, no matter what happens, don't ever give up. Take it from someone (me) who did just that after he had three jobs pulled out from under him. I left flying for a while and was totally miserable as I searched for work in a new field, looking for that real job that never appeared because my heart, my passion, has always been and always will be in aviation.

There's more to the loss of a job, too. When I say don't give up, this is only the beginning. That allows you to go out and have a drink to your dead job, but also to begin focusing your energies, not on a job, but on the *next* job because there will be one, if you don't get lazy. It's easy to find a job after you lose one, but will it be a job that pays the rent, or one that takes you up a step from where you were? That's important.

Go back to the plan I hope you've put together and look at the kind of position you've been working in. Is it really where you want to be? If you've been flying freight and you hate not being around people, perhaps in addition to the job loss, this is the boot you need to send you in a totally new direction. Why *not* apply to a regional? Or, if you are a people person, why *not* begin looking in the corporate sector?

Surviving the loss of a job boils down to planning ahead, just as you would when planning a flight. Many pilots I've interviewed recently explained why they had a degree in business or some fairly lucrative subject: "Just in case I couldn't fly." I'll grant you that never flying again is an extreme case and something I hope few of you reading this book will ever face, but it highlights the strategy again—plan ahead!

So, practically, this means you never toss away all those business cards from corporate or airline pilots you've met. You never know when you may need a reference or may want to pick up the latest details on the hiring market for corporations in your area.

Midway Airlines Revisited

When the phone rang about 11:45 P.M., I was already asleep and, now, I was quite annoyed at being disturbed. The past few weeks had not been good at Midway Airlines or its subsidiary, Midway Commuter, for whom I flew a Brasilia, as employees began to wonder when they would hear more official details about the buyout deal from Northwest Airlines. Nearly a month had passed since the buyout plans were announced, but with each passing day, the anxiety levels rose among the employees. Many never said anything specifically, but you could see it in their eyes. They were worried. And, they had good reason. We'd been sitting in Chapter 11 bankruptcy for six months, and we knew Midway could not last forever. The word had just come down that Northwest was pulling out of the buyout deal, for reasons unknown. The date was November 13, 1991 (Figure 7-9).

The voice on the end of the phone was our MEC Chairman, Ty Hackney. "Rob! Wake up! . . . It's all over, buddy! Midway ceased operation, effective at midnight tonight." So, there it was. The final struggle was over. The work ALPA had initiated many months before and taken to the MEC at Northwest that began the buyout talks in the first place had all come to nothing. I thanked Ty for the call, hung up, and went back to bed. It was not a sound sleep, though. I was unemployed.

Figure 7-9 Midway Airlines: R.I.P.

In 1979, Midway Airlines was the brainchild of Irving Tague and David Hinson (former FAA Administrator), and the darling of the deregulation set. A new airline for a new age of airlines. But, Midway Airlines filed with the now disbanded Civil Aeronautics Board (CAB) on October 13, 1976, nearly three years before the dam of airline deregulation broke open— Midway was the last airline, in fact, to file before deregulation became the law of the land.

The two entrepreneurs believed the concept of a no-frills, Southwest Airlines–type carrier with a peak and an off-peak fare structure to be a sound one. The first Midway Penny Fares promotion in 1979 would offer passengers a standard $30 one-way trip to Detroit for only 30 cents. But, first, this new airline needed airplanes, money, and an airport. Initial capitalization of Midway Airlines was a mere $5.7 million.

Finding an airport was the easiest of obstacles to overcome as Chicago's Midway Airport was then a virtual ghost town since the airlines and their runway-hungry Boeing 707s and DC-8s had pulled out in the mid '60s. Initially, employees numbered less than 200, but using three DC-9-14s purchased from TWA, Midway took the name of its home airport for its own and, in 1979, began serving Cleveland, Kansas City, and Detroit. Frank Hicks, Midway's first director of flight control, knew the airline was catching on when he relinquished his near-the-terminal parking slot for one four or five rows further away to make room for more passengers

The airline continued to grow. In 1980, Midway Airlines added five more DC-9s to the fleet, as well as four more destinations: LaGuardia, Washington National, Omaha, and St. Louis. To mark the airline's first full year in service, Midway made 850,000 shares of common stock available in the airline's first public offering. The stock sold in just two hours.

Because of the 1981 PATCO strike, Midway reversed a number of earlier route decisions and dropped service to Boston and Orlando. Many people believed this retrenchment displayed the kind of rapid-fire decision-making necessary to run an efficient airline. Captain Jerry Mugerditchian, who became the Midway pilot's Master Executive Council (MEC) Chairman in 1985 and who is now ALPA's vice president-administration and a United Airlines First Officer, said, "Midway had experienced excellent growth and expansion in 1981, but things were changing rapidly. There was little consistency to pay, work rules, and minimal training. Airline management was very short-staffed." Without clear-cut policy manuals, the airline changed the rules whenever they thought necessary. Pilots don't like lots of changes.

The first two MD-80s appeared in 1983, as Midway changed its marketing strategy to reflect its new niche—MetroLink, all business service at coach fares. In mid-1984, Midway management agreed to purchase the assets of bankrupt Air Florida. While the airline, according to Dick Pfennig, then Midway Airline's vice president of flight, was not quite ready for an expansion yet, "Air Florida was one of those opportunities that was there." This move produced a new north-south presence for the airline through the new Midway Express and the fleet of Boeing 737s, which the airline gained with its new Miami base. Midway Airlines' reach now increased to the Virgin Islands through Miami. And the airline now employed 2,000 people.

By the end of 1984, the airline carried 1.3 million passengers to 13 cities through its Midway airport hub. Also, in 1984, difficulties with management arose when it began integrating the Air Florida pilots into the Midway Airlines seniority list. "That was the straw that broke the camel's back," Mugerditchian said. By the middle of 1985, ALPA was representing Midway pilots.

Again, in 1985, the company began to reconsider its operating niche as the search for profits continued. Midway had suffered losses in both 1983 and 1984. Midway Metrolink was a failure and Midway Express was out because passengers and travel agents found them to be too confusing. John Tague, then Midway's director of airline planning, said, "It (1985) was a very difficult time for the company. But we truly believed that what may have been painful for a few would be beneficial for many."

David R. Hinson spearheaded the recovery plan. Hinson had recently been named chairman and chief executive officer for Midway Airlines. The plan was a no-holds-barred attack on excessive costs, including a 10 percent reduction in personnel, a suspension of all capital expenditures, and the elimination of Newark and Topeka from the route structure. Midway decided two-class seating was the direction to take and returned their MD-80s to the lessors.

By mid-1985, however, the airline began turning a profit and added three new cities. By December, Midway carried two million passengers in one year for the first time. By 1986, improved yields became the buzzword. Midway reported net income of nearly $7 million and annual passenger loads inched closer to three million. The carrier also established its first links with local commuters to feed the jets.

As 1987 opened, the airline's management believed Midway rated national-carrier status as they placed an order for eight MD-87s and options on 28 more. Midway President Jeffrey Erickson announced "Project New Attitude." Midway's vice president of customer service, Lois Gallo, explained, "The program recognizes that Midway employees are the experts at handling customers and gives them greater latitude to do just that. We regard our employees as responsible, capable experts, and we trust them to make proper decisions." Midway employees were overjoyed.

Midway also entered the regional market with its own aircraft in 1987, flying them under the Midway Commuter logo with Dick Pfennig, Midway's vice president in charge. Midway purchased regional Fischer

Brothers Aviation of Galion, Ohio, as a subsidiary to Midway Airlines. Pfennig made the agreement with Fischer Brothers and had the airplanes in the air just 23 days later from a new Springfield, Illinois, base. Using ten Dornier 228s—homely, noisy, but efficient aircraft—the commuter began carrying passengers to MDW from Springfield and Peoria, Illinois; Green Bay and Madison, Wisconsin; and Grand Rapids and Traverse City, Michigan.

Midway ended 1987 surpassing the three-million passengers carried mark by nearly 751,000. Annual departures totaled more than 78,000 and employees numbered about 3,500. Midway Commuter carried 433,000 passengers and Midway Airlines now accounted for 65 percent of total traffic at Midway Airport. Total airline departures numbered 123,844.

Just when most passengers and many of the employees began to wonder how the airline would top 1988, Midway, concurrent with its tenth anniversary, announced the opening of a second hub in Philadelphia, an event that was precipitated by the untimely demise of Eastern Airlines, and a glut of equipment and gates. Midway outbid USAir and planned to offer by the end of the first quarter of 1990, 70 daily departures from PHL with 16 DC-9-30s acquired from Eastern, though none of the Eastern employees came with the deal.

New destinations from Philly would include 20 U.S. cities, as well as Toronto and Montreal. Midway also established a marketing agreement with Canadian International Airlines.

Midway Commuter also saw ALPA arrive on its property in 1989, with a separate MEC structure from that of the pilots who flew the jets. One Midway employee said 1989 was "a turning point into the next decade." "Turning point" would prove an understatement. While the airline had fought its way past other economic difficulties, 1990 would see a fight that would make Daniel in the lion's den look like a walk through the park on a sunny summer afternoon.

Midway management underestimated the tenacity with which USAir would defend its East Coast turf. On October 19, 1990, a year after announcing the opening of a second hub, David Hinson told employees of the airline's intent to sell the Philadelphia hub to USAir. Hinson said "Our problem at Philadelphia is straightforward. At 60 departures per day, we are not big enough to create an efficient operation. We need approximately 100. . . . We need capital to buy airplanes, train personnel . . . financial institutions are unwilling to lend us the money."

Hinson believed the sale was a necessary one, especially after the Iraqi invasion of Kuwait sent fuel prices soaring. "If we try to continue as we are and fuel prices do not decline, we may place our company in serious financial difficulty." Hinson also admitted at the time that "economic circumstances are changing so fast, it's difficult to see too far ahead."

The plan was to reduce the number of employees, ground some of the older DC-9s, eliminate the 737s entirely, but leave the commuter in tact. The commuter had just begun flying the 29-seat Embraer Brasilias and had a fleet of 28 aircraft of its own serving 18 cities. Hinson's final words during the announcement were that "we need to back up a little and restructure our company in order to have the financial strength to fight another day." Unlike General MacArthur in the Philippines, however, Midway would never again return in such strength.

On October 23, 1990, the airline suspended its regular dividend payment. The next day, Standard & Poor's downgraded its rating on $54 million on Midway's convertible exchangeable preferred stock to single C from triple C. "Given their cash position and balance sheet, and the likelihood that they will continue to be adversely affected by high fuel costs in the weeks and months ahead, I, for one, can't see how they can last much beyond the first quarter of 1991," said one New York–based securities analyst. "In retrospect, Midway's decision to buy the Philadelphia hub from Eastern for $100 million would have been a mistake even if there had been no fuel crisis," said the analyst.

As for the possibility of another airline taking over Midway—something rumored for weeks at this point—Hinson said, "The airline is not seeking a buyer, but if the move were in the best interests of the stockholders, it would, of course, be considered." Hinson added, "If Midway was to be taken over, it would have to be a hostile takeover, because right now we have no intention of selling it."

In October 1990, ALPA helped formulate a plan to make Midway attractive to Delta. Delta, however, was in discussions with Pan Am. ALPA then approached Northwest. While both the Midway and Northwest MEC, along with ALPA national, began the initial dialogue at Northwest, efforts were swift that would bring Midway and Northwest management together to begin direct negotiations themselves.

The sand in Midway's hourglass was rapidly running out.

In March 1991, Midway Airlines filed for Chapter 11 bankruptcy protection. By June, the carrier had lost $37.1 million for the year. MD-80s were being repossessed at the gates. The airline had no money to get the lavatories serviced on the Brasilias. The swift turboprops disappeared, one by one, as pilots on both the jets and the commuter were furloughed. Morale hit rock bottom. An air of depression filled the corridors of the terminal like black smoke. Some days, passengers tried to comfort the employees aware of the state of the airline. Each day, employees would arrive at work, wondering if this would, indeed, be their last. In a desperate 11th-hour move, Midway made a deal with Fidelity Bank in Philadelphia to lease 17 old Eastern DC-9s at one-third the cost of some of the current aircraft.

This proved to be too little too late. Many ALPA pilots believed those new aircraft were worn out before they arrived. Midway pilots saw things they'd never encountered with such regularity before—Minimum Equipment List (MEL) stickers all over the cockpits, as a lack of parts pushed the capabilities of aircraft and their crews to the limits.

Then, just when many had given up, the deal with the white knight out of the Northwest appeared. On September 24th, Northwest Airlines agreed to purchase the leasehold interests of the 21 gates Midway Airlines held at MDW for $20 million. The agreement seemed to pave the way toward an agreement to purchase all Midway assets. Some analysts believed, however, that a deal would be struck only for key Midway assets, such as Midway had done with Air Florida.

Jack Hunter, with Rothschild Securities in Chicago, said, "I don't think (Northwest) would want to assume Midway's debts. Northwest has debts of their own . . . I don't think they could afford an outright purchase." Other airline analysts believed a purchase was forthcoming and this posturing was an attempt on the part of Northwest to catch up with United, American, and Delta.

Southwest and TWA were also rumored to be interested in the carrier. Southwest, indeed, put forth a $109 million offer for the airline, which the bankruptcy court turned down in favor of the $174 million offered by Northwest.

By mid-October 1991, Midway employees began to breath a bit easier. Smiles seemed to return to the terminals as rumors flourished about the results of the Northwest buyout. Quickly, initial plans were offered by Northwest for training and the overall transition to the red tail look. Of Midway's 4,300 employees, 3,800 would have jobs. Taking employees was not a part of the Southwest plan.

But by early November 1991, however, details and plans from Northwest had slowed to a trickle, and anxiety levels again began to rise. Had the deal hit a snag? No one was talking. On November 11th, Midway employees read a *Wall Street Journal* article titled, "Northwest Air Seems Hesitant On Midway Bid." "Northwest Airlines is sending signals that it is having second thoughts about its proposed purchase of certain Midway Airlines Inc. assets."

At first, it was thought as a possible ploy for wringing a bit more out of Midway for a little less money, but all were unaware of the flurry of letters being swapped between Midway and Northwest management. In the letters, Northwest denied Midway's claim that it ever had a contract to buy Midway's assets because Northwest's Board of Directors had never consummated the deal.

Northwest also claimed Midway management had provided erroneous revenue data to the Department of Transportation, which Northwest had been using to organize its original bid for Midway. A letter from Northwest's Senior Vice President and General Counsel, Richard B. Hirst, said "Midway admitted to Northwest that its DOT data are unreliable." This information only became available, Hirst continued, after Northwest had paid Midway the $20 million for the gates. Northwest was also concerned about incurring "substantial environmental liabilities."

Then, came the final paragraph. "Upon review of the transaction, Northwest's Board of Directors voted last night not to approve the Midway transaction. Accordingly, the proposed transaction between Northwest and Midway for the acquisition of the remainder of Midway's assets in terminated."

At midnight on November 13, 1991, Midway Airlines ceased operations.

On December 24th, federal bankruptcy judge John Squires approved the return of $68,000 from the Christmas party fund to employees because they'd all lost as much as three weeks pay, as well as any unused vacation time, when the airline shut down. Each employee received about $17. The judge also approved the paying of bonuses to some 57 Midway employees who hung around to shut the airline down after November 13th. President Thomas Schick, for instance, picked up $50,000, while Chief Financial Officer (CFO) Alfred Altschul and Customer Service Boss Lois Gallo each received $30,000.

In a letter dated February 24, 1992, U.S. Department of Transportation's Inspector General A. Mary Sterling told Congressman James L. Oberstar (D-MN) that, although there were certainly differences in the software used to generate passengers at Midway and Northwest, "nothing came to our attention . . . to indicate that Midway deliberately filed misrepresented sample data."

Nearly 1,000 Midway pilots read these words in the newspaper as they waited in line to collect their unemployment checks.

The Reality of Flying Today

When asked if piloting careers were less appealing in the early part of the twenty-first century, Louis Smith of FLTops.com answered, "The recent collapse of pilot compensation and pension infrastructure at the passenger airlines has removed all the glamour and much of the appeal. The pundits argue over whether it's a permanent alteration of the pilot career value, so it will take a few labor showdowns and work stoppages to determine the inherent economic value of airline pilots. "Unions have more power with profitable companies, so a profitable passenger airline sector will restore some of what has been lost. Pilot unions will become much more militant.

"Flying freight was, at one time, considered a default career for aspiring airline pilots. Now, it has become the 'crown jewel.' Business is booming for air freight because boxes are not afraid to fly and they don't mind waiting in line, even in the rain. The Internet, and the shipping it generates, has fueled much of the growth and profits for the freight carriers.

"Conversely, passenger airlines have lost much of their high-yield business passenger to the fractionals and business aviation, and we expect that trend to continue until the airlines can solve their TSA logjams. In addition, the passenger airlines have seen information technology and video conferencing cutting into the growth of their business traveler segment, which provides a significant portion of their profits."

The Moral

So, is latching on to the right flying job worth everything you're going to need to go through to turn that dream into a reality? I think it is or I wouldn't have spent months trying to organize all these ideas as a resource to help plan your journey to the left seat.

But, if this book has a moral anywhere, it's probably that a man or woman who wants to live to a ripe old age as a pilot, for an airline, a charter company, a fractional, a corporate operator as a flight instructor needs a sense of humor. Seriously. You're going to have to take a few losses along the way to a solid flying job. It simply comes with the territory. That's one of the best parts of flying in a multipilot cockpit. You have someone to tell your troubles to and you don't need to write them a check at the end of the trip.

That's why I called my blog Jetwhine (Figure 7-10). You haven't heard that joke yet?

OK. How do you tell the difference between a pilot and their jet airplane? The jet stops whining when the engines are shut down!

Make sure you visit www.jetwhine.com or e-mail me at rob@propilotbook.com with some of your hiring and flying experiences. We'll share them with everyone else because that's how this industry works.

Now, for a little fun online in Chapter 8 and a look at some great new airplanes in Chapter 9.

Did the blogs do it?

By roh on December 12th, 2006 | What do you think? »

Although there are plenty of skeptics who will think blogs, podcasts and
Word of Mouth advertising are a huge waste of time, these new
communications tactics are changing the way people think and at a rate
faster than almost anyone can keep up with.

Have you seen the redesigned elements of the Eclipse 500 VLJ? I'm not
a company insider, but I certainly wonder how much of the redesign
would have happened if the Eclipse Critic blog had not begun to field
hundreds of comments about the aircraft.

Regardless of where you stood, look at the people around the globe
who took part in talking to the Brazilian government, some through this

search

**Subscribe to Jetwhine below
via RSS or e-mail**

Subscribe Via RSS Feed

Subscribe Via Email:

enter email   go

Delivered by FeedBurner

Pages
Home
Jetwhine's Rules of Engagement
About Jetwhine's Editor
Contact

Figure 7-10 www.jetwhine.com: aviation buzz and bold opinion.

8

The Pilot
and the Internet

Throughout this book, I've made it clear that a pilot looking for a job today needs not only basic computer skills, but a good working knowledge of the Internet, as well as access to a high-speed Internet connection. As we go to press, Microsoft's Windows Vista operating system (OS) is about to appear. Be certain any software or hardware product you purchase is Vista-capable. Here are a few other ideas to remember.

Computerized Logbooks

One of the great tasks computers handle well is managing numbers like your flight times. I'm a proponent of computerized logbooks (see the accompanying article) because they take most of the drudge work out of filling out applications and keeping track of your time in the dozens of categories airlines and corporations seem to want.

One I have used for years and found easy to use is Aerolog for Windows (Figure 8-1) from Polaris Microsystems: www.polarisms.com. The benefit to me—once all the data is loaded into the software—is every time I want to update my résumé, I open the software, start up a report I organized once, and, in just a few seconds, have updated totals. But a computerized logbook is much more than simply one report. It offers the flexibility to break down pilot time in any way imaginable. Aerolog contains its own copy of an FAA Form 8710 that you need to fill out before a check ride. Click this and, in five seconds, you have all the numbers you need to transfer to the form. Because I fly for a Part 135 charter operator, my logbook also tracks the flight and duty times the Feds want to see occasionally to verify my compliance.

Other computer logbooks I've seen, though I haven't necessarily tried, are Left Seat and NC Software, a company with a variety of platforms for

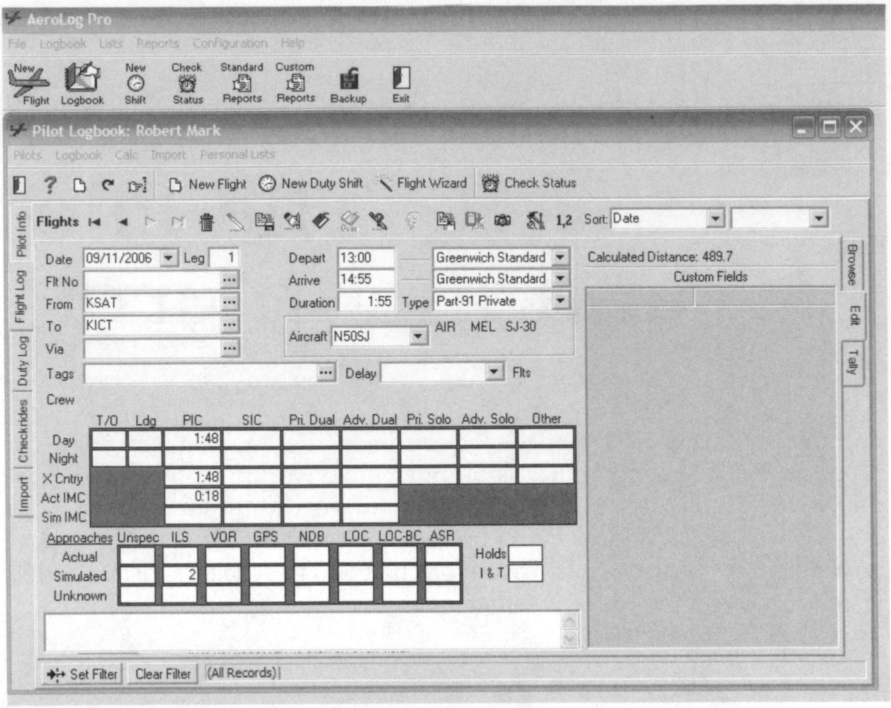

Figure 8-1 Aerolog for Windows is one of many computerized flight logbooks.

most any digital logbook—www.nc-software.com. Others are RWS Software—http://pilotlogbooks.rws-store.com, and Airlog—www. airlogsoftware.com/index.html.

One major convenience of a computerized logbook is the data needs only be entered once before a good computer can slice and dice it properly. For example, when I add a new flight by typing in N45ML, my logbook already knows this aircraft is a C-550. It also knows I have organized a special tag to track my jet time separately from my other flight time. When I ask the logbook to tell me how much jet time I have, it searches by the tag for jets and displays the result. Best of all, if someone calls me and asks how much Citation time did you have between March and September 15, 2003, I can put them on hold and be back in less than a minute with the answer.

Just like on an airplane, a computer creates nightmares for everyone. While this is a hazard when using an electronic logbook, it won't cause chaos if you regularly back up the program's data files. I make two backups of my logbook after every flight. It takes about 30 seconds to accomplish this task, and it has saved me three or four times already, such as when the hard drive on my system simply disintegrated a year ago. After the new hard drive was installed, I reinstalled the program, restored the data from the backup file, and was back in business in no time. Always, always back up your computer's data files. And, that includes your résumé and cover letters.

Blogs

Before we talk about traditional online message boards, I want to touch on a new technology that has evolved since the earlier version of this book was written ... blogging.

To the uninitiated, blogging might seem to be nothing more than another simple message board or news site. But blogging is much more than that. I like to think of blogging as a web site with personality. They're easy and inexpensive to create, so much so, that thousands are created each day all over the globe. There are dozens alone that relate to the aviation industry.

One that just happens to come to mind is the one I write at Jetwhine.com. where I offer the world my two cents on a variety of aviation news. Although some blogs only pass along the news of the day, the best do offer analysis as well. A few other essential blogs include "The Digital Aviator.com" and "Aviation Views.com." A central location of importance as well is "The Thirty Thousand Feet.com" aviation directory, which is a powerhouse of aviation resources.

So why should a future aviator care about blogging, especially if they have no desire to create one? Because blogs are not normally run by for-profit companies like CBS or CNN or Fox News. That means bloggers offer readers perspectives on a topic that they may well not see anywhere else. Some aviation bloggers evolved through the traditional journalism streams while others are simply people with a passion for aviation. Another variant of the traditional text blog is the video blog and an audio version, also know as a podcast.

Because some blogs were created by writers, some will just naturally be much more engaging than others. It's the reader's job to sort the good from the bad. And, quite honestly, there is quite a bit of junk on the Internet, so take what you read with a grain of salt until the author wins your confidence. Most blogs will also give readers a chance to offer up their opinion on a topic as well. Bloggers, in fact, welcome the dialogue. And most importantly, share the address of a good blog with your friends. Word of Mouth marketing is how good blogs can become great.

Once you find a good blog or two, or ten, you can easily subscribe through a system know as RSS feeds. These allow the reader to receive the highlights of selected blogs in one central location on their desktop—a news agrregator. This avoid the effort to constantly remind yourself to visit a web site to see what's new.

Online Message Boards

A few years ago, a discussion of online message boards focused on aviation would probably have been limited to Aviation Special Interest Group (AVSIG) on CompuServe. But, as with just about everything else that relates to computers and the Internet, last week's news is old news online. Most of these boards are free, although some do offer extra services for a price.

- **AVSIG**—www.aero-farm.com—is the granddaddy of all aviation sites, a kind of virtual water cooler where aviation people hang out, meet new people, and talk about just about anything aviation-like (and, no, I honestly have no idea how the system operator at AVSIG, Mike Overly, came up with that domain). There are also a dozen subboards to the AVSIG alone that focus on subjects such as commercial airlines chat, career chat, corporate jobs, training issues, and just plain news items people want to jaw about. A hot discussion item as I write this: Will the FAA back down after years of supporting the mandatory age 60 retirement for commercial pilots? The discussions are heated at times, but they're always interesting and informative because the talk comes from the people right there in the trenches.

  What also makes this board (and hundreds of others out there) truly valuable is the ability to post a message in search of advice about your career decisions or to find someone who just interviewed with the company you received a call from.

- **PPRuNe**—www.pprune.org/forums—I mean this in a truly nice way, but, boy, this is a crazy place. PPRuNe is an acronym for Professional Pilots Rumour Network. It was started a long time ago by a Virgin Atlantic Boeing 747 Captain in England, so the site will definitely offer you a European flavor to the aviation world. The place is absolutely swarming with professional pilots and people who want to be. PPRuNe also has a set of rules they ask you to abide by when you sign up for access. Trust me when I say they *will* let you know if you violate any of those rules (Figure 8-2).

  Most of all, though, PPRuNe, is a solid network for information about everything aviation. There are boards for Freight Dogs—that's also a nice term—military aficionados, flight training and job interviewing help, aviation history, and special boards devoted to Africa, the Middle East, Asia, and Canada. This site is a must visit.

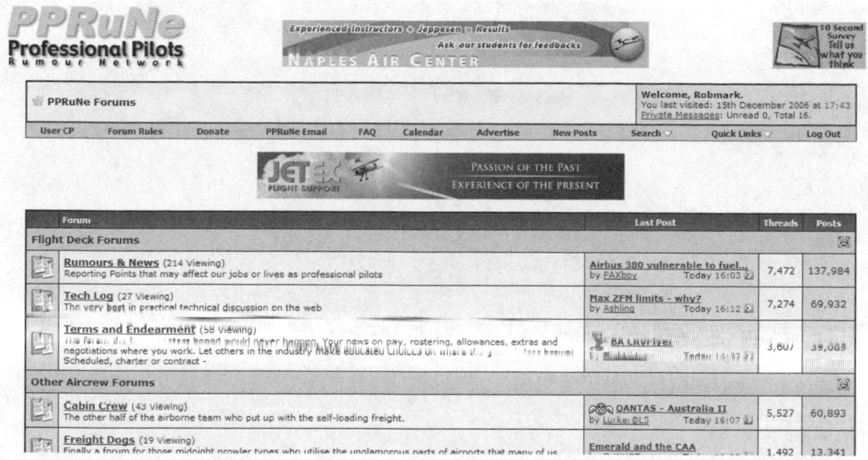

Figure 8-2 PPRuNe.org is an international pilot message board.

- **Pilot Career Center**—www.pilotcareercentre.com—I don't have a great deal of long-term experience with this board, but I like the feel of it. It's made up of pilots from Australia, the U.K., the U.S., Canada, and other parts of the European Community (EC). It posts some job information, but what I like most of all are the stories told by pilots with some of the more out-of-the-way jobs, such as Air Tanker Bird Dog pilot, Float Plane Instructor, and a C-130 Hercules pilot. I stumbled across this site the first time just by luck. I hope when you find an interesting site, you'll ship the address back to me, so I can check it out, too.

Flight Explorer

While Flight Explorer—http://www.flightexplorer.com—does not fit into the strictest of guidelines for a message board, it does offer a unique product to pilots—a real-time look at live air traffic anywhere in the United States and the world. Simply add in the airport you're after and you'll see the repeat—with data tags following the aircraft—from any region you choose (Figure 8-3).

Flight Explorer also offers a succinct listing of airline schedules that enable you not only to pick out the particular flight that grandma is on from Miami, but to see the flight symbol in real time and learn where the airplane is. This service is not free, but it is pretty neat.

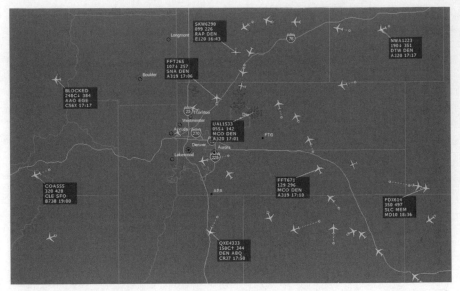

Figure 8-3 Flight Explorer allows pilots to view live air traffic.

Online Message Boards—A Taste

If you've never visited any of the online forums on CompuServe, AOL, or anywhere else or, perhaps, you have never been online before, you simply can't imagine what happens. Here's a taste of a forum from AOL that was downloaded from Greg Brown's Aviation Careers Forum.

The subject this night was law-enforcement flying, so not only will you see a bit of what online chatting looks like, you'll learn something about flying for the Drug Enforcement Agency (DEA). These are not the real IDs of the people in this session, however, except for "PaperJet," who is Greg Brown from www.paperjet.net.

PaperJet: Welcome to "Law Enforcement Flying" on the Aviation Careers Forum.

PaperJet: Our guest is "Tom," DEA special agent and pilot.

Tom: Glad to be here.

MDiprosper: Ok, what do you fly? And how long have you been with law enforcement?

Tom: I am currently flying the MD-500&BO-105 helo along with the Merlin 3B &4C.

Tom: I started my L.E. career in 1971.

Tom: After a seven-year tour as a Naval Aviator with USMC (US Marines).

PaperJet: Tom, is your primary duty intercepting drug runners?

Tom: We don't chase other A/C—that's Customs job.

PaperJet: So what are your primary flying duties?

Tom: We mainly do aerial surveillance, transport some prisoners/witnesses, and evidence.

Tom: Air-to-ground surveillance.

PaperJet: Is the aerial surveillance electronic?

Tom: We use stabilized binocs and Flirw/video (Infrared) cam.

Tom: We have infrared equipment but I can't comment on locations or plans.

PaperJet: Hi, Bill. Go ahead with your question.

Bill: Are the tethered balloons doing any good at the s/w-ern boarder?

Tom: The tethered balloons are operated by U.S. Customs—I have no knowledge of their efficacy.

Bill: Tnx.

PaperJet: Tom, because customs chases the drug runners, what is the purpose of most DEA surveillance?

Tom: Our air-to-ground surveillances are done in conjunction with ongoing criminal investigation.

Tom: Usually undercover operations.

PaperJet: I know your title is Special Agent/Pilot. What percent of your time is spent doing investigating on the ground?

Tom: Up until about six years ago, I was doing some undercover work— now most of my time is spent flying and staying current in the various a/c.

PaperJet: Jan, go ahead with your question.

Jan: What background do you have to have to get into this line of work?

Tom: A college degree and preferably some previous L.E. experience, although that is not absolutely required.

PaperJet: Taja, go ahead with your question.

Taja1983: What is the age limit?

Tom: The age limit for entry to DEA is 37.

PaperJet: Bill, go ahead with your question.

Bill: Some of the TV coverage has shown night-time surveillance. Is that done with FLIR?

Tom: That is correct.

Bill: Tnx.

Rich: Do you join the DEA, and then hope for a flying position?

Tom: We only have 110 total pilots in DEA—you must enter and qualify as a Special Agent, have two years "on the street," and then make application for entry to the air unit.

Tom: The Special Agent Basic School is in Quantico, Va., and lasts approximately five months.

Rich: Competition pretty fierce?

Tom: Yes it is.

PaperJet: Taja, go ahead with your question.

Taja1983: How many agents are in line waiting for a flying position?

Tom: Sorry, Taja I don't know—but we have a "pool" of qualified agents waiting.

Taja1983: Thanx . . . see you later.

Sue::) There are a couple of us ladies out here who were wondering how your flight suits compare to the Naval Flight Officers. And, seriously, how long did you fly in military prior to DEA?? <G!>

Tom: We wear green flight suits on helo duty and they look much better on us than NFOs!

Tom: I flew seven years in USMC.

PaperJet: Hi, GKenn! Go ahead with your question.

GKenn90806: Is a degree required prior to Quantico?

Tom: At present, that is affirmative.

GKenn90806: Two or four year?

Tom: Four year.

PaperJet: Triplebus, go ahead with your question.

Triplebus: Do DEA use txp's during intercepts and what is ATC's role?

Tom: We do not do intercepts—that's U.S. Customs.

Triplebus: Gotcha.

PaperJet: Tom, what's a "typical" day like on the job? Is it as exciting as it sounds?

Tom: I have to say that the flying is very exciting and rewarding—up until last year we were doing a lot of flying in South America. Very challenging.

PaperJet: GKenn, go ahead with your question.

GKenn90806: What type of aircraft?

Tom: In South America, I flew a CASA 212, in the early part a C-47.

PaperJet: Tom, are we making any headway against drugs in SA and Central America?

Tom: Now we are getting political.

PaperJet: Something like the C-47 used for surveillance? Or is it a gunship?

Tom: No gunship—we flew support missions for National Police of country supported.

PaperJet: Interesting! Sounds like good use of your Marine background. Scary very often?

Tom: It was great fun and exciting also very challenging.

PaperJet: Tom, are the requirements similar for most law-enforcement jobs?

Tom: I would say that probably they are—stringent background investigation.

PaperJet: How could someone prepare if that's what they'd like to do for a career?

Tom: A lot of our recent recruitment has come from state and local police agencies.

Tom: Although that does not, by any means, limit anyone who should be interested.

PaperJet: Is a law-enforcement degree beneficial?

Tom: I would have to say yes, although I have never been directly involved in recruiting.

PaperJet: How does one apply to the DEA?

Tom: We have offices in all major cities—just call and ask for the employment package (Special Agent)—it has all the forms and info.

PaperJet: I have one last question. I hear from lots of retiring mil. helo. pilots. Much demand for rotor?

Tom: It is about 50–50 in that regard.

PaperJet: Dual ratings desirable?

Tom: Yes—most of our pilots are dual-rated but that is not a requirement.

Fred: Tom . . . do y'all still do night no light ops into outlying airports in our area . . . (please don't arrest me for asking..lol).

Tom: Most of our night ops to outlying airports would be training/proficiency events.

PaperJet: SkyJock1, go ahead with your question.

John: Sorry I was late. What is the typical number of hours (helo) to be accepted into the DEA?

Tom: Just having a helo rating alone would not qualify you for entry to DEA—our helo pilots must possess a Comm helo and helo instr rating 1 250 hours 1 DEA check ride to perform operational DEA missions.

PaperJet: Welcome to "Law Enforcement Flying" on the Aviation Careers Forum.

PaperJet: Our guest is "Tom," DEA special agent and pilot.

PaperJet: Any more questions for our guest?

Tim: Tom, re Fred's earlier question, do you folks do night training ops lights out?

Tom: No we don't do that—some surveillances require it and we have waivers.

Tim: Thanks, Tom.

PaperJet: Well, Tom, thank you for joining us tonight!

Tom: You are very welcome—I enjoyed it and hope that it helped!

PaperJet: Is it OK for folks to e-mail you with questions they come up with later?

Tom: Sure—no problem.

PaperJet: Thanks and have a great evening!

PaperJet: How about a round of applause for Tom!

Tom: Same to you.

These forums take place in real time, so you'll sign on and watch all of this text scroll across your screen as you listen in—figuratively speaking. I've found these interactive forums to be of great benefit, because you can ask questions right as it is all happening—no matter where you are in the country. Unfortunately, they are bandwidth heavy and seem to be dropping in number.

Stay Sharp for the Simulator Ride on Your PC

No doubt about it, getting the call for an interview is the best news you can ever receive—short of hearing you've gotten the job, that is. The key to moving from the interview stage to the hired stage involves a number of steps, not the least of which is often a simulator ride. Again and again, companies report that one of the major reasons pilots are not hired is they are not able to walk the talk when it comes to flying instruments.

Sometimes it's simply nerves, but most of the time, pilots just haven't practiced enough, relying on the last few approaches they've flown to get them through the test. When asked about taking some simulator time before the simulator check, most pilots tell you they will, or even that they should, but let's face it, buying simulator time is not always cheap. And, then, you have to make the trip to the airport or somewhere else if no simulator is available close by.

Yadah, yadah, yadah!

If you want to pass a simulator check ride, you have to know more than the basics. You must be able to flawlessly display them under stress. When you add that to the fact that your simulator ride might be in a Boeing 747, any rusty procedures will simply be amplified.

But, now, thanks to a new Advisory Circular—AC-61-126—a pilot can use a personal computer to stay sharp on IFR procedures. And, if you're just now working on your instrument rating, you can even log up to ten hours of the time you spent on a PC toward your instrument ticket, provided the conditions of instructor endorsement, curriculum content and supporting study materials are met.

Everyone knows an airplane is a noisy, uncomfortable classroom. That's why simulators have thrived in the past for teaching the basics of flying on instruments. But, most simulators are expensive and, sometimes, out of reach for some flight schools and certainly too expensive for an individual.

Michael Thelander, director of marketing for Aviation Supplies & Academics (ASA)—www.asa2fly.com—makers of On Top, a Personal Computer Aviation Training Device (PCATD) software package, said, "There has always been a market for these devices for individuals. Long before PCATD approval by the FAA, thousands of pilots used the basic software to give them a sense of situational awareness, procedural understanding, and the ability to think ahead of the airplane before the next approach. The PCATD is a hardware/software/syllabus combination of the same basic package. While an individual may not need the entire training syllabus, the value is still there" (Figure 8-4).

Thelander discussed an issue often voiced by pilot applicants, especially unsuccessful ones. "It's a given that simulators do not fly like real aircraft. Without seat-of-the-pants sensations and, in the absence of sound and inner-ear cues, they're different. Students seriously planning for a professional pilot career should look at training facilities that start simulation training early, at the instrument student level. They should avoid an all-aircraft curriculum if they know a simulated check ride may be around the next corner." A simulator of any kind supports the training you do in an

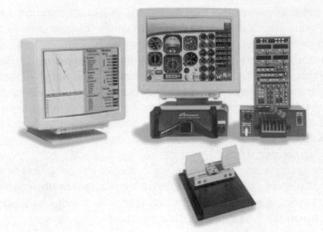

Figure 8-4 PCATDs are available in a variety of formats.

aircraft and helps prevent the information overload that often occurs when students try to accomplish everything in an airplane.

Whether a pilot can log the time on a PCATD is unimportant because there simply is no substitute for having shot 12 NDB approaches to 1 Left at Dulles before you hop into the 747 simulator for a sim check. The procedures you need to understand to fly an NDB, a full VOR, or a localizer approach will be so set in your mind, you'll easily be able to spend more time worrying about the important stuff, like how to fly the 747!

Despite my sense of humor here, you should know that seldom does a potential employer expect you to smoothly fly a large-aircraft simulator when you've been flying much smaller aircraft. But they *do* expect you to be able to keep the aircraft right side up and to notice trends off target airspeeds, altitudes, and headings. This means your scan must be fast and accurate, which only comes with practice. And, a PCATD is an inexpensive method of maintaining that proficiency, something that is often difficult, even for those of us who fly for a living. Thelander added, "Many airline pilots use the basic software at home in a nonloggable version because they never get a chance to hand-fly the aircraft any more, and also because they know, when the chips are down, they need to be prepared."

Some PCATDs—like ASAs—comes in integrated modules that begin at the least-expensive level with a simple software purchase that costs about U.S.$124. The most expensive version of ASA's PCATD costs U.S.$3,195, but it includes the software, an avionics panel that includes realistic nav/com units, a control yoke, rudder pedals, and a six-lever power-control quadrant that comes with throttles, mixture, and prop controls, and even gear and flap switches. The software emulates seven different single-engine and one twin-engine aircraft. A demo version of the ASA software can be downloaded right to your PC for evaluation at www.asa2fly.com/asa.

Another company to look into for a PCATD is Elite software at www.flyelite.com. Elite offers a number of versions of its detailed software: Core, Jet, and Premium. The Elite software ranges in price from U.S.$199 to U.S. $499, and it mimics a variety of popular single-engine aircraft. Try the demo at their site. The only thing I was not happy with here is the on-site documentation did not seem to include any sort of GPS functioning, which in this day is extremely important.

Flying a PCATD

Flying a PCATD is not as easy as flying an airplane. Like its big-brother siblings, FlightSafety's Level *C* & *D* simulators, or even the nonmotion Frasca machines, a PCATD requires a pilot's constant attention for almost the entire session because these devices lack the "feel" of a real aircraft. That's why you need regular simulator sessions. Try to pass a check ride in a simulator never having flown one before and the results could be enlightening—to say the least—and devastating to your career at worst. The question, however, is whether a PCATD can give you the real-world experience and simulator feel you need to feel confident going into a ride.

In my opinion, as a 20-year instrument instructor and line pilot in a turbojet aircraft, any IFR pilot—whether they're planning on a prehire simulator check or simply want to remain current—should own a PCATD. The experience is that real and that valuable. Much like any other kind of computer simulation system, a number of add-ons sit next to your monitor, such as a throttle quadrant, rudder pedals, or a radio stack to make your experience more realistic. The more familiar you become with the PCATD, the more realism you'll want to add. For evaluation purposes, however, I simply used ASA's basic "On Top IFR Proficiency Simulator" right out of the box and found the computer-generated panel more than adequate for my needs.

I installed the simulator on a 200 MHz Pentium machine equipped with 128 MB of RAM, a 17-inch monitor, a Kensington track ball in place of the standard mouse, and a Microsoft 3D Sidewinder Pro joystick. I had my system up and running about five minutes after I loaded the software from the CD-ROM—and with not one call to technical support necessary. I did learn that my PC lacked the video adapters necessary to run all aspects of the PCATD adequately, so the graphics outside my cockpit were rather mundane. I'm told that will improve when I update my system with a 16-bit video card.

The quality of the graphic presentation of the instrument panel on the PC monitor is astounding, however, much better than I expected. The experience was so real that I quickly lost track of the fact that the PCATD did not look like the inside of an airplane and was sitting on the desk in my office.

What makes the PCATD so valuable is simply the variety of experiences it offers—from your choice of nearly any IFR airport in the country to any of six aircraft types to fly to a wide range of weather and malfunction situations you can add to your flight, many at random for added realism. And, let's not forget the PCATD flies like a real simulator, no small challenge in itself.

Because my nonrevenue flying is often in a Piper Arrow, I chose that aircraft as the default for my system. Next, I chose the cockpit layout with a Horizontal Situation Indicator (HSI), Radio Magnetic Indicator (RMI) and even a panel-mounted Global Positioning System (GPS) to keep me on my toes. But, operation of a PCATD makes a few assumptions about the pilot. For instance, it assumes you already know how to use an RMI, an HSI, or an autopilot. If you don't, you need to work with an instructor for a bit to understand what is happening. The system does, however, offer you instruction on programming the GPS because almost all of these are a little different.

A PCATD, like the more sophisticated simulators, offers an overhead map for you to keep track of your progress across the ground throughout the flight. You can toggle out of the instrument panel, go over to the map at any point, and never lose your place, although, as in a larger simulator or an airplane, the distraction of fussing with switches and buttons can affect your instrument scan for the few seconds the sim needs to lose a hundred feet of altitude.

After setting up the cockpit and aircraft defaults of my system, I set the weather at 400 foot overcast, with a two-mile visibility, and took off for a little air work. Steep turns were just as easy—or as tough depending on your proficiency—as in other sims I've flown. I even tried a few stalls and found the PCATD to be quite realistic to both the airplane and the more sophisticated, full-motion simulators. This air work was even more challenging because it was performed in the clouds.

I decided the best test of my PCATD was to take it out for a little cross-country work. I planned a flight from PWK (Chicago-Pal Waukee) to RFD (Rockford, IL) to shoot some approaches. After setting the weather parameters for takeoff, I added in a few variables, such as the clouds being near minimums on all the approaches and only a slight northeast wind. The variability factor the PCATD offers meant, during some letdowns, the clouds might be below minimums calling for a missed approach, while at other times, I could land or make a low approach and return to the system—but I'd never know for certain when it was going to occur. I also set the forecast weather to deteriorate some during the flight but, again, I would not know exactly when. The one personal limitation I set for myself was to begin the session and hand fly it continuously to a final landing at RFD, provided the weather remained above minimums. I made certain I had enough fuel on board to get back to Chicago if I needed it.

The takeoff—as well as much of the Arrow's performance throughout the rest of the flight—was agonizingly slow because I made the aircraft weight equivalent to maximum while I set the outside air temperature to 96 degrees. During the enroute phase, I tried some more steep turns, slow flight, and stalls for proficiency. I continued on airways to RFD where I cleared myself to the NDB for the full ILS 1 approach. If you haven't played with the NDB in your airplane or a simulator recently, this is the place to shake out the cobwebs. Many prehire simulator checks add an NDB approach as a starting point, simply because they know a sharp pilot who performs well here will most likely do well on the ILS or VOR approaches.

I saw the runway at minimums on the first approach, but executed a low approach and flew the published miss out for a few turns around the holding pattern before shooting a B/C 19. At minimums on the back course, I saw nothing, so I went around again and headed back for an NDB 1 approach. I was pretty sure I would see nothing at minimums, but the variability of the simulator's cloud layering gave me just enough of a glimpse of the runway that I knew it was there, but not enough to land. I missed again and vectored myself to the RFD VOR (thanks to a quick toggle to the overhead map for a heading) and set up for a full ILS 7. Thankfully, I landed to a full stop on this one. Total time from takeoff to landing was an hour and 25 minutes. By the time I applied the parking brake at RFD and took a deep breath, my hands were sweaty. It had been quite a realistic workout!

A major benefit of the PCATD is the system's evaluation function. During lunch, a little later, I sat back down in front of my computer and punched up the instant replay. There, I was able to see just what my holding patterns, as well as all of my approaches, looked like not only from the overhead perspective, but also in a profile view. The PCATD even displayed my ability—or lack of it, at times—to stay on the glide slope during my two ILS approaches.

My only word of caution to an IFR pilot using one of these devices is not to get too dependent on the fancy gadgets installed, such as the autopilot and flight director, or even two nav systems. Let the simulator do its job by failing some things along the way. Better to find out how well you cope from the comfort of your desk than watching things fall apart when you're in the soup. You'll also never stay in touch with the "feel" of the simulator unless you hand fly it quite a bit.

Do you need a PCATD? Any pilot who doesn't have one installed in their desktop computer is missing out on one of the most exciting and least-expensive methods of staying IFR current, whether it is to prepare for a prehire simulator check or just to be ready for the real world of IFR flying.

Flight Simulator X

At the 25-year point, *Microsoft Flight Simulator X* is the newest of the Flight Simulator series that offers many new aircraft, thousands of airports around the world, and the capability of connecting any pilot online with another playing the game 10,000 miles away through Free Flight. Pilots can chat with one another in real time using the Flight Simulator headset. Flight Simulator X offers pilots to chance to learn the Garmin G-1000 series glass cockpit avionics, like the one used in the Cessna Citation Mustang (see Chapter 9's Pilot Report).

Never played with Flight Sim? You only need a decent speed PC, a graphics card, and a sound system to experience flying that is as close to the real thing as you can find on the ground. Download the demo of Flight Simulator X (includes two airports, three missions, and three different aircraft at St. Maarten in the Caribbean): http://www.microsoft.com/games/flightsimulatorx/downloads.html.

Trust me, though, if you see an FS X demo available anywhere—they are often running at most aviation exhibitions—you won't be able to leave without buying a copy.

Flight Simulator X includes 20 new and legacy aircraft to carry your flights of fancy across the globe by putting you in control of the aircraft you've dreamed of flying. Whether it's an ultralight, a DC-3, a Beech Baron, or a glass cockpit Boeing 747, you'll find it's easy to spend a lot of time playing at being a pilot. Just don't forget you're supposed to be out there looking for a full-time flying position. See Table 8-1. These aircraft come standard with FS X.

Table 8-1 Flight Simulator X Aircraft

Airbus A321	DG 808S 18 Meter Sailplane
Air Creation SL450 Ultralight	Douglas DC-3
Bell 206B JetRanger III	Extra 300S
Boeing 737-800	Grumman G-21A Goose
Boeing 747-400	Maule M7-260C Super Rocket
Bombardier CRJ 700	Mooney Bravo
Bombardier Learjet 45	Piper J3 Cub

Cessna 208B Caravan   Raytheon Baron 58
Cessna 172 Skyhawk SP   Raytheon King Air 350
deHavilland DHC-2 Floatplane   Robinson R22 Beta 2

System Requirements:

- Windows XP SP2/Windows Vista

- Processor: 1.0 GHz

- RAM: Windows XP SP2—256MB, Windows Vista—512MB

- Hard Drive: 14GB

- DVD Drive

- Video Card: 32MB DirectX 9–compatible

- Other: DX9 hardware compatibility and audio board with speaker/ headphones

- Online/Multiplayer Requirements: 56.9 kbps or better for online

And, once you get drawn in to Flight Simulator as many of us have, you won't want to miss the magazine that covers this piece of the industry— *Computer Pilot* magazine—www.computerpilot.com—where you'll find dozens of articles about the varied bits and pieces you can add to your version of flight sim, from hundreds of airline, business, and private aircraft to dozens and dozens of specific scenery to enhance the entire trip. There's even a motion-filled simulator chair with a 50" flat screen in front of it to play Flight Simulator between trips. It retails for a cool $24,000, so it might be best to visit some of the other Flight Simulator sites such as www.avsim.com, www.pcaviator.com, and www.microwings.com, where things are a bit less expensive.

Virtual Airlines

OK, maybe you're not quite ready to run off to sleep yet, because you're not finished dreaming about that cockpit waiting for you somewhere down the road. Into every pilot's life, a little fun must fall—if you're going to remain sane during your hunt for just the right flying job—and here it is. Your opportunity to become an airline pilot without a written exam, without sweating through an interview, or even without worrying about which way to enter the hold when you miss the approach on the ILS 1 to Dulles in the simulator. Joining a virtual airline can deliver a great deal of fun and enjoyment, with little aggravation. And, best of all, joining a virtual airline will cost you nothing—virtually.

But, here to explain more about what a virtual airline is, how it works, and why you should consider not filling out one more application for the rest of the afternoon until you're tried one, is Sean Reilly, the dean of the virtual airline set.

Virtual Airlines 101: A Crash Course

By Sean "Crash" Reilly, formerly of WestWind Airlines

In the later part of the evening and, occasionally, into the wee hours of the morning, a hearty group of individuals—most of them seemingly rational, grown men—sit perched in front of computer monitors with sweaty palms tightly clenching flight yokes. Distant cries of "Honey, come to bed" have long since fallen on deaf ears as, with razor-sharp concentration, these airmen skillfully guide their aircraft down glide slopes to airports across the world. The late night silence is shattered by loud screeches of rubber on runway, immediately followed by the deafening whine of reverse engine thrusters and, finally, by sighs of relief from the flight deck. The equipment, crew, and thousands of virtual passengers have safely arrived at their destinations. Just another routine day in the life of a pilot flying for a virtual airline (Figure 8-5).

Are you one of those people whose eyes are constantly cast skyward, watching aircraft pass overhead? Ever dreamed of flying the heavy iron or wondered what it would be like to be a pilot for an airline? Want to develop a greater understanding of, and/or appreciation for, real-world aviation? Looking for more structure from your flight simming than booting up to the trusty Cessna at Meigs, and then having to figure out where you want to fly? If you answered yes to any of these questions, I highly recommend you consider pursuing a "virtual career" with a virtual airline.

Virtual Airlines (VAs) are not a new concept—they've been around for several years. Although I'm not certain of their exact origins, I believe their birthplace can be tracked to Flight Sim forums of online services like AOL or CompuServe. VAs were started by a small group of individuals with a passion for flight simulation, who decided to create the virtual equivalent to a real-world airline. They created a structure where flight sim enthusiasts could fly specified routes in aircraft other than the Cessna 182. With the introduction of a software program called Flight Shop, by the now-defunct BAO Software, it was possible for these enthusiasts to create custom-designed aircraft of all types and colors. This further stimulated the growth of the VA indus-

Figure 8-5 WestWind conducts new pilot training in an Embraer E-120.

try. Now it was possible to fly a much wider variety of aircraft not included in the default Flight Sim program.

While VAs based online grew in popularity and number, it wasn't until the boom of the Internet that—if you'll pardon the pun—VAs really took off. As a conservative estimate, 1,001 virtual airlines are operating either on the Net or online today. They employ (virtually speaking) thousands—probably tens of thousands—of pilots of varying skill levels and aviation knowledge. These numbers will only increase as Microsoft continues to improve on its versions of Flight Simulator, as computer hardware improves, and as outstanding add-on software and programs designed to enhance the virtual-flying experience (like Squawk Box and Pro-Controller, which enable you to fly with real-time ATC!) are introduced. The future of flight simulation, and of the virtual airline industry, is extremely bright.

To back up a step, to give you a better sense of how a virtual airline operates, let's compare the structure of a typical VA to that of a real-world commercial airline. Say, one day, you wake up and decide you want to pursue a career in aviation. The first thing you do is get over to your local airport and sign up for flying lessons. You begin in a small, single-engine aircraft and, eventually, you work your way up to more complex, multiengine aircraft. During the process, you earn various ratings. After you accumulated the necessary hours and secured the required ratings (and shelled out tens of thousands of dollars in the process) you race down to the personnel office of your favorite airline, fill out an application and—with some luck—are issued a polyester uniform and are hired on as an entry-level pilot. Most entry-level pilots begin by flying twin turboprops.

Switching to the virtual world, this is the jump-in point for most virtual airlines. Most assume you have secured the necessary training and accumulated the minimum number of hours to step into a twin turboprop aircraft, such as a Beech 1900D or Brasilia. Your first step, once you decide which VA you want to fly for, is to log on to its web site and follow the links to the pilot application form. Fill out the form and wait for a letter of acceptance. Once accepted, many VAs require you to take basic training in their entry-level aircraft. Training varies from one VA to another. If you are looking to increase your knowledge of, and appreciation for, flying, I highly recommend you select a VA with the most realistic training program possible. A few VAs, including the one I fly for (www.flywestwind.com), offer training programs crafted by real-world certified flight instructors (CFIs). Some even go so far as to provide downloadable, custom-designed, training center scenery and training adventure files! Our training file introduces some "hairy" weather conditions that makes for quite a challenging landing.

Much of what you learn in a virtual training program mirrors what you would learn in a nonvirtual program. The more realistic the program, the better. If you survive the training exercises and return the training aircraft in the same condition you received it (wheels round on all sides, no scratches on the bottom, wings attached, and props unbent), you are issued virtual wings and your career begins. Congratulations and welcome aboard! An important caveat for those who may be wondering—there is no charge to fly for any of the VAs I'm familiar with. Even the virtual jet fuel is free!

OK, back to the real world for a moment—when you fly for an airline, you fly aircraft painted in your airline's colors. In the virtual world, VAs offer a livery of aircraft—ranging from small turboprops all the way up to heavies, such as the Boeing 747 and 777—painted in their unique colors. When choosing a VA, I recommend you not only consider the fancy paint job of the fleet, but also how well each aircraft flies. Better VAs offer both attractive color schemes and highly accurate flight models of every aircraft in their fleet.

Some VAs also offer pilot-operating handbooks (POHs) for each aircraft. In case you are wondering, VA aircraft can be downloaded, free of charge, from each VA's respective web site. This process is not complicated and neither is the process of loading those aircraft into your Flight Simulator program. Most aircraft come with readme.txt files that walk you through the process. Some VAs, including ours, offer self-installing aircraft files for maximum pilot convenience. A real plus.

In the real world, commercial airlines fly predetermined regional, domestic, and/or international routes. Larger airlines fly to destinations throughout the world, using a wide variety of aircraft, while smaller airlines may fly only regional or domestic routes with a limited fleet. It goes without saying that the types of aircraft each airline flies is determined by the nature of the routes they fly. The same applies to the virtual world. Some VAs concentrate solely on regional or domestic routes with few aircraft types, while others maintain diverse fleets of aircraft that fly a multitude of routes to destinations throughout the entire virtual world. When looking for a VA, consider the type of flying you most enjoy (short hops, long hauls, or a combination) and the type of aircraft you like to fly. VAs with small fleets concentrating on regional routes can be every bit as fun to fly for as larger VAs with diverse fleets flying a full-blown assortment of routes. This is a matter of personal taste. There is even one VA I know of that flies only classic (vintage) airliners. Quite specialized and clever.

One area of confusion, for those unfamiliar with the VA concept, that should be touched on is where the flying actually takes place. VA web sites are similar to FBOs at your local airport. This is "home base" where you come to sign up, and check out (or download) aircraft, scenery, adventure files, and other add-ons. In addition, training materials, POHs, and so forth can be read and printed out from the web site. You can also communicate with other members of the VA from the web site. Some VA sites offer bulletin boards or forums for their members. All flying is done using a Flight Simulator program (99.9 percent are of the Microsoft variety). To be clear, you don't fly from within the web site of a VA. The world you fly in is determined by the Flight Sim software you are using.

A word about career advancement. In the real world, most entry-level pilots flying twin turbos want to rapidly log as many hours as they can to advance to aircraft of more substance—the latter promises a pilot greater challenge and a far more attractive salary. Career advancement in most VAs works much the same way. You begin your career in smaller aircraft, build hours (which you should log in your

Flight Sim log book), and advance in aircraft type-ratings over time. Most VAs require you to, on regular intervals, report your hours to management. This is usually done via an online PIREP (pilot report) form at your VA's web site or by e-mail. Most VAs have an "honor system" for flight-time reporting. A few require you to submit your Flight Sim log book occasionally. As you accumulate hours, you are promoted in aircraft type and gain increased bragging rights. Sorry to say, however, that in the virtual world, salaries are the same for small turboprops as they are for triple sevens—zip, nada, zero. But, consider the bright side—you didn't need to spend tens of thousands of dollars for flight training and you don't need to wear those polyester uniforms.

A few other things to consider as you search for a virtual airline: first off, please take time to look at many VAs before signing on with one. Download and test fly various aircraft from at least a few VAs to see if they fly as good as they look. Look critically at the management structure/team of a VA—is it well organized? Do the managers seem knowledgeable? Consider the longevity of the airline. Many VAs disappear as fast as they appear, which is a real frustration for pilots who then have to sign on with another VA and, in many cases, begin their career at the bottom again. Take a good look at the training program, the route structure, and the minimum requirements to maintain active pilot status.

Are they to your liking? You might also look critically at the web site of any VA you are considering. Look for depth and organization of content. Does the site offer interesting and informative information? Does it offer a "pilot help" section? Has it been updated recently or is the content dated? Does the site contain a pilot and/or visitor comments page? If it does, what are people saying there? Does the VA promote and support real-world aviation? Ours is a proud sponsor of GA Team 2000's "Stop Dreaming/Start Flying" campaign to increase real-world, general-aviation pilot starts. Has the site and/or airline received any awards and, if so, for what and by whom? Another great resource to consider are the flight sim magazines—like this one [Computer Pilot]. Has the VA received any favorable press? Does the VA offer specialty divisions, such as a Cargo or Charter Division? Does the VA host special events such as fly-ins? While you might not want to choose one VA over another for any single thing previously listed, you should consider the bigger picture—a combination of the previous—as an indicator of how well run and stable a VA is, and how seriously the management team takes its operation. Just because it is a virtual operation doesn't mean it shouldn't be a professional one.

Another thought regarding career advancement in the virtual world: if you would like to do more than just fly for a virtual airline, after accumulating some hours and experience with your VA, inquire about a management position within the organization. The single most important ingredient of any successful virtual airline—large or small—is its management team. It takes more than a few talented people to run a successful virtual airline—it takes many, all working in sync toward a common goal. A virtual airline is only as good as the sum of its leaders. VAs are always in need of creative and talented individuals to fill positions, such as hub managers, aircraft and scenery designers, personnel directors, route creators, training directors, special project directors, web designers, and others. Although the pay isn't great—"virtually nothing," in fact—a management career in a virtual airline can be highly gratifying.

Perhaps the most enjoyable thing I have found about flying for a virtual airline is the creativity and camaraderie that exists among our members—and the flight sim community as a whole. While our airmen come from literally all over the world, are of different ages, cultural backgrounds, and so forth we all have one thing very much in common—a passion for flying, be it simulated, real-world, or both. The sense of "community" that exists within the structure of a successful virtual airline, coupled with the common interest we share, is second-to-none. Consider a career with a virtual airline—it's a blast! (Figure 8-6)

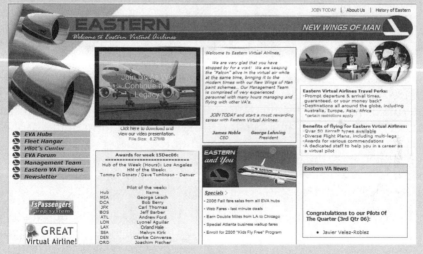

Figure 8-6 Eastern Virtual Airlines offers enthusiasts a way to learn about the airline industry.

An additional list of Virtual Airline and Flight Sim resources is available in Appendix C.

Sean "Crash" Reilly is the "Virtual" Executive VP of Marketing & New Biz Development and cofounder of WestWind Airlines. Reilly can be reached by e-mail at wan2fly@sbcglobal.net or via the WestWind Airlines web site at http://www.flywestwind.com.

Web Sites No Pilot Can Do Without

You bought this book was as a resource. Because I've been discussing how important the new electronic community is to your job search, this review of some unique aviation web sites should help you on your way toward finding the ideal job. If you have not had the fun of whiling away an hour here and there surfing aviation web sites, this list—as well as the additional lists stored on most sites themselves—should prove valuable. But, be careful. Sometimes you can start clicking one link after another and lose track of where you are—or where you were—before you know it.

This section includes just a few of the web sites I've looked at that have value. But they are hardly all of the sites available. Let me know if you find some additional aviation sites that prove valuable and I'll include it in the next edition of this book. E-mail the site address to me at: rob@propilotbook.com. You can also track me down via this book's web site: www.propilotbook.com.

A word of caution, too—web sites come and web sites go. Many contain information that is difficult to independently verify because a web site might be run from someone's basement or an office in the Mile High City, and you'd never know which it was. Information found on the Web should normally be checked with similar or contrasting viewpoints from other sources before you make any career decisions. Prior to sending money or offering a credit card to any company on the Internet, call their telephone number and talk to a live person. Some, of course, don't offer a phone option. But a reputable company should quickly answer e-mails sent their way. If not, pass them by. This message posted on one web site I visited explained web site concerns perfectly: "Disclaimer: This database may not be complete or up-to-date. Do not rely on it in any way."

- http://www.smilinjack.com/—Smilin Jack Site—Interesting little page produced by a guy who liked the Smilin' Jack cartoon from years ago. It contains one of the largest lists of worldwide airline web sites around—about 150—from all points on the Earth. This also has an extensive list of weather-related web sites and even a long list of airport web sites from the United States, Europe, and the Far East. Great place to add research info before the interview.

- http://flighttraining.aopa.org/ft_magazine/—*AOPA Flight Training* magazine—If a good pilot is always learning, they need a magazine focused on just those concerns. *Flight Training* magazine does that. In *FT* mag's own words, they are "Aviation's How-To, Back-To-Basics magazine for new pilots, their instructors, and those who own and operate flight training schools." Many experienced pilots also find *Flight Training* magazine's monthly issues provide an excellent up-to-date review of aviation-training information and piloting techniques.

- www.icao.int—The International Civil Aviation Organization, the world aviation organization to which 189 contracting countries around the world come together to formulate airspace and regulation policy.

- http://interactive.wsj.com/pj/personal.cgi—The *Wall Street Journal* Interactive Edition—If you're in search of work as a pilot, you already know I recommend the *Wall Street Journal (WSJ)* as an important source of current information about the airline or other company you might be interviewing with. But, for just $29.95 extra per year, *WSJ* print subscribers can build their own page—I've made mine the home page I log into each day—and track anything they find of interest by adding as many keywords as they like to their profile. Then, when you log on each day, the *WSJ* brings you stories that match your keywords. Want to know what the Midwest Express, Continental, and Coca-Cola are doing before your airline or corporate interview? Find out here. Then, print the piece and save it to your folder for future reference. This site is a must-have!

- http://www.pilotshop-usa.com/index.html—The Pilot Shop—A Midwest-based pilot supply shop that offers great products at great prices, as well as an online ordering form.

- http://www.avweb.com/—AV web site—AvWeb bills itself as the World's Premier Independent Aviation News Resource. In addition to a twice-weekly e-mail newsletter covering a range of general aviation topics, AvWeb also produces a regular podcast series available on the site.

- http://www.landings.com/—The Landings—Lots and lots of stuff is here—regulations, airworthiness directives, NTSB briefs, the Airman's Information Manual, and a listing of aircraft country prefixes. Now, when you see an aircraft whose tail number begins with YV, you'll know its from Venezuela. Also includes a solid list of aviation search engines.

- http://www.airnav.com/—Air Nav—Provides free, detailed aeronautical information on airports and navigational aids in the USA. Type ORD in the navaid database and you'll learn more than you probably ever wanted to about the VOR/DME that serves the world's busiest airport.

- www.af.mil/—Air Force Link—This is the official site of the U.S. Air Force and a jumping-off spot to tons of information about places you might see if you join the Air Force, such as Officer Training School (OTS).

- www.nbaa.org/—National Business Aircraft Association—This site does not post job information, but if your goal is a career as a corporate pilot, this is a source of information about what is happening in this segment of the industry. The information, however, is somewhat limited if you are not an NBAA member.

- www.aviation.org/—The Aviation Safety Connection—This is a site I would not pass up. Focused on human factors issues in flight safety, it offers an informative online newsletter—Cockpit Leadership—as well as access to an online discussion forum about safety issues. It also includes analysis of National Transportation Safety Board (NTSB) and Aviation Safety Reporting System (ASRS) reports.

- www.aopa.org/—Aircraft Owners and Pilots Association—Although AOPA is not necessarily a career-focused organization, the site does offer terrific message boards for aviators of all experience levels and a well-stocked library, as well as impressive database access. If you're not an AOPA member—join. It's the best $39 per year you can ever spend. One of the best extras I've found for my $39 is the capability to download electronic copies of practically any instrument-approach plate in the United States. Try picking out a few interesting ones and sit down with a friend to ask questions. This is a great way to learn how much you do or do not know.

- http://www.atwonline.com—*Air Transport World (ATW)* magazine—*ATW* is a business magazine focused on airline issues from the majors to the regionals, but all from a management standpoint. Not a great deal of original content here, but it does offer you a chance to search back issues of the magazine for information.

- http://www.bts.gov/oai/—The U.S. DOT's Office of Airline Information—More statistical info than you'll ever have time to digest. Choose from such show stoppers as On-Time Statistics, FAA Statistical Handbook of Aviation, or Sources of Air Carrier Data. Seriously worth it.

- http://www.airlines.org/—Air Transport Association—This is the trade organization that represents most of the major airlines in the U.S. You'll find policy speeches here, as well as much airline statistical information.

- http://ww2010.atmos.uiuc.edu/(gh)/guides/mtr/home.rxml—University of Illinois Online Meteorology Guides—These are pretty nifty online workbooks about all facets of weather, including a great section on severe storms, weather forecasting, cloud types, and air mass explanations.

- www.beapilot.com/—Stop Dreaming, Start Flying—This is the official site of the new Learn to Fly program sponsored by most of the major aviation organizations in the country. Great information source for those new to flying. The site includes great articles, such as Susan Paul's "Twenty Life Lessons I Learned From Becoming a Pilot," that makes great reading, no matter what your experience level.

- www.corporatepilot.com—Corporate Pilot.com—Designed to help business and corporate aviation-flight departments and pilots find each other for full-time employment or just more contract work. Some actual job listings are posted here.

- http://www.airforce.com/ —The U.S. Air Force Career Forum—This site is organized much like an actual Air Force base and offers career information to various people, in addition to pilots. Archived version of an online career seminar, held in late 1998, is worth reviewing. A little light on information, but worth reviewing if you're considering the Air Force as an option.

- www.aerospace.bombardier.com/—Bombardier Aerospace—Company site that includes lots of information about Bombardier regional and corporate aircraft in both photos and news stories.

- www.jeppesen.com/—Jeppesen corporate site—Information on lots of Jeppesen training materials, not the least of which is info on the FS-200 flight simulator.

- http://www.aviationjobsonline.com/—Aviation Jobs Online—It's tough to review this site because you must pay the fee to gain access to the actual job listings. Might be worth trying for a month or two to learn more.

- http://findapilot.com/—Find A Pilot—Posts lots of job openings and also offers pilots an opportunity to post their résumé online. Résumés are only on a list, however, and are not searchable through any special parameter by a potential employer.

- www.boeing.com/—The Boeing corporate site—Information on Boeing products and lots and lots of great downloadable photos, as well as a great screensaver of Boeing aircraft.

- www.airbus.com/—The Airbus corporate site—Airbus aircraft family site with information and photos.

- http://www.weather.com/aviation/—The Weather Channel Aviation site—The title says it all. It's a gotta have.

- http://www.webring.org/ringworld/rec/aviation.html—The Web Ring—There's just no way to describe The Ring, except to say you must try it. This is a continual link from one aviation site to another. Simply too many are linked together to review. Let me know—via e-mail—which ones you find useful.

- www.microwings.com/—Micro Wings, the International Association of Aerospace Simulations—Not particularly attractive graphically, but if you've turned into a flight sim crazie while you search for that dream flying job, you need to bookmark this site. You'll find software reviews, links to other sites—like the flywestwind.com mentioned elsewhere—and even links to software add-ons for your simulator. Take a break from your job search for a little while and have some fun.

- www.polarisms.com—Polaris Microsystems—This is the corporate site for the company that developed and sells the electronic pilot logbook—Aerolog for Windows. The site includes a downloadable demo of its electronic logbook program for evaluation.

- www.ntsb.gov/—National Transportation Safety Board—Official site of the NTSB and simply loaded with information of interest to pilots. Search a database of some 41,000 aircraft accidents reports to read and learn from. Also, plenty of aviation statistics that might be valuable during an interview.

- www.wiai.org/—Women in Aviation International—Dedicated to the encouragement and advancement of women in all aviation career fields and interests. Women in Aviation provides year-round resources to assist women in aviation and to encourage young women to consider aviation as a career.

- www.trade-a-plane.com/—Trade A Plane—This is the yellow tabloid of just about anything anyone could ever want in aviation—from aircraft parts to pilot jobs. It contains a searchable classified site for employment that sets you back $2.95 per month, or $29 if you purchase a year's worth.

- www.raa.org/—The Regional Airline Association site—This site includes photos and stats on regional airline aircraft, as well as the latest details about how the various companies in this segment of the industry are performing.

- www.aerolink.com —Aerolink.com—In their own words, "Aerolink is an aviation-specific farm. No eye candy, no page counters and no propaganda. Nothing to crash your browser. Instead, just links. Probably more links than any other aviation web site." At last count, Aerolink listed over 7,000 aviation-related sites.

Employment (Reviewed in Chapter 4)

- http://www.aeps.com/aeps/aepshm.html—Airline Employee Placement Service
- http://www.airapps.com//index.html—Air Inc.
- http://www.fltops.com/—FLTops.com site
- http://www.airlineapps.com—Airline Applications online
- http://www.aeps.com/—Airline Employee Personnel Service
- http://www.avcrew.com—AvCrew.com site
- http://www.pilotjobs.com/
- http://www2.allatps.com/AirlinePlacements/
- http://www.pilotcareercentre.com/CareerHelpResumes.htm
- http://bizjetjobs.com/
- http://www.avjobs.com/index.asp
- www.netjets.com
- http://www.satsair.com/index.html
- www.corporatepilot.com
- http://www.dayjet.com/
- www.jet-jobs.com
- http://pilot.fedex.com/
- https://ups.managehr.com/screening/professional/
- http://www.raa.org/airline_directory/
- http://www.qantas.com.au/infodetail/about/employment/pilotCareerInformationBooklet.pdf

Official Union Web Sites

- http://www.alpa.org/ The National Air Line Pilots Association site
- http://www.alliedpilots.org/—The Allied Pilots Association (American Airlines Pilots)
- http://www.ipapilot.org/—Independent Pilots Association (UPS Pilots)
- http://www.usairwayspilots.org/homepage.htm—USAirways ALPA site

Simulator Training

- http://www.simulator.com/—Simcom Training Centers site
- http://www.flightsafety.com/—FlightSafety site with searchable database for aircraft type and training locations
- http://www.simuflite.com/—SimuFlite Training International site

Aeronautical Academia

- http://cid.unomaha.edu/~unoai/uaa.html—The University Aviation Association
- http://www.aero.und.edu/Academics/Aviation/index.html—University of North Dakota, Aviation Department
- http://www.aviation.uiuc.edu/—University of Illinois at Champaign, Institute of Aviation
- http://www.tech.purdue.edu/at/—Purdue University, Aviation Technology
- http://raptor.db.erau.edu/—Embry-Riddle University
- http://www.spartanaero.com/—Spartan School of Aeronautics
- http://www.amerflyers.com/—American Flyers
- http://www.flightsafetyacademy.com/
- http://www.gibill.va.gov
- http://www.43airschool.com/
- www.deltaconnectionacademy.com

One of my favorites for those interested in simply listening to ATC chit-chat, but those who want to watch as well:

- http://atcmonitor.com/
- http://www.liveatc.net/

Here's one that offers a look at all the registration information for U.S. air carrier companies.

Aviation Message Boards

- http://www.aero-farm.com/avsig.htm
- http://news.airwise.com/airline_news.html
- http://www.pprune.org
- http://www.aia-aerospace.org/

Valuable If You Fly the Airlines a Bunch

- http://www.seatguru.com/

Photography

- http://www.airliners.net/
- http://www.jetphotos.net/members/contact.php

FAA

- www.faa.gov/—The Federal Aviation Administration site—Besides the career database that pulled up a recently announced GS-14 pilot position (salary $66,138 to $85,978) the site also houses considerable information about some of FAA's major components—Air Traffic Service, Regulation and Certification, Airports and Security. Also includes links to tons of FAA statistical information.
- http://www.faa.gov/other_visit/aviation_industry/airline_operators/training/
- http://ostpxweb.dot.gov/aviation/index.htm
- http://www.faa.gov/pilots/training/
- http://av-info.faa.gov/OpCert.asp
- http://www.ntsb.gov/Pressrel/pressrel.htm

Other Pubs . . .

- http://www.ainalerts.com/ainalerts/
- http://www.trade-a-plane.com/index.shtml
- http://transportation.northwestern.edu/programs/icarus/index.html (The Icarus Society aviation discussion group)

Aircraft Manufacturers

- www.cirrusdesign.com
- www.eclipseaviation.com
- www.cessna.com
- www.gulfstream.com
- www.piper.com

- www.falconjet.com

- www.embraer.com

So, there you have a glimpse at what's happening electronically in aviation. But, again, this is just a glimpse. Spend some time and you're bound to find other tools for your PC that will be of value to you in your job search. Be sure and share them with the rest of us.

The End . . . or Just the Beginning?

o here you are at the end or, hopefully, the beginning. You've had a glimpse of what lies in store for you if you decide that a career as a professional pilot is for you. I've spoken about some of the jobs, as well as how to find them. We've looked at the ratings you'll need and how to pick them up. I've discussed some of the schools available and how to finance your career. Best of all, I've spoken with people who have made it—people who developed a plan and followed through until they reached their goal (Figure 8.7).

A plan is important, but the most important part is you must keep moving toward your goal. Don't let anything (or anyone) get in the way. If someone says no, realize that means no . . . today. Tomorrow it could be a maybe. If you try something and you fail along the way, realize you really didn't fail. You learned something. You learned what doesn't work. If you learned something, you didn't fail. You never fail until you stop trying.

Here are some final words from Lt. Michael Fick from my interview with him about the Air Force. These thoughts sum up a great many of the feelings of the other pilots I've spoken to. I asked Lt. Fick how much he enjoyed what he was doing. He said, "You can't beat getting paid to fly. There's no other fun like it." Notice he didn't use the word "work" in that sentence. Flying is hardly like work. "The real work comes in those years when you're learning to become a rated pilot."

Computers are a must-have if you want to stay on top of the latest information about who is hiring and how to connect with them. I hope you'll keep me abreast of your progress. To make that a little easier, I'd again like to offer my e-mail and web site address for this book. Please make it a point to let me know what portions of this book worked well for you and which ones might be improved. If for no other reason, e-mail me when you get that next job.

This book's web site is www.propilotbook.com.

My e-mail address is rob@propilotbook.com.

The best blog around is www.jetwhine.com.

Figure 8-7 FedEx MD-10s and MD-11s (Courtesy Tim Wagenknecht: www.jetphotos.net.)

And one more tip . . . buy a digital still or digital video camera and take it with you throughout your career. Send us copies of your best shots and we'll post them here at www.propilotbook.com. Digital information sharing is where the world is headed.

Good luck!

Robert Mark, Evanston, IL
May, 2007

9

Pilot Reports

When you began reading this book, I mentioned one of the things I personally wished for most was a mentor, someone who might have been able to share a little of the aviation world before I experienced it all on my own. Consider this a chance for me to offer you a glimpse at some of the aircraft I've been lucky enough to have been given the keys—so to speak—in my career as a pilot and journalist. My thanks to the manufacturers who took the time to let me experience these machines and to *Aviation International News* magazine—www.ainonline.com—for trusting me with these assignments for their readers.

Here are a few I think you might enjoy. Many more can be found sitting at this book's web site—www.propilotbook.com—so please stop by.

Enjoy your flying career.

Rob Mark

Pilot Report: Citation Mustang

Reprinted by *Courtesy Aviation International News*—
www.ainonline.com
By Robert P. Mark/October 2006

The champagne corks were surely popping in Wichita on September 8 when Cessna Aircraft announced it had earned full type certification of its newest jet, the Mustang. The paperwork was signed just short of four years after the company announced the project at the 2002 NBAA Convention in Orlando, Fla.

In the middle of last month, AIN visited Cessna's Wichita headquarters and Independence, Kan. manufacturing plant where the Mustang is built, to fly the airplane and find out whether the Mustang is a true entry-level jet or simply a turbine-powered airplane hurriedly developed to keep Cessna a step ahead of its competitors.

Figure 9-1 Figure 9-1: Cessna Citation Mustang (Courtesy Cessna Aircraft Company.)

Cessna began cutting metal on the Mustang, designated the 510 (Figure 9-1), less than a year after the project won company approval. The company made it clear in 2002 that the Mustang would not be simply a stripped-down light jet, but a Part 23 airplane that would offer the true versatility of load, speed, and range customers demand.

And while the Mustang promised to raise the stakes for entry-level jet makers, it also delivered a klaxon message to competitors lacking the resources of Cessna and Textron: when this company sees a lucrative marketplace opportunity, get out of the way.

At press time, the Mustang was certified for day and night VFR and IFR, as well as RVSM. Only one item remained outstanding: known-icing paperwork, expected shortly, but certainly in time for first scheduled aircraft deliveries in the first quarter of next year. Cessna's current backlog covers 250 Mustangs and extends to the fourth quarter of 2009.

Cessna's first business jet, the 500, was priced at nearly $590,000 more than 30 years ago. In today's dollars, that 500 would cost $2.3 million. Billed as Cessna's true entry-level Citation, the Mustang is priced at $2.6 million.

The Mustang price and performance numbers are certain to make turboprop salespeople plenty nervous, too. A Beech King Air C90GT, a derivative of a 42-year-old design, offers a comfortable size cabin, but costs $2.95 million, and its cruise speed trails by more than 100 knots. The 270-knot Pilatus PC-12 turboprop single rings in at $3.3 million, typically equipped. The TBM 850 turboprop single sells for slightly more than a Mustang, but flies 20 knots slower.

At the heart of the Mustang's avionics is the Garmin G1000 suite, similar to that found in Cessna's current piston-single line, but with the addition of a center multifunction display. The functionality of the G1000 is easy to spot once the pilot spends the time learning the system. Much of the system is intuitive enough that, by the second or third flight, many of the entries will be second nature.

Russ Meyer III, Cessna's Mustang program manager, plowed right through a number of company scheduling issues just weeks before NBAA 2006 to make an airplane available to AIN. Cessna engineering test pilot Don Alexander, a former Air Force C-141 commander, served as my instructor.

The flight began at Cessna's Wichita Service Center aboard N510KS, the second production-conforming aircraft. The ramp was a toasty 32 degrees C on the cockpit gauges, while the ATIS called it 28 degrees C with a gusty southeast wind and only a few high clouds. The Wichita field elevation is 1,333 feet msl.

Entering the cockpit of the Mustang requires only minor gymnastics, thanks to a well-placed handle to grab and a short throttle pedestal. The control wheel on the Mustang is connected through the instrument panel and is reminiscent of a single- or light twin-engine aircraft, rather than the typical long arm of a jet control wheel sticking up from the floor. The seat is comfortable and easy to move vertically and horizontally.

I didn't care for the traditional automobile-type shoulder harness, and a five-point harness is not offered as an option. The Mustang has an electrically heated glass windshield that eliminates the once-deafening bleed-air blowers, used on earlier Citations, to keep the pilot screens clear of ice and condensation.

Because the ground power unit was plugged in, the pilot's 10.4-inch flat primary flight display screen was already lit up and showing the basic HSI configuration that combined the engine and other system gauge readings into what Alexander called the "compressed mode," or the configuration that would be seen during a total electrical failure. While the display is certainly crowded in an emergency, nearly every piece of information a pilot might need is available in one place.

The copilot side uses the same type of flat screen as the left seat. The multifunction display (MFD) is a single 15-inch flat screen that holds the engine indication and crew alert system.

The plan was to depart at maximum takeoff weight and climb directly to FL410, while recording some of the pertinent data along the way. With full fuel—193 gallons usable per side—100 pounds of ballast in the nose, 120 in the rear compartment and a passenger, we were just a few pounds short of the 8,750-pound max ramp weight. Because mtow is 8,700 pounds, we'd need to burn a bit of fuel during taxi. Be warned: the Mustang does not burn fuel very quickly on the ground.

With the nose facing into the sun on the ground, the air conditioners—running because the aircraft was plugged into the GPU—did a pretty nice job of keeping the cockpit and cabin cool. The fans have three positions for cockpit and cabin—high, low and off. In the high position, a necessity in 28 degree C heat, the fan noise made it difficult to talk to Alexander in the right seat, or even to hear the radios. Low-fan speed would be a necessity on the ground for most operations in high-traffic areas, I think.

The instrument panel will be familiar to previous Citation pilots, with engine start buttons to the left, as well as easy-to-read generators and fuel boost pump switches that need only be left in normal. The aileron and rudder trim are a bit hidden from the pilot's view at the base of the center pedestal. Although they are not often used, it might be tough to find them in the dark.

Alexander filed a flight plan to the Cessna Test Area west of Wichita for our climb and a later route that would take us back south of town for a landing at Independence. Because the field elevation at our first landing point—Independence—was 825 feet, the only work necessary to run the pressurization system was to load that airport data into the Garmin before takeoff. The Mustang's maximum pressure differential is 8.4 psi, which delivers a sea-level cabin to 21,000 feet and will hold an 8,000-foot cabin to 41,000 feet.

Engine start with GPU means being able to run the avionics full-time, which is very handy. A battery start would require turning off the avionics master switch, as well as the generators. The Mustangs are equipped with lead acid batteries. The air conditioning is also turned off for the start. With a clearance in hand, we started the left engine with a single touch of the start button. At 8 percent N2, the throttle was moved over the gate to begin fuel flow to the engine.

The nice thing about FADEC is the automation. Alexander said that the throttle could actually have been moved over the gate at the beginning of the start sequence. The FADEC is smart enough not to allow fuel into the combustion chamber until the proper time.

Both motors spun quickly and reached idle speeds within 15 seconds. There are no time limits between starts using the battery. Temperatures never climbed much past 700 degrees C during the sequence. As always, it is the rate of temperature increase that will be the first tip-off to a hot start. A rapid rise toward the 800 degrees C mark would be of concern to Alexander. At idle, fuel flow settled at 120 pounds per side.

After engine start, the generator check is a matter of turning off one operating generator to check that the remaining unit has picked up the load, followed by the same check on the opposite side. The pilot's indication that the GPU has been disconnected is the drop from 28 volts back to the approximate battery voltage of 24 volts. Windshield anti-ice is normally turned on after the start because there are no overheating concerns. While the Mustang uses a number of annunciator tones, there is only a red light to indicate a possible engine fire. There is no stick shaker or stick pusher before a stall.

Taxi control of the Mustang is accomplished through rudder pedal steering. I initially called the brakes sensitive, but my passenger more accurately labeled them responsive. And indeed they are. When I needed stopping power during the taxi, I had it.

Because the small Pratts don't offer a lot of residual thrust, it takes a bit of power to get the aircraft rolling, but little to keep it rolling at a comfortable taxi speed. The pilot needs to stay ahead of taxi control to avoid tapping the brakes to steer, which can produce an annoying pull to one side and announce to everyone on board you're a newbie. Taxiing would later prove much easier. Because the emergency gear extension releases the uplocks when needed, the gear freefalls once the handle is pulled. A blow-down bottle is installed as a backup.

Flight controls are all manually actuated. The hydraulic system is an electrically powered pack located in the left-hand side of the nose area that provides for the landing gear extension, retraction, and braking. The Mustang is equipped with two 300-amp starter generators. One is capable of powering everything on the airplane and also provides automatic load shedding in case of an emergency.

The flight plan called for outlining the test area on the center multifunction display to track our progress over the ground. With the Garmin G1000, we could have simply departed and eyeballed the map to remain within the confines of the test area. One particularly useful aspect of the avionics system is its capability to record and later play back a clearance by pressing the record button on the audio panel.

This keeps readbacks and radio chatter to a minimum, while making it easier to review an original clearance well after it was first received. For some reason, the com 1 function on N510KS was not usable on the copilot's side, so we operated the rest of the flight from com 2. Had it been IFR, I would have wanted both panels to be fully functional.

Getting Under Way

ATC sent us to the 8,000-foot Runway 19 Left. Another large-aircraft tool on the Mustang is Garmin's Safe Taxi system, which uses the 15-inch MFD to provide pilots with an exact graphical position in relation to labeled taxiways, runways, and buildings during taxi at more than 680 airports across the U.S. The G1000 also uses the XM satellite system to provide weather overlaid on the moving-map display.

Takeoff numbers pulled from the tab data were calculated at V1 of 90, Vr of 90 and V2 of 97 knots. We would use a cruise climb of 170 knots until reaching Mach 0.44 through FL410. The numbers were entered in the Garmin by toggling through an on-off function that I did not particularly care for. Without turning the takeoff numbers "on," they would not appear on the speed ribbon during the roll. It would seem a simpler system would be to allow the pilot to turn on "takeoff," or "landing," at one time, rather than to click through each V-speed individually.

A final takeoff check confirmed that flaps were set at takeoff, the speed brakes were retracted, and the HSI heading was the same as runway heading. Establishing takeoff power once on the runway was easy, thanks to the FADEC , but still began with easing the throttles up to 35 percent or so to confirm smooth acceleration before pushing the levers to the firewall.

Acceleration was brisk and we quickly reached takeoff speeds. Rotation forces were light as I retracted the gear and climbed west toward our first altitude limit of 10,000 feet. Flaps were retracted at V2 plus 10, while climb power was set in the detent at about 1,000 feet agl. Alexander said the engines can be held at takeoff power for ten minutes.

My goal was to hand fly the airplane all the way to FL410. The only issue I noted with the 170-knot climb was a nose attitude most pilots might not feel too comfortable with. While a higher speed would translate into a longer climb to altitude, I'd probably use it just to have better visibility over the nose. Despite the nose attitude, the visibility from the Mustang cockpit can only be described as excellent. The Mustang does not include TCAS but, instead, uses Traffic Information System (TIS), which works only with certain local ATC radars.

Although other initial climb restrictions included plans to stop us at 15,000 feet and FL230, we found ourselves cleared to FL340 within ten minutes of brake release, with a clearance to FL410 following soon after. As we passed 13,000 feet, our climb rate showed 1,650 fpm, despite the heat.

At FL190, we were climbing at 1,100 fpm with a temperature of ISA+17. We were hoping the temperatures would cool down a bit to be certain we could make Flight Level 410. At FL230, it was ISA+14, with the true airspeed settling in at 251 knots.

Out of FL270, we were climbing at 260 ktas. ISA dropped to +10 at FL300, with a climb rate of about 1,000 fpm and a climb speed of Mach 0.44. Out of FL360, we were burning 280 pounds per side and climbing at 600 fpm with ISA back up to +12. Approximately 39 minutes after brake release, 510KS leveled out at 41,000 feet at the miserly cost of 500 pounds of fuel, about 40 gallons per side. I removed my headset out of FL370 and found the noise level quite comfortable. Once level at FL410, it took about ten minutes to accelerate to 319 ktas at ISA+4.

At the temperatures we saw throughout the climb, the choice of FL410 might have been questionable, especially because the Mustang seemed to be struggling at 500 fpm out of FL390. I'm glad we hung in there, though, because the rate picked up again for the last 1,500 feet. Long-range cruise speed is 298 ktas and would be needed to make the full 1,150 nm.

Before we left FL410, I stepped into the back of the cabin to check the noise level. From the rearmost right seat, I could carry on a conversation with the other pilot, who now sat in the left seat. The nearly 15-foot-long cabin in the back is comfortable, with a 54-inch width and 55-inch height.

Cabin seats include shoulder harness belts. Cabin outfitting is nice—not lush, but comfortable. The interior has folding armrests on the reclining, aft-facing seats. There's a 110-volt electrical outlet in the console. Cabin windows are tinted, which Cessna believes will reduce the need to lower the shades, an important move because the Mustang cabin looked small with all the shades down.

The potty seat, located between the cockpit and the cabin, is unbelted and really for emergencies only. The cockpit curtain pulls shut for privacy on that side while an expandable visor that attaches with magnets closes off the area from the rest of the cabin.

After the acceleration run at FL410, we descended to FL350 to check cruise speed, which turned out to be 338 ktas. Alexander said the best altitude on flights of 600 to 800 nm is FL350, where the true airspeed will be about 340 knots.

We headed toward Independence, a nontower airport that is home to the Mustang production line. The Vnav function on the Garmin is robust enough for most any pilot. Alexander and I told the system to plan on our reaching an altitude about 3,500 feet agl a few miles away from the airport, so we could test how well the system would couple to the GPS 17 approach there.

Within a few minutes of the top of descent (TOD) warning from the computer-generated "Vertical Track message," the TOD appeared on the MFD. The autopilot captured the slope and my only job was to keep the power in line. Because there is no radar coverage at Independence, we made visual vectors to the final and watched the system fly the approach.

The Landing Learning Curve
Initial approach flap setting is 184 kcas, while landing flap can be dropped only when the speed is below 148 knots. The gear can be dropped at up to 250 knots. Beginning the base-leg turn, I dropped approach flaps. Approaching the final approach fix, I slowed to 180 knots and dropped the gear. Final landing checks were only three green down.

Passing PUWES final approach fix inbound put us on a five-mile final for the 5,500-foot-long Runway 17. I popped off the autopilot, called for final flaps and maintained the airspeed of about 150 until two miles out. Our ref was indeed low, about 94, even on the first approach, for which we were just under the max landing weight of 8,000 pounds. I should have slowed the aircraft a little earlier on final to fit into the traffic pattern better.

Crossing the end of the Independence runway at about 50 feet, I pulled the power to idle. My first landing taught me one lesson about the Mustang: don't flare. Even though I floated somewhat in the gusty wind, the touchdown was firm on the left main first and settled quickly to the other two gear. I extended the speed brakes with my right thumb on the button and pressed heavily on the brakes. We made the first turn, which we estimated to be about a 3,000-foot ground run. The brakes are an excellent combination of power and ease of use. Cessna says we could have landed the airplane in as little as 2,600 feet at sea level. Alexander said on a dry runway, 3,000 feet made him feel much more comfortable.

When we left Independence, we headed south toward Tulsa to avoid an MOA to the west. It took just under eight minutes to climb to 16,500 feet, where we set up for VFR steep turns. I rolled into a number of 45-degree banks and found that, with only a slight boost to the throttles from Alexander and a small crank of trim, I could hold the bank attitude with a speed near 200 knots.

Rudder pressures rolling from left to right were not light, but just enough to make the airplane feel solid. The visibility throughout the turns was good, if only somewhat restricted out of the side windows during very steep turns.

I tried racing along at 220 to 230 knots to see how quickly the aircraft would slow for a simulated traffic pattern. With power back to flight idle, I dropped the gear and within five seconds found myself below the 200-knot limit for approach flaps. With those down, speed dropped to less than the 161-knot limit for landing flaps. With the second notch down and the gear out, the airplane slowed down to the 130-knot range within ten seconds.

I punched the go-around button on the throttle to prepare for a simulated balked landing. I brought the levers to the takeoff detent and found only a modest amount of nose-down trim was needed to keep the nose under control. Now weighing in at somewhere around 7,650 pounds, the Mustang acceleration was quick.

Still at 16,500 feet, we headed for Wichita to try some VFR approaches. ICT actually snuck up on us a little faster than expected and we found ourselves at 16,500 feet, 17 miles southeast of the airport. Perfect. I pulled the power to idle, extended the speed brakes, and pitched down toward a base-leg entry to Runway 19 Left.

With an indicated airspeed of approximately 200 to 210 knots, the rate of descent easily held around 3,500 fpm and we comfortably reached a point five or six miles from the traffic pattern at 2,500 feet. I noticed no perceptible pressure bump on the way down, but the speed brake rumble was quite pronounced.

The wind at Wichita was gustier than when we'd left a few hours before, with peaks to 28 knots. Knowing what was coming, I was ready to finesse this next approach. With landing flaps, I pulled the power to idle at 40 or 50 feet agl and held the same pitch attitude I'd had on final.

The Mustang settled on the mains nicely and I quickly lowered the nose gear as my right thumb commanded the speed brakes to extend. This time, I also pressed hard on the brakes. The amazing part of the Mustang brakes, again, was their responsiveness. I used only moderate pressure to produce large results. We almost made the Runway 19 Left M5 turnoff, which would have indicated about a 3,500-foot landing distance. But so as not to cook the brakes, I let the aircraft roll to M6. Because the best landing is most likely the next one, we taxied back to 19L for just one more.

The Mustang displayed its rocket-like performance down the runway as we departed at 8,000 pounds. The crab angle on downwind meant that the strong crosswind was still there. On final, as the PAPI showed the wind was trying to best me, I added power, then more, and then more as more of the red lights became visible than white.

Finally arresting my descent, I dragged the airplane the last half mile or so with a lot of power. Once I had the runway made, I eased off the power and made a, well, solid touchdown, connected to another short ground run, indicating I should have quit earlier while I was ahead. After three hours of flight time, we had burned 1,370 pounds of fuel.

Cessna hopes the Mustang will fill the long-empty entry slot of its popular business jet line that began with the Citation 500 just over 30 years ago, and that it will be the aircraft of choice for people thinking about an attractively priced, easy-to-maintain alternative to piston-engine or small turboprop aircraft, as well as those ready to move up to the speed and comfort offered by high-altitude jet travel.

Having flown the early 500s and now the 510, I think the Mustang is an entry-level jet that far surpasses the company's first footsteps into the business jet market more than three decades ago. It is sure to be a match for many competitors because the move up to the left seat in a Mustang will be relatively simple for most piston and turboprop pilots, whether they fly the aircraft single pilot or as part of a crew.

After an evaluation of slightly more than three hours, we'd give the Citation Mustang an A. When the company makes a five-point cockpit shoulder harness standard in front, we'll give it an A+. The Mustang is, indeed, an impressive machine.

Boeing Business Jet

by Robert P. Mark
Reprinted by *Courtesy Aviation International News—*
www.ainonline.com, June 2004

The pinstripe 737 is a "space" ship compared with other business jets, and a joy to fly, but its very size is restricting its market in the current economic climate.

Size may actually matter. It seems that BBJ sales to corporations have not materialized as planned due to the "royal barge" image of the big business jet, with many entering service in head-of-state, wealthy individual, or charter applications.

Figure 9-2 Sales of the Boeing BBJ have been almost double of the numbers expected in 1996.

When General Electric's then CEO Jack Welch decided it was time to shop for a new business aircraft in 1995, he knew he needed more of what every executive wants in a corporate airplane–space (Figure 9-2). The Gulfstream V, then under development, was expected to have the range, but on a 12-hr trip, the GE head thought the cabin was too claustrophobic. Because GE was already negotiating a deal for a number of Boeing 737s, Welch called his friend, Boeing chairman and CEO Phil Condit, and asked if he could deliver a 737 with more range. Condit asked Borge Boeskov, Boeing's vice president of product strategy at that time, the same question. Boeskov said, "Yes."

The obvious airframe choice was a member of Boeing's Next Generation (NG) series, the 737-600. But while the -600's cabin was right for the job, the aircraft still did not have enough range. Boeing then looked at the 737–700, but, again, while the cabin offered the room business owners wanted, range was limited. By combining the 737–700 fuselage with the -800 wing and landing gear and extra fuel tanks, Boeing developed the combination that would become the BBJ, officially titled the Boeing 737–700 IGW (increased gross weight), offering a 6,200-nm range.

In early 1997, Boeing began accepting deposits for the Boeing Business Jet (BBJ), which is powered by two 27,300-lb-thrust CFM56-7 engines produced by CFM International (Figure 9-3), a 50/50 joint company between Snecma of France and GE.

In addition to the necessary range, the BBJ offers space—lots and lots of cabin space. Although the BBJ's ramp footprint is not significantly larger than that of a Bombardier Global Express or the GV, the cabin is massive, topping out at 807 sq ft, with a ceiling more than seven feet high. Another, even larger version of the BBJ, called the BBJ2 and based on the 737–800 cabin, offers 1,000 sq ft of interior cabin area. Typical interiors include seating for 14 to 27 people, conference tables, additional conference rooms, sleeping quarters, bathrooms with showers, and gourmet galleys.

One BBJ operator, who wished to remain anonymous, compared regular trips to Europe from the West Coast in his company's Challenger 601 with those in its new Boeing. "You buy the aircraft for the boss, not the pilot, but the long days were killing our crews. From Van Nuys, for example, it was six hours to Gander, an hour on the

Figure 9-3 Winglets are most effective on long-range aircraft like the BBJ.

ground to fuel, and six and a half more to Paris. Add in the two-hour [before takeoff] show times and we were easily seeing 15-hour days going over and 17 coming back. Now that we go nonstop, we knock off four hours going over and six coming back. We looked at a Global Express, and then did the analysis on a BBJ. I don't think there is any comparison. The BBJ does the job hands down." On April 7, Boeing flew a BBJ nonstop from Seattle's Boeing Field to Jeddah, Saudi Arabia. The 6,854-nm flight, which took 14 hr 12 min, represents the farthest distance flown in a BBJ.

Distinctive Aviation Partners winglets helped to improve the airplane's payload-range equation.

Another Dallas-based operator said, "The aircraft burns 20 percent more fuel than some smaller jets, but I can save money by tankering fuel. On a trip from the West Coast to New York, I saved $7,600 because I didn't buy any of the high-priced fuel in the Big Apple. Buying this airplane was the easiest decision I've ever made." Kevin Russell, executive vice president of NetJets, said, "It is one of the most reliable aircraft ever built for corporate aviation. Most other aircraft require more maintenance. Our dispatch reliability is over 99 percent. It has been a wonderful aircraft that has exceeded our expectations." NetJets currently flies seven BBJs with 22 more on order (sales of BBJ shares, however, have reportedly not met NetJets' original expectations).

The Dallas operator explained the value of the BBJ this way: "My boss and his wife climb aboard the aircraft in the late afternoon, have plenty of room to get nice and comfortable as we head off to Europe. Our flight attendant can cook them a gourmet meal, they can take a shower if they'd like, and then climb into bed at their normal time. They wake up about an hour before landing, have breakfast, and are pretty much ready to go when we land in Paris. It's like they take a part of their home with them."

An owner can do a considerable amount of personalization with an 800-sq-ft-cabin, which is precisely why some customers choose the BBJ over the GV and the Global Express. But while this personalization works well for many of the individuals who own and operate BBJs, the public visibility issues that accompany owning an aircraft the size of a 737 might be slowing sales of the Boeing. In fact, some experts believe it could be one of the most significant marketing problems Boeing faces when competing with the GV and the Global–two aircraft, that despite their own size, have become almost normal in size for a large-cabin business jet.

A Marketing Dilemma

Clay Lacy, president of Clay Lacy Aviation and a BBJ operator, said, "I think the BBJ is a great airplane. But most of the BBJs are owned by wealthy individuals and only a few by publicly held companies. There are only so many of those wealthy people to go around, so the BBJ might be reaching a saturation point." Lacy, a long-time proponent of large aircraft for business aviation, added, "I thought [corporate] people would get used to these big airplanes the way wealthy individuals have. I don't know how Boeing is going to overcome that issue. Perhaps the marketers should educate people that Boeing pays for almost everything the first five years and that these aircraft don't cost much more to operate than a GV or a Global. We lease the GV for $8,000 a flight hour and the Boeing for $9,500 per hour."

Figure 9-4 Boeing Business Jets chief pilot Mike Hewett contends that the BBJ can take a licking and still keep ticking.

Chuck Colburn, Boeing Business Jets' director of marketing, said, "One of the reasons the BBJ costs more to operate is that it burns about 20 percent more fuel. But check out how seldom our aircraft is ever down for maintenance." The Dallas operator said, "I can operate the BBJ cheaper than a GV or the Global because my maintenance costs are so low." Russell agreed: NetJets flies its BBJs about 1,000 hr per year. "There is not a great differential between the occupied hourly costs on the GV and the BBJ," he said.

A green BBJ now runs $39 million, up from $32 million in 1996. Operators can expect to spend anywhere from $10 million to $12 million more to turn the aircraft into a home away from home. "All the things we've done to make the 737NG a great airliner make it a no-brainer for a business jet," said Mike Hewett, Boeing Business Jets chief pilot (Figure 9-4) "But since the downturn in our nation's economy, our challenge at Boeing Business Jets is to convince corporate boards of directors that the BBJ is a better economic investment and a more effective tool than our competitors." One industry expert explained the issues he faces selling time in an airplane as large as a 737: "One corporation I called asked who else in the neighborhood we'd sold one to. He loved the aircraft, but did not want to be the first one in his area to operate such a large aircraft."

Airbus, with its Airbus Corporate Jetliner (ACJ), is hot on Boeing's heels. "We cannot afford to dismiss Airbus as a competitor," said Colburn. "They make a fine aircraft. But I do think that until recently, they simply viewed the ACJ as an extension of the A319 and not a true business aircraft. They have since changed that view and are being a lot more aggressive with our potential customers, too. But walking through our aircraft and theirs, most people could not tell the difference. The ACJ is a foot shorter and five inches wider than the BBJ."

Not Just Another 737

While many people might think the DDJ is simply a 737 dressed up in pinstripes and wingtips, Hewett said that's hardly the case. "There is only 9 percent of the BBJ that is common to the 737 Next Generation. We kept the architecture of the flight control systems and the human interface, but everything behind the scenes was redesigned to the latest standards or upgraded to what Boeing offers on other new aircraft. There have been hundreds of changes to FAA certification regulations since the 737 was certified. We chose to certify the BBJ as a derivative, which means you don't need to meet all of those changes. But we stepped up to all but about six or seven of them."

Because the BBJ incorporates much of the same safety architecture as the original Boeings, a pilot can safely control the aircraft to landing through the aircraft's cable-boosted flight control systems, even after a complete hydraulic failure, which no 737 has ever had. The Boeing 737 has 120 million flight hours behind it, 100 million on the 737 classic (–100 thru –500 models) and 20 million on the Next Generation aircraft (–600 models and later).

Operationally, Boeing sells the BBJ as an airplane independent of most ground support anywhere it travels, with a stairway that folds neatly into the fuselage and an APU and air conditioning packs "that could make it snow in 110-degree heat on the ground in any Middle East airport," according to Hewett. Access to the baggage compartments from the ground is easy for anyone more than five feet tall, with a simple one-handed turn of the door's handle. The BBJ floors are approximately 51 to 55 in. above the ground, depending on aircraft weight. The ACJ, by comparison, has baggage compartment floors considerably higher at 78 to 83 in.

Certainly there are a few things Boeing operators don't like. For example, not all business airport runways can accept a 171,000-lb aircraft. By contrast, the GV-SP tips the scales at 91,500 lb and the Global Express weighs in at 96,000 lb. Colburn said, "The BBJ's weight has been an issue only at a small number of airports, such as Teterboro and Aspen. And, in both cases, it is not necessarily a runway capability issue as much as it is a local political one. There is a 100,000-pound arbitrary weight limit at Teterboro designed to keep out airline traffic. At Aspen, the airport manager is concerned that the BBJ may clutter up the ramp and has imposed a wingspan limit on aircraft. The Gulfstreams just make it in, but not the BBJ. We are not interested in a legal battle over the issue and are continuing to work with the communities." He added, however, that some BBJ customers might be able to lodge legal challenges at airports that take federal money while excluding the BBJ.

The Speed Issue

Also at issue is the BBJ's speed limit of Mach 0.82 and service ceiling of FL 410. Hewett would like to see the aircraft fly faster and knows his airplane can easily do it. "I've flown the aircraft to Mach 0.92, which means I could clear it to 0.84, but because we'd be up against the barber poll, we'd have to go back and reconfigure the computers to accommodate 0.84. Every time we want to make a tiny change anywhere, we have to prove to the FAA we haven't affected any of the other systems, which can take thousands of hours. So we tend to make modifications when we are going to open the boxes anyway for a major change. But I'm still pushing for Mach 0.84." Bruce Marsteller, a BBJ captain and special projects manager at Huizenga Holdings in Fort Lauderdale, Fla., said, "Mach 0.80 is pretty much it in the BBJ. We'd like to go 0.82, but with the auto throttles on, the engines surge as they keep looking for the right power setting. But I like everything about this airplane. It will easily out-climb our GV at altitude."

Buddy Rogers, a Houston-based BBJ pilot, said, "I thought the BBJ was much slower operationally than I'd expected. On long trips you had to back off to Mach 0.78. Above FL 350, you're against the barber pole if you push it above 0.80. With that said, though, I found the BBJ to be simple and very reliable. I'd recommend it over the Gulfstreams or the Globals to anyone." Jerry Sonderstrom, a BBJ captain for Kevin Air in Van Nuys, said, "We fly at 0.80 for trips of ten hours or less and at 0.78 on a trip like Van Nuys to Paris. It is a great airplane."

Hewett added, "We're lowering the cabin to 6,500 feet at 41,000 feet. To accomplish that, though, we had to change all the software in the digital controller and do a year's worth of structural analysis that defines how many life cycles—the BBJ is good for 75,000—we would have to give up when we blow more pressure into the cabin. It will cost us 20 percent of the useful life of the BBJ, but a business jet customer will never see that kind of time anyway. They'll never even see 10,000 cycles. If it were an airliner, it would be an issue."

But the Dallas flight department manager said, "I don't like the fact that the BBJ only goes to 41,000 feet. I'd rather lose cycles on the airframe and be able to go to FL 450. On a long trip that could save me fuel, which could mean making a legal alternate or not." Colburn countered, "Although the aircraft wing and engines are more than capable of flying the aircraft safely at FL 450, there is the matter of an FAA restriction to contend with that is slowing the approval process." This problem may, in the end, completely prevent the BBJ from legally flying above FL 410.

An FAA document, "Amendment 2587, Standards for Approval of High Altitude Operations of Subsonic Transport Airplanes," imposed a new standard when it was released in July 1996—ironically just a month after the BBJ was announced—for the size of a hole in the cabin an aircraft must be able to withstand and still not completely depressurize before reaching the magic 14,000-ft emergency descent altitude. "Right now, none of the Boeings can comply with this amendment, except the 777, which is certified to FL 430," said Colburn. Interestingly, the A319, of which the ACJ is a derivative, is certified to only FL 390 as an airliner, but Airbus obtained FL 410 as a service ceiling on the ACJ itself.

BBJ Maintenance
Simple maintainability is one of the highlights of a BBJ. The *BBJ Gold Card* entitles purchasers to hotline access at Delta Air Lines' technical operations center, fuel discounts, and flight support through Air Routing, not to mention advice through a connection with Boeing's trouble desk. The Gold Card also allows purchasers to pick up a badly needed part at a local participating airline parts department.

Despite Boeing's claims of an inexhaustible supply of parts, one operator complained, "We're the little fish in the Boeing pond. AOG for a part from Boeing is 24 hours, and that is simply too long. They should have them to us in four to eight hours maximum." But another expressed joy at the fact that recapped tires for a BBJ list out at just $400 each. Boeing even has items in the catalog for a dollar, he said.

Another operator compared windshield replacement procedures between his new BBJ and his old Challenger. "Replacing a windscreen on the Challenger costs about $35,000 and takes the airplane out of service for three days. On the Boeing, it's $5,000 and takes about seven hours." Admittedly, though, no one at Boeing could even recall an instance where any 737 had an in-flight windscreen failure that required replacement. None of the operators AIN spoke to reported more than a few minor delays due to mechanical issues.

If a buyer uses the aircraft 1,000 hr or less a year, Boeing puts the airplane on a low-utilization CAMPS maintenance program. The first hot section would be five years after the aircraft goes into service and ten years before the first major overhaul. "We're running at about 99.9-percent dispatch reliability with about 30,000 hours of BBJ operations," said Hewett. "Everything that is easily repairable is in the electronics and equipment (E&E) bay near the nose wheel. Hydraulic and electrical pumps, and much of the aircraft's plumbing, all meet in the wheel well where they can quickly be repaired. The wheel well has been designed to withstand a complete tire explosion and keep everything intact."

While airlines choose from a huge option list before they purchase a 737–700, BBJs have only a few extras. "We went through the option list and hand picked the ones we believed were most important to business aviation customers," said Colburn. Other than the choice of either Honeywell or Goodrich for brakes—both certified for about 1,500 landings—the only significant option is Category 3A auto land capability. "We believe that most corporate flight departments flying 400 hours a year would not be able to stay current on the Cat 3A system," said Colburn. "But we also learned that most company principals don't want to be out trying to land in Cat 3A conditions anyway."

The BBJ standard equipment includes the Smiths FMS that has been a standard on the 737 for many years. Boeing chose the Smiths unit for its superb navigational capabilities, recently demonstrated by the RNP approaches Alaska Airlines has been flying with the units into remote mountain airports in the 49th state. Boeing gives each customer a new Dell Laptop Computer Tool with proprietary weight and balance and performance software that makes flight planning easy.

The numbers must, however, be manually transferred into the FMS before takeoff because the Smiths unit does not store performance data. "We simply didn't want to partition off a part of the FMS computer to add the performance data," Colburn explained. The Dell laptop also holds all of the aircraft's maintenance documents, including the aircraft maintenance manual, fault isolation manual, illustrated parts catalog, wiring diagram equipment list, dispatch deviation list, service letters, and in-service activity reports.

The possibilities of a rudder problem have been plaguing the 737 ever since two high-profile crashes, one in Pittsburgh and another in Colorado Springs, Colo. Although nothing was ever conclusively proven to indict the 737 rudder system, Boeing is being forced to upgrade all of the rudder power control units (PCU) on all 737s still in service, including the BBJs, at Boeing's cost. The Dallas operator said, "All this hoopla on the rudder just irritates me because they are not sure there is even a problem with the system."

Hewett added his personal thoughts: "I don't believe the 737 had a rudder problem. We probably ran that Pittsburgh scenario in the simulator a hundred times. We did everything but drop the PCU off a 50-story building and it didn't fail. They did try freezing the unit and they managed to get it to fail when a Boeing engineer jammed on the rudder after we'd shot boiling hydraulic fluid through the PCU. The NTSB thought they had the solution, although this kind of scenario is impossible in a 737NG. But a later test we did measured the temperature of the hydraulic fluid in actual flight and found it was near zero in the Pittsburgh aircraft, not -35 like the test. And the Pittsburgh aircraft crashed in the summer from 9,000 feet, where it was much warmer. I believe a rotor cloud toppled the Colorado Springs aircraft, not the rudder." None of this speculation on the rudder PCU applies to the PCU in the BBJ, however.

Most privately owned BBJs are operated under Part 91, while those available for lease are operated under Part 125. Left over from another era when rules were set in stone to restrict competition with the regularly scheduled airlines, Part 125 BBJ operators cannot advertise their aircraft as being available for charter at all. In fact, they cannot even use the word "charter" in any form. "I'd like to advertise my BBJ in the *Wall Street Journal*, but I can't," said Lacy. "I'm sure that's why our use of the BBJ has not been nearly as much as I initially expected."

But one solution may be operating the BBJ under Part 135, a concept both the FAA and DOT blocked until recently. "There was the perception that this airplane is too big and more complicated than some smaller ones and, hence, it needs more oversight [like Part 125]," Colburn said. "We disagree." One BBJ customer who intends to use the aircraft under Part 135 filed a petition, knowing it would become a test case. Top-level FAA officials recently determined a suitably configured BBJ will be approved for Part 135 operation and were expected to issue pending Part 135 certificates by the middle of last month.

It's What's Inside That Counts

Boeing salespeople had to negotiate a learning curve as they made the switch from marketing airliners to marketing business aircraft. Although the company sold the aircraft to their customers in a green

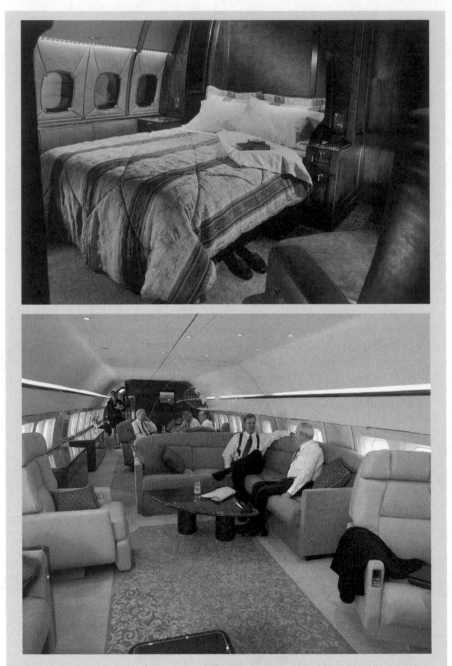

Figure 9-5 TTypical BBJ interior offers roomy quarters for both work and rest.

configuration, it shied away from making recommendations about available completion centers.

"When we started out, Borge Boeskov and I visited all the centers capable of working on BBJs," Colburn recalled. "When we first visited Raytheon E-Systems in Waco, Texas, they didn't seem very interested until we told them about the size of the market. They told us they could finish 12 aircraft a year."

But the relationship between Boeing and Raytheon E-Systems soured quickly when "it became obvious that Raytheon did not have the resources to complete the planned work and got progressively further behind schedule." Hewett said, "We're a lot smarter now." Boeing now reviews the business plans of all BBJ completion centers and audits their work processes.

A fairly typical BBJ interior adds 12,000 lb to the aircraft. "Airbus claims its aircraft can do exactly what ours can," Hewett said. "But since it never got its aircraft's gross weight increased, it can only do what a BBJ can if it has a 5,000-pound interior. Try and find one of those." Hewett also compared landing performance. "I'll take this aircraft into a 5,000-foot runway all day long. I can also get it down and stopped in 2,000 feet."

Boeing's market expectations were originally small, estimating six to eight aircraft per year. The market demand turned out to be greater, with 83 aircraft sold between 1996 and this year, with 50 already in service. But BBJ sales have stagnated this year, with no sales recorded during the first two quarters. Production rates for the BBJs have also fallen from a peak of 24 per year in 1997 to currently about 12 per year. Hewett said, "There are a number of sales about ready to go, but they just haven't happened yet." Colburn said Boeing plans to announce some orders at this month's NBAA show. While Hewett said there are no white-tail aircraft sitting on the ramp at Boeing Field, he admitted that this situation might be rapidly approaching.

"I know a large majority of the BBJs in service in the U.S. is currently with private individuals," said Colburn, "but we hope to expand our base. The most significant thing that will help would be to get more BBJs in service." But Colburn is also aware of the Catch 22 aspect of Boeing's marketing challenges. Big corporations are wary of being seen by shareholders flying around in an airliner, and that makes them tougher to sell. "We were hoping our link with General Electric would help put more aircraft into the field [with corporations]. Then, too, there is the product-loyalty issue. Many potential customers have been flying Gulfstreams for a long time and are reluctant to change."

But a strong advantage to the BBJ is its lineage. The 737 is the most successful airliner in the world today, with more than 3,500 aircraft flying. Boeing officials say a 737 lands and takes off somewhere in the world every six seconds. No other aircraft manufacturer can make that claim. And that lineage goes backwards and forward, with such aircraft as the Boeing 707, 727, 747, 757, 767, and the 777, some 10,750 airframes in all.

The Flight

Walking through the security doors at Boeing's delivery center at Seattle Boeing Field (BFI) made it abundantly clear that, at 171,000 lb gross weight, this was not going to be an ordinary airplane for a flight report. The ground rules set by Boeing were that this AIN reporter would fly left seat on a typical trip, with an approach or two at the destination, but not much time for air work in between. The aircraft was being repositioned for a new customer demonstration in Raleigh-Durham, N.C., about four hours east. Boeing Business Jets said no flight simulator time was available before my flight.

It was a partly sunny day in Seattle, with light winds and a temperature of 20 deg C, as BBJ chief pilot Hewett and I approached N130QS, a BBJ originally destined for fractional provider NetJets. After climbing on board, Hewett took me on a tour of the aircraft from the cockpit back. Returning to the cockpit, Hewett offered me the left seat, which was certainly easy enough to climb into because the subliminal message everywhere inside and outside the BBJ is space. Once the pilot adjusts the seat, the field of view out all six windows is superb.

Avionics are arranged around five 8 by 8-in. Honeywell LCD flat-panel displays—the same as those used in Boeing's 777—connected through dual Smiths FMS units. The BBJ comes standard with a Honeywell Enhanced Ground Proximity Warning System, a Flight Dynamics HGS 4000 Head-Up Guidance system, dual ADIRUs, 120-min cockpit voice recorder, triple VHF, dual HF, and TCAS II. The airplane is also certified for RVSM airspace, 8.33-kHz spacing, and required navigation performance (RNP) 0.5 nm. The BBJ also uses a Smiths auto throttle system.

Like all other Boeings, the BBJ incorporates a traditional control wheel. "We were asked why we didn't move to a fly-by-wire side stick," said Hewett. "For our airline customers, that meant nothing except additional expense and they are already in trouble right now. We had to provide them with an airplane that flew faster, further, higher, with higher reliability, and maintain the same type rating, while requiring no more than two days of FAA training."

With only two passengers and a crew of five, N130QS was well under its maximum gross takeoff weight at 147,000 lb with 42,000 lb of fuel. The BBJ can carry as much as 64,000 lb. Hewett estimated we'd burn about 5,000 lb per hour for the 4.5-hour trip to the East Coast. Boeing's Laptop Tool quickly calculated V1 at 133 kt, Vr 136 kt, and V2 142 kt. Required runway was 5,401 ft if we needed to stop before V1. The available pavement at BFI's Runway 31R is 10,000 ft. The nice part of using the Boeing Laptop Tool is that all the numbers consider necessary obtain ls clearance entorno. Not all competing systems do.

Before doing the external walk around, Hewett described what's needed to bring the airplane online: turn on the battery switch, run the fire tests and start the APU. The APU can, incidentally, be started all the way up to 41,000 ft, with a generator capable of supplying the entire aircraft. Next, the avionics come on to begin the IRU alignment process while the first officer handles the preflight. Outside, Hewett pointed out the easy access to the E&E bays in front of and behind the nose wheel. Easy access to the forward cargo hold from ramp level offers up 300 cu ft of baggage space. Despite the large cargo bay, many buyers opt for an in-cabin closet that offers passengers in-flight access to their suitcases.

The lowest point of the engine nacelles of the CFM56s is just 19 in. off the ground, raising questions about FOD. "The velocity of the air going in is so low, because of the size of the opening, that small objects simply get tossed aside," Hewett explained. The oil service door is at chest height for most people, making that verification a snap. The landing gear can easily be checked from the same point.

As one walks in front of the wings, the seven-foot-tall winglets are prominent. Boeing put them on initially to enhance the airplane's image, but soon learned the benefits were tremendous, according to Hewett. "The performance improvement from the winglets allowed us to pull out the 10th fuel tank, as well as enhance the aircraft's hot-and-high performance."

Moving to the rear cargo area, a look inside revealed little cargo room on 130QS, with available space taken up by fuel tanks. The BBJ2 offers an additional 20 cu ft of cargo space here. Looking up from the cargo bay allows an easy view of the separate APU intake and exhaust stacks.

The first officer can now walk forward toward the left main gear well, duck down, and emerge standing in a Rube-Goldberg-like place with miles of piping and electrical bundles all available for quick checks. After checking the left main landing gear, left wing, and engine, it is time to climb the stairs to the cockpit.

Hewett said, "We do flows and verify with the checklist. The first officer has an assigned part of the overhead panel and normally loads the FMS. Essentially, everything goes to 'on' or 'auto' and all the lights go out. It is pretty simple." Hewett also pointed out the emergency galley power isolation switch, now required on all new transport aircraft since the Swissair crash.

Hewett attempted to upload the ATC flight plan into the FMS through Universal Data's system. But, after three or four attempts, it became clear a manual entry would be necessary. Although the Smiths system is similar to the Honeywell or Universal systems I'm familiar with, I needed a few minutes of training from Hewett before I put the flight plan in.

Although the start sequence on the big CFMs is simple, the right engine took two tries for reasons unknown to any of us. At the beginning of the taxi out, it was apparent that having marshallers, while important, was not absolutely necessary because the pilots can easily see the tips of those big winglets from the cockpit. Hewett said the added airfoils have actually made getting the Boeing to descend a bit more difficult than originally planned. The tiller was necessary for the tight turns as we taxied between the dozens of new 737s awaiting completion and delivery, but the pilot can otherwise easily steer the BBJ with the interconnected link between the rudder pedals and the nose wheel steering, something that made the takeoff a few minutes later quite easy.

We discussed the takeoff procedures out of BFI, normally planned to miss traffic departing and arriving SeaTac, just a few miles south. As Hewett turned the auto-brake system to "RTO," he discussed engine failure before V1. The procedure would be speed brakes out, full reverse thrust, and hang on while the auto brakes pump 3,000 psi into the system. He added that the auto-brake system is so effective that a rejected takeoff below V1 will find the airplane slowing nicely by the time the flying pilot even gets the reversers out. But, he also cautioned against stepping on the brake pedals during the rejected takeoff, something that pilots coming from smaller jets might normally do. "Once you add more than 90 pounds of pressure to the pedals, the auto braking kicks off and you're on your own." There are really not many emergencies in the 737 to get excited about. The only two requiring memory items are an engine fire and an emergency descent.

The HUD was left down for the flight, so all attitude and speed information was readily available. Lining up for takeoff, I brought the throttles up a few inches and pressed the auto throttle button to engage them. As a Hawker pilot, watching the throttles move on their own seemed a bit odd to me. They moved smoothly and the airplane accelerated quickly through V1 with only minimal rudder steering inputs. On rotation, I pitched the aircraft to about 20 deg nose up. We'd used about 4,300 ft of runway. The light feeling of the elevators was a precursor to how the big Boeing would fly later. At positive climb, the gear was retracted and the aircraft quickly climbed toward our first target altitude of 3,000 ft. Because the initial rate of climb can easily reach 3,500 fpm soon after gear up, the level-off needed to begin almost immediately. Even though the speed bug had been set at 200 kt, a pilot not paying attention could easily have blown through 3,000 ft when hand flying the aircraft as I was.

Cleared to 11,000 ft, we were given a turn east toward our first point, Moses Lake, Wash. Still hand flying the aircraft, the response for a 150,000-lb vehicle was impressive. It was quickly trimmed to fly hands off as we passed through 10,000 ft and the speed increased to 280 kias. The climb rate settled at about 3,000 fpm. By 28,000 ft, the rate had fallen to 2,000 fpm, which held through level off at FL 370. From takeoff to level off took about 19 min for an average rate of 2,000 fpm. Still hand flying the aircraft, I found the effort minimal.

Once we accelerated to cruise at M0.80, I tried some steep turns at 37,000 ft. Initially I used 30 deg of bank so as not to frighten the folks in back. Hewett tried a few at nearly 60 deg and gave it back over to me for another attempt. It felt like the BBJ could easily have completed a great aileron roll from my 60-deg banks, and that is not surprising considering the BBJ's grandfather—the Boeing's 707—was rolled by Tex Johnson on its first flight 45 years ago. The HGS system makes steep turns a breeze. After just a few minutes of practice, I was able to easily hold altitude within 20 ft. Fuel flows at Mach 0.80—our cruise speed—settled down to about 2,600 pph per side. The book showed that at 41,000 ft and Mach 0.80, the flows would have saved about 800 pph total.

Discussing handling qualities again, Hewett said the BBJ, like all 737s, has no stick-shaker system. "When I did the flight test on this aircraft, I took it up to 30 degrees pitch. I was coming down at 7,000 feet per minute and if I'd let go of the wheel, it would have recovered on its own. But the FAA wants you to hold the stick back and have the nose drop by itself. To make that happen, they put stick pushers on airplanes, which I think are the stupidest things in the world. So we put a stall strip in, but we also gave it an artificial feel. When it gets close to a stall, we ramped up the stick force so high, a pilot can't hold it back any longer."

During the next 3.5 hr, the conversations about rudder issues and other marketing perspectives on the BBJ were easily handled because the cockpit noise level was so low. After two hours, I handed the controls to our extra Boeing pilot, Jim Ratley, and Hewett and I moved aft to sample the noise level in the cabin. The room in the back would quickly spoil any corporate pilot, once you pass the galley and the crew rest quarters. Five people could easily walk abreast in the cabin and never interfere with each other, which is easily one of the largest selling points to the Boeing.

As the noise check proceeded, I noted there was some noticeable, but tolerable, vibration noise right near the wing connect points. Hewett said Boeing is looking at a noise-suppression system with actuators that measure the vibration in the cabin and reduce it by exciting the pylons. It will be retrofittable to earlier aircraft, but is probably still a year off.

About halfway through the flight, Ratley climbed to FL 410, where N130QS accelerated to Mach 0.80, about 460 ktas. Fuel flows became about 2,500 pph per side. By the time the BBJ is light, the flows can reach down to less than 4,000 pph total at this altitude. About 150 mi west of Raleigh-Durham as I slid back into the left seat just before we began our descent, a number of respectable thunderstorms began to pop up. A great practical benefit of the HGS is navigating weather in daylight hours. If the tops of the thunderstorms penetrate the wings-level pitch attitude of the aircraft on the head-up display, you won't clear the weather. If you need to deviate, rather than guessing how far, simply turn the heading bug until it clears the weather visually on the HGS screen, hit nav, and you're guaranteed to miss it all.

Skies were partly cloudy and winds light out of the northeast at Raleigh with a temperature of 35 deg C. The first approach was to be a flaps 40, full auto land affair to a touch and go. I flew the aircraft by hand out of 10,000 ft, slowing to 250 kt. Vref was 126 kt at our landing weight of 121,000 lb. All there was left to do was lower the flaps incrementally and turn the speed bug to the required number. Flaps one begins around Vref+70 kt, or 196 kt, which occurred as we turned downwind. Abeam the airport, it was flaps five and I selected 176 kt on the airspeed selector. Because it looked as if the turn might be close, we selected flaps 10 and slowed to 156 kt. As we were vectored to a 90 deg heading for the ILS 7L approach, I reengaged the autopilot and punched approach. The system quickly coupled. At glide slope alive on autopilot, I called for flaps 15 and gear down as one dot beneath appeared. As we joined the slope, flaps went to 25 and speed slowed to 136 kt. Just inside the marker, we went flaps 40 and selected 131 kt or Vref+5.

At 50 ft the throttles moved to idle and the aircraft began to flare. The touchdown was so gentle that I wondered whether I could duplicate it next time. As the nose gear touched down, Hewett set the flaps for 10 deg and I gently brought the throttles back up. The aircraft required only a mild rotation and we were quickly airborne within 4,000 ft of touchdown. The final landing would be a hand-flown version and I was determined to equal the auto land I'd just witnessed, although I thought that would be well high on my first attempt at landing a BBJ.

The second approach was set up much like the first and I used the flight director all the way to minimums. There was very little trimming necessary until flaps 25. Surprisingly, at 121,000 lb, the approach and flare were a piece of cake, with a touchdown so gentle, I wondered if Hewett had a hand in it, something he denied completely. After touchdown it was speed brakes and reversers out. The massive brakes brought the aircraft to a stop to make the turn off for our ramp at Million Air after using just under 5,000 ft of runway. Hewett handled the after landing checks.

The BBJ is an airplane with a heritage as broad as it is deep and millions of flight hours behind it. It flies like a pussycat, yet has the strength to take much more abuse than most corporate flight departments might hand it. It is also an airplane that can, indeed, be a chariot for kings. As Colburn said, "It will take a while for the BBJ to be accepted as a regular member of the business aviation community, but we are seeing new markets develop all the time. We maintain a high profile at all the trade shows and we're always on the lookout for that next opportunity."

Hewett concluded, "There are only two reasons you should buy a BBJ. One is for the space and the other is because it's a Boeing airplane, which is designed to have the heck beat out of it 7 days a week, 12 hours a day. And every day you get ready to go fly, this airplane will be ready to go. When you buy a BBJ, you get an unbelievably reliable aircraft."

Epilogue

Throughout this book, I've shared the thoughts and ideas of many professional pilots, including what I've learned during 35 years of flying (Epilog Figure 1). The question eventually became how to squelch the spigot because there simply was not enough space to log all these ideas and perspectives.

My goal as a pilot and a writer has been, and always will be, to share as much knowledge and experience with people good enough to ask by placing their money on the table to purchase one of my books. But there is also a fine line between teaching and pontificating.

That's why when I ran across this note from a Gulfstream 4 pilot on the West Coast—I'll call him Tom—to a harried flight department manager in Virginia, it seemed to sum up the reason for books like this, as well as web sites, message boards, and blogs to share our experiences.

True professional pilots share their ideas and thoughts with others on the way up because we all wore those very same shoes years before (Epilog

Epilogue Figure 1

Epilogue Figure 2

Figure 2). So, what you'll read here and, I hope what you've read in this entire volume, are stories about conviction and dedication to a profession we all care about deeply. That's why Tom's advice to his comrade here is so important. It focuses a beam on one of the most important aspects of a flying career and the one that is also the most difficult to get your hands around . . . a gut feeling.

You know what this is. You've experienced it before in your life, even though you might not have heeded its advice. It's that little voice in your head that begins silently calling to you when something may not yet look wrong, but definitely feels wrong.

It could be the sense that you're drifting too close to a set of thunderstorms as you try to find a smooth route home at the end of a long day. As a flight instructor, you just somehow know your instrument student is not well-enough prepared for the check ride. As an airline pilot, a cabin attendant tells you about a passenger who makes them feel uncomfortable and yet hasn't quite asked for your help.

They all mean the same thing. Take action, now. Surprisingly, you'll likely know exactly what to do when you're confronted with the situation.

Fly safely, my friends. And do whatever you can throughout your career to help someone else along the way. That's why we're all here.

Rob Mark
Evanston, IL, May 2007
rob@propilotbook.com

A Flying Job Is Not Always Easy

Here's the question one chief pilot asked Tom in late December 2006. The chief pilot is obviously frustrated, but read the sage advice Tom offers.

"It's been one year now since my company acquired a light jet and asked me to come on as the 'Chief Pilot.' Since that time, the company has experienced the growing pains of owning and operating an aircraft that I cautioned them about.

"The one that is really getting me, though, is their expectation that I be available at their beck and call on a 24/7 basis. It hasn't been as much of a problem with actual flying (although they are pushing that now, too), but they seem to think I should always be on the other end of the phone. For 8 out of the past 12 months, I have operated a one-man 'flight department' where I am the Chief & Only pilot.

"Maybe that's the way it is in a small 'flight department.' But I am beginning to feel very stressed by these unrealistic expectations. To be on a 24/7 leash should not be an issue for a company whose business is not life-threatening.

"What does the group think?"

Tom's Response

"I feel bad for you. I'm fortunate enough to be in a single-airplane flight department that is well staffed with both pilots and maintenance techs. Our owner treats us as professionals—and trusts in our judgment. He is a smart guy. He has earned our loyalty through his leadership.

"I think, though, that we all tend to drift down to the least favorable working conditions that are acceptable by the least choosy among us. In other words, if one of us is willing to work for nothing and fly 24 hours a day, then that is what the rest of us will drift towards. The 25-year-old who is building time comes immediately to mind.

"I'm not saying that we need to have 'old-style' airline working conditions, however, we need to be able to count on rest and reasonable schedules to do those things that all people live for—family, work, recreation, and even the routine chores in life. While we may not be the CEO of their company, we are the Chief Operating Officer of their airplane, and they need to not only understand that, but to embrace it.

"We have legal and moral authority to ensure that we operate with a safety-first mindset, NOT A MISSION-FIRST mindset. Too many of us are afraid to do anything other than knuckle under. We feel that because we work for them, we must unquestionably do what they want. That is never the case in life.

"As their Chief Pilot, it is reasonable for them to expect to be able to get hold of you via phone. However, instant beck and call is a bit much. They may view you as a chauffeur, and if so, that needs to be changed. It sounds as if they either have never truly understood the realities of aviation or they have chosen to reject them. It seems as if they don't respect you, your position, or the value of your talents.

"If it's the respect issue, then run, run away quickly!

"If it's the former, you need to find a way to convince them your method of running their flight department is better than theirs. If they aren't willing to operate within your 'envelope,' then you have to question whether you are willing to stay. If safety is not a concern to them, do you really want to be there?

"There WILL come a day when they push you to do something you know is not right—something that might get you killed.

"These are tough decisions, and I don't envy you.

"But, ultimately, you must be true to yourself and to the profession. I am certain you know what the right answer is already. Will they accept this when you present it to them? And, if they choose not to accept the right answer, will you stay?

"THAT is the ultimate question."

A

AOPA Flight Training 2007 College Directory

Special thanks to the *Aircraft Owners and Pilots Association (AOPA) Flight Training* magazine for sharing this valuable resource and allowing us to reprint it here.

AOPA Flight Training is pleased to present its 2007 College Directory. This directory lists colleges and universities in the United States that provide associate's, bachelor's, or master's degrees in aviation fields. You can access *AOPA Flight Training*'s college database online (http://ft.aopa.org/colleges).

If you're interested in learning to fly or to upgrade your pilot certificate and you already have a degree, or if college isn't the route for you right now, check out our online flight school directory (http://ft.aopa.org/schools).

A school identified by a plus sign (+) following its name is accredited by the Aviation Accreditation Board International (AABI); for more information, see the AABI web site (www.aabi.aero). Schools identified with an asterisk (*) are members of the University Aviation Association (UAA). To learn more about UAA, see the web site (www.uaa.aero).

Alabama

Auburn University Aviation Management and Logistics+*
Auburn University, 334/844-4908, 334/844-6848
http://www.business.auburn.edu/academicdepartments/aviation.cfm

Enterprise Ozark Community College
Ozark, 334/774-5113
www.eocc.edu/divisions/avn_div.htm

Tuskegee University*
Tuskegee, 334/727-8761
www.tuskegee.edu

Wallace State Community College
Vinemont, 256/737-3040, 800/238-6681
www.wallacestate.edu/aviation/index.html

Alaska

University of Alaska-Anchorage*
Anchorage, 907/264-7430, 907/264-7462
www.uaa.alaska.edu/aviation

University of Alaska-Fairbanks, Aviation Technology*
Fairbanks, 907/455-2889, 907/455-2809
www.uaf.edu

Arizona

Arizona State University+*
Mesa, 480/727-1381, 480/727-1775
www.poly.asu.edu/aviation

Chandler-Gilbert Community College
Mesa, 480/988-8000, 480/988-8103
www.cgc.maricopa.edu/aviation

Cochise College
Douglas, 800/966-7943, 520/417-4114
www.cochise.edu

Coconino Community College
Flagstaff, 928/527-1222, 928/526-7696
www.coconino.edu

Embry-Riddle Aeronautical University+*
Prescott, 800/888-3728, 928/777-6601
www.erau.edu

Mohave Community College
Lake Havasu City, 928/855-7812
www.mohave.edu/pages

Pima Community College
Tucson, 520/206-5910, 520/206-5906
www.pima.edu

Arkansas
Henderson State University*
Arkadelphia, 870/230-5012, 870/230-5584
www.hsu.edu/aviation

Northwest Arkansas Community College*
Bentonville, 479/629-4159
www.nwacc.edu/academics/aviation

California
California State University
Los Angeles, 323/343-4568
www.calstatela.edu

City College of San Francisco Aircraft Maintenance
San Francisco, 415/452-5001, 415/239-3000
www.ccsf.edu

Cypress College
Cypress, 714/484-7411, 714/484-7253
www.pilotage.com/vendor/cypress

Glendale Community College
Glendale, 818/240-1000
www.glendale.edu

Long Beach City College
Long Beach, 562/938-4567, 562/938-4387
www.lbcc.edu

Mendocino College
Ukiah, 707/468-3102, 707/468-3453
www.mendocinocollege.edu

Miramar College
San Diego, 619/388-7660
www.miramarcollege.net/programs/avim/

Mount San Antonio College
Walnut, 909/593-3351, 909/594-5611
http://aeronautics.mtsac.edu

Orange Coast College
Costa Mesa, 714/432-5987, 714/432-0202
www.occ.cccd.edu

Pacific Union College Flight Center
Angwin, 707/965-6219
www.puc.edu/angwinairport/

Palomar Community College
San Marcos, 760/744-1150
www.palomar.edu/aeronautics

Sacramento City College
Sacramento, 916/750-2721
www.scc.losrios.edu

San Bernardino Valley College
San Bernardino, 909/384-4451
www.valleycollege.edu

San Diego Miramar College
San Diego, 619/388-7659
www.miramar.sdccd.net/programs/avia/

San Joaquin Valley College
Fresno, 559/453-0123
www.sjvc.edu

San Jose State University*
San Jose, 408/924-3190
www.engr.sjsu.edu/avtech

Santa Rosa Junior College
Santa Rosa, 707/527-4475, 707/527-4757
www.santarosa.edu

Solano Community College
Fairfield, 707/447-4578
www.solano.edu

Westwood College of Aviation Technology Inglewood
Inglewood, 310/642-5440
www.westwood.edu

Colorado
Aims Community College Flight Training Center*
Greeley, 970/356-0790, 800/677-2467
www.aims.edu

Colorado Northwestern Community College
Rangely, 970/675-3284, 800/562-1105
www.cncc.edu

Metropolitan State College of Denver*
Denver, 303/556-2983, 303/556-2982
www.mscd.edu/~aviation

Westwood College of Aviation Technology Denver
Broomfield, 303/466-1714, 303/464-2320
www.westwood.edu

Connecticut
Naugatuck Valley Community College*
Waterbury, 203/575-8191
www.nvcc.commnet.edu

Delaware
Delaware State University*
Dover, 302/857-6710, 302/730-5075
www.desu.edu/som/airway_science

District of Columbia
University of District of Columbia
Washington, 202/274-6205, 240/460-6740
www.udc.edu

Florida
Broward Community College Aviation Institute*
Pembroke Pines, 954/201-8084, 954/201-8087
www.broward.edu

Embry-Riddle Aeronautical University+*
Daytona Beach, 386/226-6525, 800/862-2416
www.erau.edu

Everglades University*
Boca Raton, 561/912-1211, 888/772-6077
www.evergladesuniversity.edu

Florida Community College"
Jacksonville, 904/997-2811, 904/997-2800
www.fccj.org

Florida Institute of Technology, College of Aeronautics+*
Melbourne, 321/674-6500, 321/674-6501
http://coa.fit.edu

Florida Memorial University
Miami Gardens, 305/623-1440
www.fmuniv.edu

Hillsborough Community College
Tampa, 813/503-3322
www.hccfl.edu

Jacksonville University*
Jacksonville, 904/256-7894
www.ju.edu

Lynn University*
Boca Raton, 561/237-7333
www.lynn.edu

Miami-Dade College, Eig-Watson School of Aviation*
Homestead, 305/237-5060, 305/237-5260
www.mdc.edu/homestead/aviation/

National Aviation Academy
Clearwater, 727/531-2080, 800/659-2080
www.naa.edu

Palm Beach Community College
Lake Worth, 561/868-3474
www.pbcc.edu

Seminole Community College
Sanford, 407/328-2121
www.scc-fl.edu

Georgia
Clayton College and State University*
Morrow, 770/961-3569, 770/961-3400
www.clayton.edu

Georgia Aviation Technical College*
Eastman, 478/374-6980, 478/374-6402
www.gavtc.org

Georgia State University
Atlanta, 404/651-1099, 404/651-3533
www.gsu.edu/~wwwavi

Middle Georgia Technical College
Warner Robins, 478/988-6924, 800/474-1031
www.middlegatech.edu

South Georgia Technical College
Americus, 229/931-2590, 229/931-2583
www.southgatech.edu

Hawaii
University of Hawaii/Honolulu Community College*
Honolulu, 808/837-8098, 808/837-8099
www.tech.honolulu.hawaii.edu/avit/

Illinois
Harper College
Palatine, 847/925-6065, 847/925-6000
www.harpercollege.edu

Lewis University*
Romeoville, 815/836 5511
www.lewlsu.edu

Lincoln Land Community College
Springfield, 217/786-2406
www.llcc.edu

Quincy University
Quincy, 217/228-5200, 800/688-4295
www.quincy.edu

Rock Valley College Aviation Division
Rockford, 815/921-3016, 815/921-3014
www.rockvalleycollege.edu

Southern Illinois University-Carbondale*
Murphysboro, 618/453-1147, 618/453-8898
www.aviation.siu.edu

Southwestern Illinois College
Belleville, 618/235-2700
www.southwestern.cc.il.us

University of Illinois*
Savoy, 217/244-8606, 217/244-8601
www.aviation.uiuc.edu

Indiana
Indiana State University*
Terre Haute, 812/237-2641
http://aerospace.indstate.edu

Ivy Tech State College
Bloomington, 812/332-1559
www.ivytech.edu

Purdue University+*
West Lafayette, 765/494-7070
www.tech.purdue.edu

Vincennes University
Vincennes, 800/262-9077, 812/888-4500
www.vinu.edu

Iowa
Indian Hills Community College
Ottumwa, 641/683-5214
www.indianhills.edu

Iowa Central Community College
Webster City, 800/362-2793, 515/832-1632
www.iccc.cc.ia.us/business/programs/aviation/index.htm

Iowa Lakes Community College*
Estherville, 800/242-5106, 712/362-7961
www.iowalakes.edu

Iowa Western Community College
Council Bluffs, 402/325-3414, 800/432-5852
www.iwcc.edu

University of Dubuque+*
Dubuque, 563/589-3180
www.dbq.edu

Kansas
Central Christian College of Kansas
McPherson, 620/241-0723
www.centralchristian.edu

Hesston College
Hesston, 316/282-8978, 800/995-2757
www.hesston.edu

Kansas State University-Salina+*
Salina, 785/826-2679
www.salina.k-state.edu

Wichita Area Technical College
Wichita, 316/677-9400
www.watc.edu/

Wichita State University
Wichita, 316/978-5597, 800/642-7978
www.niar.wichita.edu

Kentucky
Eastern Kentucky University*
Richmond, 859/622-1014
www.eku.edu

Northern Kentucky University
Highland Heights, 859/572-5469
www.nku.edu/~technology

Louisiana
Louisiana Tech University+*
Ruston, 318/257-2691
www.aviation.latech.edu

Northwestern State University
Natchitoches, 800/256-5822, 318/357-3209
www.nsula.edu/aviation

Southern University at Shreveport-Boissier City
Shreveport, 318/676-5590
www.susla.edu

University of Louisiana at Monroe
Monroe, 318/342-1780
www.ulm.edu/aviation

Maryland
Community College of Baltimore County*
Catonsville, 410/455-4157
www.ccbcmd.edu

University of Maryland Eastern Shore*
Princess Anne, 410/651-6489, 410/651-6365
www.umes.edu

Massachusetts
Bridgewater State College*
Bridgewater, 508/531-1779
www.collegeaviation.com

Holyoke Community College
Holyoke, 413/538-7000, 413/552-2276
www.hcc.mass.edu

North Shore Community College+*
Danvers, 978/739-5592
www.northshore.edu

Michigan
Andrews University Air Park*
Berrien Springs, 269/471-3547
www.andrews.edu/academics/cot/aviation

Baker College
Muskegon, 231/777-5282, 231/798-2126
www.baker.edu

Delta College
University Center, 989/686-9249
www.delta.edu/aviation

Eastern Michigan University*
Ypsilanti, 734/487-1161
www.emich.edu

Jackson Community College
Jackson, 517/263-1351, 888/522-7344
www.jackson.cc.mi.us

Lansing Community College Aviation Center
Lansing, 517/483-1406, 517/267-5944
www.lcc.edu

Northern Michigan University
Marquette, 906/227-2070, 800/682-9797
www.nmu.edu

Northwestern Michigan College Aviation Division
Traverse City, 231/995-2914, 231/995-1220
www.nmc.edu/aviation

Western Michigan University+*
Battle Creek, 269/964-6993, 269/964-6375
www.wmich.edu/aviation

Minnesota
Academy College of Aviation*
Bloomington, 952/851-0066
www.academycollege.edu

Anoka Technical College
Anoka, 763/785-5940
www.anokatech.edu

Central Lakes College
Brainerd, 218/829-3398, 800/933-0346
www.clc.mnscu.edu

Inver Hills Community College
Inver Grove Hgts, 651/450-8564, 651/227-8981
www.inverhills.edu

Lake Superior College
Duluth, 218/723-4880, 218/878-4731
www.lsc.edu

Metropolitan State University
St. Paul, 651/793-1780
www.metrostate.edu

Minneapolis College-Aviation Center
Eden Prairie, 612/659-6142
www.mctc.mnscu.edu/aviation

Minnesota State University-Mankato*
Mankato, 507/389-6116, 507/389-6371
www.coled.mnsu.edu/departments/aviation

Northland Community and Technical College
Thief River Fall, 800/959-6282, 218/681-0829
www.northlandcollege.edu

St. Cloud State University+*
St. Cloud, 320/308-2107, 320/308-2978
www.stcloudstate.edu/aviation

University of Minnesota-Crookston
Crookston, 218/281-8114, 218/281-8128
www.umcrookston.edu

Winona State University
Winona, 507/452-7220, 800/342-5978
www.winona.edu

Mississippi
Delta State University*
Cleveland, 662/846-6083
www.deltastate.edu

Hinds Community College*
Raymond, 601/857-3300
www.hindscc.edu

Jackson State University
Jackson, 601/979-2471
www.jsums.edu

Missouri
Central Missouri State University+*
Warrensburg, 660/543-4975
www.cmsu.edu/index.xml

Ozark Technical Community College
Springfield, 417/447-7500
www.otc.edu

Parks College of Engineering and Aviation+*
St. Louis, 314/977-8251
http://parks.slu.edu/aviation

Montana
Montana State University/Great Falls College of Technology-Bozeman
Bozeman, 406/994-6151
www.msugf.edu

Rocky Mountain College*
Billings, 406/657-1060, 406/628-2860
www.aviation.rocky.edu

Nebraska
University of Nebraska-Kearney*
Kearney, 308/865-8976
www.unk.edu/aviation

University of Nebraska-Omaha Aviation Institute+*
Omaha, 402/554-3424, 800/335-9866
http://ai.unomaha.edu

Western Nebraska Community College
Sidney, 308/254-7448, 308/254-7443
www.wncc.net

Nevada
Community College of Southern Nevada*
Boulder City, 702/651-4040, 702/651-4053
www.ccsn.nevada.edu

Great Basin College
Winnemucca, 775/623-4824
www.gbcnv.edu

New Hampshire
Daniel Webster College+*
Nashua, 603/577-6000, 603/577-6403
www.dwc.edu

New Jersey
Mercer County Community College+*
Trenton, 609/586-4800, 609/530-1178, 856/261-2606
www.mccc.edu/aviation

Ocean County College*
Toms River, 732/255-0390
www.ocean.edu

New Mexico
San Juan College*
Farmington, 800/232-6327, 505/566-3348
www.flightcareers.com

New York
Columbia-Greene Community College
Hudson, 518/828-4181
www.sunycgcc.edu/aviation

Dowling College/Brookhaven Center*
Shirley, 631/244-1314
www.dowling.edu

Dutches Community College
Poughkeepsie, 845/431-8411
www.sunydutchess.edu

Farmingdale State University of New York
Farmingdale, 631/420-2308, 631/420-2070
www.farmingdale.edu

Jamestown Community College
Jamestown, 716/338-1322, 800/388-8557
www.sunyjcc.edu

Mohawk Valley Community College
Utica, 315/792-5509
www.mvcc.edu

Schenectady County Community College
Schenectady, 518/381-1266
www.sunysccc.edu

Vaughn College of Aeronautics*
Flushing, 718/429-6600
www.vaughn.edu/

York College The City University of New York*
Jamaica, 718/262-2707, 718/262-2829
www.york.cuny.edu/aviation

North Carolina
Caldwell Community College and Technical Institute
Hudson, 828/726-2387, 828/322-6044
www.cccti.edu

Cape Fear Community College
Wilmington, 910/362-7326, 910/362-7189
www.cfcc.edu

Elizabeth City State University*
Elizabeth City, 252/335-3290
www.ecsu.edu/

Guilford Tech Community College
Greensboro, 336/334-4822, 336/334-4822
www.gtcc.edu/transportation/aviation

Lenoir Community College*
Kinston, 252/522-1735
www.lenoircc.edu

Missionary Aviation Institute
Mocksville, 336/998-3971
www.pbc.edu

Wayne Community College
Goldsboro, 919/735-5152, 919/735-5151
www.waynecc.edu

North Dakota
University of North Dakota-Aerospace+*
Grand Forks, 800/258-1525, 701/777-2791
www.aero.und.edu

Williston State College
Williston, 701/572-9901
www.wsc.nodak.edu

Ohio
Bowling Green State University*
Bowling Green, 419/372-2870
www.bgsu.edu

Kent State University School of Technology/Aeronautics Division+*
Stow, 330/672-1933, 330/672-9476
www.tech.kent.edu/pages/academicdivisions.asp

Ohio State University*
Columbus, 614/292-5286, 614/292-5580
www.osuairport.org

Ohio University Department of Aviation*
Albany, 740/597-2626, 740/597-2623
www.ent.ohiou.edu/avt

Sinclair Community College
Dayton, 937/512-2242, 937/512-4134
www.sinclair.edu/academics/egr/departments/avt

University of Cincinnati/Clermont College*
Batavia, 513/735-9100
www.ucclermont.edu

Oklahoma
Northern Oklahoma College
Tonkawa, 580/628-6200
www.north-ok.edu

Oklahoma State University-Stillwater*
Stillwater, 405/624-8022
www.okstate.edu/aviation

Southeastern Oklahoma State University*
Durant, 580/745-3252, 580/745-3252, 580/745-2000
http://aviation.sosu.edu

Southern Nazarene University
Bethany, 405/789-6400, 405/491-6639
www.snu.edu

Spartan College of Aeronautics and Technology
Tulsa, 918/836-6886, 800/331-1204
www.spartan.edu

Tulsa Community College*
Tulsa, 918/828-4270, 918/828-4041
www.tulsacc.edu

University of Oklahoma+*
Norman, 405/325-7231, 405/325-7232
www.aviation.ou.edu

Western Oklahoma State College, Maffry Aviation Center
Altus, 580/477-7825, 580/678-8603
www.western.cc.ok.us/aviation/

Oregon
Central Oregon Community College
Bend, 541/318-3736
http://aviation.cocc.edu

Lane Community College
Eugene, 541/463-4195
www.lanecc.edu/flight

Portland Community College
Portland, 503/614-7457
www.pcc.edu/fly

Pennsylvania
Community College of Allegheny County
West Mifflin, 412/469-6222, 412/498-1084
www.ccac.edu

Community College of Beaver County*
Beaver Falls, 724/847-7000, 800/335-0222
www.ccbc.edu

Lehigh Carbon Community College*
Allentown, 610/264-7089
www.lccc.edu

Marywood University
Scranton, 570/348-6274
www.marywood.edu

Pennsylvania College of Technology
Williamsport, 570/326-3761
www.pct.edu

Pittsburgh Institute of Aeronautics
Pittsburgh, 412/346-2100
www.pia.edu

Reading Area Community College
Reading, 610/372-4721, 800/626-1665
www.racc.edu/

Puerto Rico
Interamerican University School of Aeronautics*
San Juan, 787/724-1912, 787/724-1912
http://bc.inter.edu

South Carolina
Bob Jones University Aviation
Greenville, 864/987-9330, 864/987-9331
www.bju.edu

South Dakota
South Dakota State University*
Brookings, 605/688-5126
http://learn.sdstate.edu/aviation

Tennessee

Middle Tennessee State University+*
Murfreesboro, 615/898-2788, 615/898-2054
www.mtsu.edu/

Tennessee State University*
Nashville, 615/963-5371, 615/963-5378
www.tnstate.edu

Tennessee Technology Center-Nashville
Nashville, 615/425-5600
www.nashville.tech.tn.us

Texas

Baylor University*
Waco, 254/710-3563
www.baylor.edu/bias

Central Texas College*
Killeen, 800/792-3348
www.ctcd.cc.tx.us

LeTourneau University*
Longview, 903/233-4260, 800/759-8811
www.letu.edu

Midland College
Midland, 866/749-2376, 432/685-4799
www.midland.edu

Mountain View College*
Dallas, 214/860-8763, 214/860-8852
http://aviation.mvc.dcccd.edu

Palo Alto College
San Antonio, 210/921-5173, 210/921-5171
www.accd.edu/pac/avt/avt.htm

San Jacinto College*
Pasadena, 281/478-2789
www.sjcd.edu

Southwest Texas Junior College
Uvalde, 830/591-2939, 830/278-4401
www.swtjc.net

Tarleton State University*
Killeen, 254/519-5469
www.tarleton.edu/centraltexas/departments/aviation

Texas Southern University
Houston, 713/313-1841, 713/313-1846
www.tsu.edu

Texas State Technical College-Waco*
Waco, 800/792-8784, 254/867-2070
www.theflightcollege.com

Westwood College of Aviation Technology Houston
Houston, 800/776-7423, 713/847-1503
www.westwood.edu

Utah
Salt Lake Community College*
Salt Lake City, 801/957-3598, 801/957-2359
www.slcc.edu/

Utah State University-College of Engineering+*
Logan, 435/753-4289, 435/787-1346
www.engineering.usu.edu/ete/

Utah Valley State College
Provo, 801/863-7784, 801/863-7771
www.uvscaviation.com

Westminster College*
Salt Lake City, 801/321-0345, 801/321-0343
www.westminstercollege.edu

Virginia
Averett University
Danville, 434/791-5733
www.averett.edu/academics/aeronautics/index.html

Hampton University+*
Hampton, 757/727-5418, 757/727-5519
www.hamptonu.edu

Liberty University*
Lynchburg, 434/582-2183
www.liberty.edu/aviation

The George Washington University Aviation Institute
Ashburn, 703/726-8334, 800/822-6717
www.gwu.edu/~aviation

Washington
Big Bend Community College
Moses Lake, 509/793-2241
www.bigbend.edu/aviation

Central Washington University*
Ellensburg, 509/963-2364
www.cwu.edu/~flight

Clover Park Technical College
Puyallup, 253/583-8903, 253/589-5744
www.cptc.edu/

Everett Community College Aviation Department
Everett, 425/388-9533
www.everettcc.edu

Green River Community College
Auburn, 253/833 9111
www.ivygreen.ctc.edu/aviation

Northwest Aviation College
Auburn, 253/854-4960, 800/246-4960
www.northwestaviationcollege.edu

South Seattle Community College
Seattle, 206/763-5133, 206/768-6629
www.southseattle.edu

Spokane Falls Community College
Spokane, 509/533-3370
www.spokanefalls.edu/academic/aviation

Walla Walla College*
Walla Walla, 509/527-2323, 509/527-2712
www.wwc.edu/academics/departments/technology/avi

West Virginia
Fairmont State University*
Bridgeport, 304/842-8300
www.fairmontstate.edu

Mountain State University
Beckley, 800/766 6067, 304/253-7351
www.aviation.mountainstate.edu

Wisconsin
Blackhawk Technical College
Janesville, 608/757-7743, 608/757-7742
www.blackhawk.edu

Fox Valley Technical College*
Oshkosh, 920/232-6001
www.fvtc.edu

Gateway Technical College*
Kenosha, 262/564-3900, 262/564-3902
www.gtc.edu

Wyoming
Casper College
Casper, 800/442-2963, 307-268-2459
www.caspercollege.edu/

B

200 of the Best Interview Questions for *Any* Flying Job

The following questions were compiled from multiple interviews:

1. Describe a time where you were subordinate, but assertive.
2. Describe a time where you disagreed with a company policy.
3. Describe a time where you were disappointed as a leader Certified Flight Instructor (CFI) or captain.
4. Have you ever failed a check ride? If so, why?
5. Why do you think you never failed a check ride?
6. What was the most challenging part of your career and why?
7. Are you happy about your college grades?
8. Do you think you could have done better?
9. Have you applied to any other company?
10. Why aren't you receiving job offers now?
11. Have you ever had a difficult time with a crewmember?
12. Have you ever changed a company policy or procedure?
13. What will you do if you do not get hired by our company?
14. Where have you been applying?
15. Are you willing to accept employment as a CFI again, just to keep yourself current?
16. Tell me a little about yourself.
17. How did you become interested in being a commercial pilot?
18. How did you get to where you are today?

19. Why do you want to work for this company?

20. Tell me about your flight training.

21. Tell me about the biggest work decision you ever had to make.

22. What was the most difficult part of your flight training?

23. Tell me about a time when you felt a company policy was personally distasteful.

24. Tell me about a situation where you broke company policy.

25. Give me an example of a situation in which a company policy was unfair to you. How did you cope with this problem?

26. Tell me about a time when you were fairly criticized?

27. Tell me about a time when you were unfairly criticized?

28. Tell me about a time you became involved with a problem faced by a peer or a subordinate.

29. What was the most difficult situation you experienced in establishing a rapport with a crewmember?

30. Have you ever flown with anyone who talks too little or too much?

31. Describe the perfect work environment.

32. What is the best decision you ever made?

33. What was the worst decision you ever made?

34. What leadership role are you most proud of?

35. What is your greatest failure as a leader? As a student?

36. When have you gone above and beyond the call of duty?

37. Have you ever diffused a situation between yourself and a crew member or between two other crewmembers?

38. Have you ever flown with a crewmember who was dissatisfied with your performance?

39. Tell me about a time your flying was criticized, other than on a check ride.

40. What was the most negative event in your life?

41. What was the most positive event in your life?

42. Has a subordinate ever questioned you?

43. Who have you respected for their leadership qualities? What were these qualities?

44. What are the most important qualities of a practical leadership philosophy?

45. Give an example of a time you were part of the problem, instead of part of the solution.

46. Do you think of Flight Engineer as a leadership position?

47. Have you ever had to diffuse an argument between a supervisor and another crewmember?

48. Did you ever disagree with a decision made by a supervisor?

49. Have you ever worked with someone you disliked? How did it affect your performance?

50. Tell me about a major success you've had.

51. Has anyone ever disliked you? How did you handle it?

52. Name a time you reversed a decision.

53. Tell me about your college years.

54. Tell me about a conflict in the cockpit and how you handled it.

55. Tell me about a person you worked with who was hostile. How did you deal with that person?

56. Tell me about a company policy you thought was completely outrageous.

57. Share with me a situation where you had to ignore company policy/procedures/checklist and you had to improvise.

58. Tell me about a time where you made up your mind, and then changed it.

59. What projects have you accomplished in the past year?

60. How do you make decisions?

61. Have you ever disobeyed company decisions?

62. Have you ever made a bad decision?

63. Have you ever been described as "hard headed"? By whom?

64. In what kind of social situations do you "freeze up"?

65. How skillful do you think you are in sizing up people?

66. What has been the most difficult situation you have experienced in trying to establish rapport with a crewmember?

67. Tell me what you know about wake turbulence.

68. Tell me about a time you disagreed with a captain.

69. Tell me about your flying career.

70. I am interested in knowing how you became interested in becoming a pilot and what steps you took to achieve this goal.

71. Tell me about the biggest work decision you had to make.

72. Would you describe yourself as being more logical or more intuitive in making decisions?

73. Have you been tempted to break company policy to get the job done?

74. Give a brief summary of all the leadership positions you have held.

75. Have you ever had to take over the leadership role unexpectedly?

76. Give me an example of when you feel you failed in a leadership position.

77. Why should I hire you over the other candidates?

78. Describe a time when you made multiple decisions with little information.

79. When did you arrive in town?

80. Where are you staying?

81. Explain how fog forms.

82. Which hydraulic system on the 737 uses engine-driven pumps?

83. What items are located on this system?

84. Explain how to read a METAR (don't make it too easy).

85. What is the trend of today's weather?

86. Can you land at the airport with the reported weather? (Choose a few difficult locations, for example, ASE.)

87. What kind of weather conditions would you need to be able to land?

88. What would the weather need to be at your filed alternate (both precision and nonprecision requirements)?

89. And what if the applicant is handed the LDA/DME to runway 25 at EGE?

90. What is the lowest reported weather when a Part 135 captain can initiate the approach?

91. Explain how you'd enter the hold at Jesie when you're inbound from the north.

92. Explain the latest point you can legally execute a missed approach at EGE?

93. Draw a holding pattern on an en-route chart.

94. What is the radio call on entering holding?

95. What is the max-holding airspeed for a turboprop?

96. What is outbound timing for holding at 14,000?

97. How would you enter holding on the missed?
98. How soon would you slow down?
99. Give the definition of MEA and MOCA.
100. What is Class A airspace? What is the max speed in Class B and what is max speed below lateral limits?
101. What is the standard day temperature in Class C at 20,000?
102. Numerous time/airspeed/distance problems.
103. Determine the visual descent point on an approach.
104. Why do you have to continually monitor an ADF, but not an ILS?
105. What is V1 and how is it used?
106. What is the critical engine in the Beech King Air and why?
107. What is P-factor?
108. Why do you need 5 degrees into the good engine?
109. What kind of A/C and D/C power do the T-44 generators produce?
110. Read a weather line.
111. Numerous time/distance to descend problems.
112. Heading versus radial questions.
113. Explain FAA weather minimums.
114. What are the FAA fuel requirements?
115. Numerous time to fuel dump problems.
116. Define a visual descent point. What kind of obstacle clearance does it provide?
117. How is an MSA defined?
118. Can a controller vector you below MSA?
119. What is the formula for hydroplaning? What are the considerations?
120. What is the difference between type 1 and type 2 deicing fluid?
121. What is an inversion layer and what is associated with it?
122. What weather is associated with a cold front/warm front?
123. What kind of front is associated with a snowstorm?
124. Do snow flurries indicate a stable or an unstable air mass?
125. What is the significance of the temperature/dew-point spread?
126. What is advection fog and what causes it?
127. What is an ILS hold line and when is it activated?

128. What is the criteria to continue an approach at the MAP?

129. What does it mean if you hear "PAN, PAN, PAN"?

130. What are minimum and emergency fuel?

131. What would you do if you received a windshear report of a 25-knot gain in airspeed from a Cessna 152 inside the FAF?

132. What are outside signs of windshear?

133. What would the weather brief say that would indicate windshear?

134. What are the considerations for landing in a wet or slick runway?

135. What would you do if you went NORDO in IMC right after takeoff? At what altitude would you fly?

136. What if the weather is above takeoff rain, but below landing mins?

137. You have a sick passenger and the nearest landing minimums are two hours away. What do you do?

138. What are the considerations for takeoff/landing behind a "heavy jet"?

139. Define accelerate/stop and accelerate/go.

140. What is a balanced field takeoff?

141. Where is the final approach fix for a precision approach?

142. What is VIRGA and what is it associated with?

143. What if the weather goes below mins while you are outside/inside the FAF?

144. If you want to track 330 degrees, with a 30-kt crosswind, what crab angle would you need?

145. If you are cleared for the ILS, are you cleared for the localizer?

146. What is mountainous versus nonmountainous terrain?

147. Numerous time to climb problems.

148. What is the touchdown zone and the requirements?

149. Lost comm scenarios.

150. Talk me through a normal approach/landing from 100 feet, 3/4 mile out.

151. What do you do with an engine failure after you rotate?

152. What is Mach buffet?

153. What are the conditions for the use of anti-ice?

154. What is a Northeaster, an Alberta Clipper, and an occluded front?

155. What is your fuel planning if the weather is below minimums?

156. What is clear air turbulence and where would you find it?

157. What is the MAP for a Cat III approach to Rwy 27 Right at Chicago O'Hare?

158. In what state is the highest mountain, and how high is it?

159. What would you do for rapid decompression? What altitude would you descend to if you were over the Rockies?

160. What does the arrow in the plan view indicate?

161. How close would you fly to a thunderstorm?

162. What is the transponder code for hijack?

163. When do you need to file an alternate?

164. What are the alternate minimums?

165. How do you check the VORs?

166. What factors are involved in determining V1 and Vr?

167. What do you do if your engine fails before/at V1?

168. The tower issues a windshear alert as you are #2 for takeoff. Explain if this is significant to you.

169. How long does a windshear alert last?

170. What does it mean to climb out at Vx or Vy?

171. What is the lowest altitude you would fly on a route with no published?

172. When would you slow down to enter holding?

173. Which leg of the hold will be shorter/longer if you have a headwind/tailwind in the inbound leg?

174. When can you descend below the minimums?

175. Why is nitrogen used to inflate most aviation tires?

176. Where do you see yourself in five years?

177. What was the lowest grade you received in college?

178. Why do you believe you received this grade?

179. Have you ever flown with someone you didn't like?

180. You are flying as captain and discover that your aircraft is going to be grounded by a mechanical discrepancy. A spare aircraft is available, but it will take approximately 30 minutes to get the aircraft to the gate. How will you prepare your crew for the aircraft swap? What will you tell the passengers? What will you ask the flight attendants to do?

181. What qualities make a good captain?

182. What would your friends say is your biggest weakness?

183. Have you ever been arrested?

184. Have you ever been the subject of a FAA accident/incident investigation?

185. How many moving violations have you been issued since you got your driver's license?

186. Would you rather fly with a good friend or someone who is extremely competent, but difficult to get along with?

187. When do you have to declare a departure alternate?

188. Can your departure alternate also be your destination airport?

189. What other airlines have you applied with?

190. How did you choose these airlines?

191. How great a factor is salary in your decision-making process?

192. How far apart are runway centerline lights?

193. How far apart are runway edge lights?

194. Alternating white and yellow runway edge lights are an indication of what?

195. What is the scariest thing you have experienced as a pilot?

196. What are your three strongest personality traits?

197. What would you do if you couldn't fly?

198. What was the biggest sacrifice you made to become an airline pilot?

199. Is a four-year degree important for a career as a professional pilot?

200. What are the criteria to question a controller's instructions above FL240?

C
Virtual Airline Resource Guide

Hundreds of great, and not so great, web sites containing Flight Sim and Virtual Airline–related content are on the Net. Here are just a few of the better web sites I recommend to all flight simmers—especially those who fly for a virtual airline—to enhance their virtual flying experience. Enjoy!

Virtual Airline Listings/Links

Here are a few places to go if you're searching for a virtual airline (VA) to fly for. It's easier to provide online resources containing direct links to the 1001 VAs on the Net than it is to list URLs for all of them here. Good hunting!

- **MicroWINGS VA Directory:** http://www.microwings.com/CoolWebs.HTML#VA.
- **Virtual Pilots Association:** www.virtualpilots.org.
- **Flight Sim Corner:** http://www.fscorner.com/. This place is simply loaded with a variety of aircraft, instrument panels, and necessary goodies to enjoy Flight Sim.
- **VA News and Information:** Virtual Airlines have become so popular that news services were developed to track the "latest and greatest" goings-on at many. This is a great way to keep abreast of what is happening in the VA industry.
- **Free Add-On Panels:** If you're going to fly the advanced aircraft in any VA's fleet, you're going to need better panels than the default ones in Flight Simulator.
- **Navigation Software:** If you'd like to navigate with greater ease and accuracy, consider using a GPS and/or flight-planning software. Several programs are out there. Here are a few.

- **Airport and Navigation Data/Information:** Here are additional real-world navigation resources to help you get where you want to go in the virtual world.

- **How Far Is It?** http://www.escapeartist.com/travel/howfar.htm. This site enables you to plug in From and To information, and then determine the distance from one point to another. It also provides you with specific latitude/longitude and elevation information, which can be used in GPS and other navigation.

• **Weather:** Those serious about flight simming like to fly in weather conditions other than the default Flight Sim weather. The following are three weather-related links. The first two provide weather conditions that you can then manually plug in to your Flight Sim's weather dialog. The third is a freeware software program that imports real-world weather condition data into your Sim.

  - **Aviation Weather Center:** http://www.noaa.gov/. Aviation weather galore including winds aloft forecasts, National Weather Service information, AIRMETS, and SIGMETS. Everything but the NY-METS<g>.

  - **NOAA's National Weather Service:** http://weather.noaa.gov/weather/ccus.html. Additional, very detailed weather information for your flight planning. Searchable by region and specific areas. The server is a bit slow (government web site) but it's worth the wait.

  - **Real Weather 5.0:** http://www.avsim.com/mike/weather/index.html. This is a freeware program that lets you use weather data downloaded from the Internet in your actual Flight Simulator flights. The program works in FS98 and FS6.0, and is a Windows 95–based program.

• **Flying With Real-Time ATC:** If you want to enhance your virtual flying by a hundredfold, you absolutely must check out the following links. The following programs are hugely responsible for the continued growth and success of the VA industry.

• **Simulated ATC Organization (SATCO):** http://www.satco.org/. The official organization of "virtual" air traffic controllers (many of whom are real-world air traffic controllers) who direct you through all phases of your Squawk Box flight—from push back to shut down. They provide clearance, vectors, and even put you on SIDs and STARs! These are the guys on the Pro Controller end of things.

• **Sites with Great Links Pages:** A few additional sites that contain links to all kinds of flight sim–related sites.

• **Preflight Checklist:** http://www.geocities.com/TimesSquare/Arcade/6721/links.html.

- **Mike's Flight Sim Homepage:** http://www.geocities.com/Cape Canaveral/7090/.

- **Essential Web Sites:** The following web sites contain more information related to flight simming than you can view in a single sitting. Bookmark and surf both. These are outstanding resources.

  - **Top Ten Virtual Airlines Rated:** http://www.flightsimaviation.com/ _virtual_airlines_top10.html.

  - **Virtual Airlines:** Partial List (more follow) http://flyawaysimulation .com/article299.html.

  - http://flyaow.com/virtual.htm.

  - **Avsim.com:** http://www.avsim.com/.

Flight Sim and industry news, product reviews, free downloads, online surveys, links to some of the best flight sim sights, flight simmers' guides, help files and tutorials, patches and fixes, a listing of many VAs, and much, much more. Plan to spend considerable time surfing the site, and loving every minute of it.

- **Flightsim.com:** http://www.flightsim.com. Another must-visit site. This site has been built over several years and is jam-packed with everything related to simming, much like the Avsim sight. It also hosts online forms for many of the virtual airlines, which makes it unique.

Reprinted by permission of *Full Throttle* magazine, 5376 52nd Street SE, Grand Rapids, MI 49512, 1-800-821-1707, www.ftmagazine.com.

Some of the Best Virtual Airlines

- WestWind—http://www.flywestwind.com/
- Brussels—http://snba.svagroup.org/
- Eastern—http://www.evair.com/
- Singapore—http://www.cpavirtual.org/singapore2/
- Delta Virtual Airlines—http://www.deltava.org/
- Northwest Virtual Airlines—http://www.nwva.org/nwva/
- Altair—http://www.altairva.com/
- Canadian—http://canadianva.com/
- Sun Air—http://www.sunairexpress.com/
- American—http://www.joinava.org/
- Continental—http://www.aussiepeter.com/cva/cva.htm

- Transload—http://www.transloadairlines.com/
- Monarch—http://www.monarchvirtualairlines.com/
- American Flight—http://www.flyafa.com/

D
Small Cargo Pilot Pay Rates

One cargo shipping company that employs a considerable number of pilots is AirNet. Here's a look at the system that company uses to pay those pilots. More information is available at http://www.airnet.com/.

In 2007, AirNet Systems, Inc. provides a unique pay system that allows all team members to share in the success of the organization. This philosophy serves to strengthen our team commitment and emphasizes serving our customers.

The *Pilot Pay system* utilizes a duty rig component as the base with which work is measured. *Duty rig* is accrued based on the flight hours and time spent on duty for each qualifying assignment. The duty rig accrues at accelerated rates for assignments worked in excess of the minimum guarantee of ten accrued rig hours per duty assignment and for working more than eight duty assignments in a two-week pay period. This approach assures this: the more a pilot works, the more they are compensated! Each full-time pilot is also guaranteed a minimum 80 rig hours per two-week pay period. Pay is calculated by multiplying the accrued rig for each assignment by the assigned Duty Rig Rate.

Table D-1 shows the three[1] ranges for Duty Rig Rates based on a pilot's Duty Position.

---

[1] Prop SICs receive a Duty Rig Rate of $7.69 per hour with no guaranteed minimums.

Table D-1 Duty Rig Rate Table

Duty Position	Minimum Duty Rig Rate	Maximum Duty Rig Rate
Jet Pic	$19.20	$40.96
Jet Sic	$13.64	$18.39
Prop Pic	$11.12	$14.39

Pilots receive yearly merit adjustments on their anniversary date. Merit adjustments for all AirNet Team Members are made in accordance with company guidelines, which are based on overall corporate performance. A standard adjustment is applied when pilots upgrade duty position. The bottom of each scale is determined to assure no pilot in that position falls below a minimum compensation level. Most pilots upgrade into a position above the minimum established.

Table D-2 demonstrates Annual Pilot Pay based on the guaranteed minimum and the average accrued rate of the current route structure:

Table D-2 Annual Income Table

Duty Position	@ Minimum Rig Rate		@ Maximum Rig Rate	
	Min. Accrued	Ave. Accrued[2]	Min. Accrued	Ave. Accrued[2]
Jet Pic	$39,936	$50,569	$85,197	$107,880
Jet Sic	$28,371	$35,925	$38,251	$48,436
Prop Pic	$23,130	$29,288	$29,931	$37,900

[2] The average assignment accrues 101.3 rig hours per two-week pay period.

Additional compensation is available to pilots for working on holidays, for traveling on a day that they would not normally have worked, and for working on normally scheduled off days. A generous per diem system is also available to pilots for rest periods spent out of base.

E

Pilot-in-Command/ Second in Command Defined

The terms Pilot-in-Command (PIC) and Second-in-Command (SIC) are used on a regular basis in every aspect of flying. Here's a look at not only the strict definitions of the terms, but also the roles of these two jobs. In cases where large aircraft used, or still do employ, a flight engineer, the term Second Officer is used to refer to that position, which carries its own set of responsibilities.

PICs are responsible for all aspects of the flight. The PIC does not necessarily need to be occupying the left command seat. The flight's PIC could be seated in the right seat, as in the case of training a new captain.

The PIC ensures that Company policy, procedures, Client/Supplier Contract provisions, Company Insurance provisions, and Federal Aviation Regulations will be complied with. The PIC ensures that all elements of flight planning, aircraft airworthiness, crew requirements, and equipment requirements are completed.

Duties

PIC(s):

1. Are responsible for the safe and efficient conduct of the flight assignment.

2. Shall plan each flight (VFR and IFR) to the extent required to ensure maximum safety and efficiency.

3. Shall, by whatever appropriate means available, determine the status of enroute and destination weather, and possible delays.

4. Shall ensure the aircraft is ready for flight prior to departure.

5. Shall determine the flight can be completed without exceeding any maintenance due time or date on the Aircraft Airworthiness Status Sheet, as well as if all maintenance squawks on the MEL Deferral Log have been properly deferred or corrected.

6. Shall ensure all flying time is properly logged on the flight record.

7. Shall utilize Crew Resource Management (CRM) to work together as a team (crew coordination).

8. Shall ensure the securing of the aircraft at the end of the flight.

9. Shall be responsible for awareness of current information and policy, issued by the Company through verbal briefings or written pilot notices.

10. Shall coordinate flight activities with the CP and DO.

11. Shall not depart on any flight without adequate charts, handbooks, and other references appropriate to the nature and characteristics of the flight.

12. Shall not depart on any flight without proper authorization.

13. Shall perform other appropriate duties as assigned.

Second-in-Command (SIC/First Officer)

Area of Responsibility

First officers or pilots acting as SIC perform duties directed by the Captain or PIC. The primary function of the SIC is to assist the PIC in the operation of the flight. Usually, the SIC performs such duties as aircraft preflight, some or all aspects of flight planning, weight and balance computation, aircraft loading, aircraft cleanliness and security, passenger comfort and requirements, and generating required paperwork.

Duties

SIC(s):

1. Are responsible for the safe and efficient conduct of the flight assignment.

2. Shall plan each flight (IFR and VFR) to the extent required to ensure maximum safety and efficiency.

3. Shall, by whatever means available, determine the status of en route and destination weather.

4. Shall preflight the aircraft prior to each initial flight.

5. Shall review the Airworthiness Status Sheet prior to each flight to determine the aircraft is in an airworthy condition, and also to determine if maintenance squawks have been addressed or deferred.

6. Shall be responsible for logging flying time.

7. Shall be responsible for securing the aircraft at the end of the flight day.

8. Shall utilize Crew Resource Management (CRM) to work together as a team (crew coordination).

9. Shall be responsible for awareness of current information, policy, and so forth issued by Waste Connections, Inc. through verbal briefings or written pilot notices.

10. Shall coordinate flight activities with the PIC, CP, and DO.

11. Shall not depart on any flight without adequate charts, handbooks, and other references appropriate to the nature and characteristics of the flight, and portable equipment, as necessary.

12. Shall perform other appropriate duties as assigned.

Index